Confronting Consumption

Confronting Consumption

edited by
Thomas Princen, Michael Maniates, and Ken Conca

The MIT Press
Cambridge, Massachusetts
London, England

Chapter 5, with revisions, is reprinted from *Ecological Economics*, vol. 20, Thomas Princen, "The Shading and Distancing of Commerce," pp. 235–253, 1997, with permission from Elsevier Science. Chapter 11 is reprinted from *Global Society*, vol. 14, no. 1, E. Helleiner, "Think Globally, Transact Locally," 2000, with permission. http://www.tandf.co.uk

This book was set in Sabon on 3B2 by Asco Typesetters, Hong Kong and was printed and bound in the United States of America.

Library of Congress Cataloging-in-Publication Data

Confronting consumption / edited by Thomas Princen, Michael Maniates, and Ken Conca.
p. cm.
Includes index.
ISBN 0-262-16208-3 (hc. : alk. paper) — ISBN 0-262-66128-4 (pbk. : alk. paper)
1. Consumption (Economics)—Environmental aspects. 2. Environmental policy. 3. Substainable development. I. Princen, Thomas, 1951– II. Maniates, Michael. III. Conca, Ken.
HB801 .C623 2002
339.4′7—dc21 2001059640

10 9 8 7 6 5 4 3 2

To the memory of Donella Meadows

Contents

Preface

This project began with a simple idea: if consumption is self-evidently a major driver of environmental change, consumption itself is not self-evident. The book derives from our deep-felt concern about unsustainable trends in consumption, especially the overconsumption of the North, and all that it entails socially and environmentally. But it was also inspired by our dissatisfaction with limited ways of thinking and talking about consumption that predominate in both policy and academic circles. To equate consumption with what economists call the *demand function* or with what sociologists and others take as a quest for meaning through shopping is, in our view, to ignore structural elements of the problem grounded in political and economic systems.

This book, therefore, develops an "ecological political economy" of consuming—or, probably more accurately, it points analysis and intervention in that direction. One facet we emphasize is the ecological: the "using up" of natural capital in ways that diminish the current and, especially, future ability of individuals and societies to develop. The other is the social and political, especially the organized exercise of influence and power to skew benefits to some and harm to others. Whatever success this book may have in making analytic connections and identifying promising social movements, there remains much left to do, in the realm of thought as well as in practice.

We intend that each chapter stand on its own. A reader interested in conceptual issues surrounding consumption could turn to any of the chapters in part I and find substantive analysis, for example, while those interested in specific aspects such as the effects of globalization or social movements that respond to consumption could turn to the respective chapters and gain insight. But we have also sought to lay out in the introduction and conclusion—by way of overview, synthesis, and projection—

some of the broad themes of overconsumption and responses thereto that link the separate chapters. If anything, we have erred on the side of being provocative, as, indeed, confrontations must be.

The ideas found in these pages are a product of many peoples' thoughts and concerns. Over the course of five years and with stops in Toronto, Minneapolis, Meadville, Ann Arbor, Washington, DC, Eugene, Chicago, Lancaster and Oxford, England, and Stockholm, we have become indebted to many. We extend a special thanks to all those who attended our conference presentations, engaged in our workshops, and sat through our courses. Also, we deeply appreciate the helpful comments and frank criticism of numerous readers, including Aaron Ahuvia, Mark Amen, Maurie Cohen, Edward Comor, Jim Crowfoot, Karen DeGannes, Raymond De Young, Maya Fischhoff, Tim Kasser, Anu Kumar, Ronnie Lipschutz, Dierdre Lord, Donald Mayer, Daniel Mazmanian, Laurie Michaels, Ron Mitchell, Michael Moore, Norman Myers, Jason Park, Rodger Payne, Christer Sanne, Elizabeth Shove, Paul Stern, Karl Steyaert, Richard Wilk, and three anonymous manuscript reviewers. Clay Morgan, senior environmental studies editor at The MIT Press, production editor, Sandra Minkkinen, and copyeditor Elizabeth Judd were valuable developers of this work and are greatly appreciated. Finally, we owe perhaps the largest debt to our contributors, whose dedication to task and collegial spirit made this a wonderfully collaborative experience.

For generous financial support we thank the John D. and Catherine T. MacArthur Foundation (grant 96-34311). We are indebted as well to Allegheny College, the School of Natural Resources and Environment at the University of Michigan, the University of Maryland's Harrison Program on the Future Global Agenda, and the Environmental Studies Section of the International Studies Association for institutional support.

An earlier version of chapter 5 appeared in the journal *Ecological Economics* in 1997 and a version of chapter 11 in *Global Society* in 2000. Chapters 2, 3, 6, and a condensed version of chapter 1 appeared as a special section of the summer 2001 edition of *Global Environmental Politics*.

Finally, a note about the dedication. As we were bringing this book to a conclusion, the world lost Donella Meadows, a leader in environmental thought and activism. In a remarkable life of study and public engagement, she combined rigorous scientific analysis with a deep concern for humans and nature. In many ways, she was a true pioneer in confronting consumption.

1 Confronting Consumption

Thomas Princen, Michael Maniates, and Ken Conca

Consumption and consumerism have long been consigned to the edges of polite talk among North Americans concerned about environmental degradation and the prospects for sustainability. How much, and what, do we consume? Why? Are we made happier in the process? How much is enough? How much is too much for the social fabric or health of the planet? Small wonder that these questions are addressed only obliquely, if at all. They are hard to answer, and when answers emerge they can be problematic, for they have an awkward tendency to challenge deeply held assumptions about progress and the "good life"; they call into question the very idea of consumer sovereignty, a cornerstone of mainstream economic thinking. They also challenge prevailing distributions of power and influence and smack of hypocrisy, coming as they so often do from those who consume the most. To confront such questions is to bite off, in one chunk, a large and vexing body of social, political, and cultural thought and controversy. It is no exercise—intellectual or practical—for the timid.

Perhaps it is no surprise, then, that comforting terms like *sustainable development* have come to frame the dominant environmental discourse in North America, where the contributors to this volume live and work. Those who developed the term—a concept that suffused the 1992 Earth Summit in Rio de Janeiro and, to this day, reverberates powerfully through the environmental debate—defined sustainable practice as actions that meet the needs of current populations without endangering the prospects and livelihoods of future generations.[1] Just what constitutes the needs of today's people remains blurred, out of focus, even usefully ambiguous: everyone has become adept at talking about sustainability without having to wade into the treacherous waters of consumption.

Consequently, much that is said today in the name of sustainability continues to stress the familiar environmental themes of population (too large), technology (not green enough), and economic growth (not enough of it in the right places). Consumption occasionally enters the discussion, but only in nonthreatening ways, and most often in the form of calls for "green consumption" or in support of some moral imperative to consume recycled or recyclable products. Much of this sustainable development talk steers clear of escalating consumption levels and, especially, the roots of such escalation. In the United States, for example, conventional wisdom casts recycling as a primary mechanism for mass publics to "save the planet" without confronting the hard truth that recycling can be a reward for ever-increasing consumption. Questions about driving forces and the impact of consumption continue to hang there, unaddressed. They are like the proverbial 800-pound gorilla in the living room that almost everyone chooses to ignore.

Until now. In the elusive search for a truly sustainable pathway for modern industrial society, frank talk about what this book calls "the consumption problem" is beginning to surface. Across North America—a place arguably viewed as the economic model by much of the rest of the world and undoubtedly a major source of global stress on environmental systems—people are examining their everyday lives. Books such as Juliet Schor's *The Overspent American* and Joe Dominguez and Vicki Robin's *Your Money or Your Life* have proven surprisingly popular. *Affluenza*, a film produced for American public television, chronicles what it labels the "epidemic of stress, overwork, waste, and indebtedness caused by dogged pursuit of the American Dream"; it has become something of a cult classic. Remarq, a major provider of public access to USENET discussion lists, reports that the USENET list on "responsible consumption" and frugal living ranks twentieth in terms of activity, this out of more than 15,000 discussion lists it monitors.[2]

Indeed, aided by the Internet, practices such as green consuming, socially responsible investing, and "responsible shopping" are enjoying a resurgence. Organizations such as the Center for Civic Renewal and the Center for the New American Dream have seen heavy demand for their print- and Web-based materials on American lifestyles that illuminate ways of reducing consumption and exiting the "rat race." Even the distinguished pages of *Science* magazine hosted a debate between a prominent biologist and two leading economists on the existence and nature of a consumption problem.[3] Scores of local communities have launched initiatives under Agenda 21, the action plan that emerged from the Earth Summit. They examine questions of growth, planning, sustainability, and the quality of life.[4] A growing cultural backlash against gas-guzzling sport utility vehicles and intrusive cellular phones—symbols of the frenetic, overconsuming society—is evident. Antisprawl movements are springing up all over.

Ironically, the seeds of this potential groundswell were sown by the very same 1992 Earth Summit that advanced the obscuring principle of sustainable development. At that unprecedented global gathering, government officials and activists from across the global South challenged their Northern counterparts to confront Northerners' own complicity in global environmental damage. Southerners argued that the South's underdevelopment and overpopulation have not been nearly as important for global environmental degradation as the North's overconsumption. In the wake of the Earth Summit, many individuals, community groups, and municipalities throughout the industrialized world have taken up this challenge, examining their own society's patterns of resource consumption and exploring ways of reducing such consumption.[5] One result, evident in many pockets and corners of these societies, has been growing interest in environmental issues that conventional debates over population, technology, green production, and recycling tend to miss: throughput (the overall flow of material and energy in the human system), growth (increasing economic activity or throughput or both), scale (the relationship of the scope and speed of economic or "material provisioning" activity to human and ecological capacity), and patterns of resource use (the quantities and qualities of products used, their meanings, and their changes per capita over time).

Although these emerging concerns about consumption have an environmental dimension, they transcend narrowly biophysical considerations to embrace issues of community, work, meaning, freedom, and the overall quality of life. For some the concern is consumerism, the crass elevation of material acquisition to the status of a dominant social paradigm. For others it is commoditization, the substitution of marketable goods and services for personal relationships, self-provisioning, culture, artistic expression, and other sources of human well-being. For still others it is overconsumption, in the popularly understood sense of using more than is necessary. For most, these themes converge in a troubled, intuitive understanding that tinkering at the margins of production processes and purchasing behavior will not get society on an ecologically and socially sustainable path.

Such intuition is even making its way, albeit slowly, into scholarly circles, where recognition is mounting that ever-increasing pressures on ecosystems, life-supporting environmental services, and critical natural cycles are driven not only by the sheer numbers of resource users and the inefficiencies of their resource use, but also by the patterns of resource use themselves.[6] In global environmental policymaking arenas, it is becoming more and more difficult to ignore the fact that the overdeveloped North must restrain its consumption if it expects the underdeveloped South to embrace a more sustainable trajectory.[7] And while global population growth still remains a huge issue in many regions of the world—both rich and poor—per-capita growth in consumption is, for many resources, expanding eight to twelve times faster than population growth (see box 1.1). Given current forces of economic globalization, these ratios are likely to endure, if not rise. It is little wonder that one leading nongovernmental organization (NGO) recently placed "The Consumption Juggernaut" at the top of its list of "ten hot sustainable development issues for the millennium."[8]

How might ordinary people living in high-consumption societies begin to clarify and act on these unsettling intuitions? Where can they turn for insight, systematic analysis, support, intervention strategies, or hope of effective action? Certainly not to the policymaking arena. There one finds processes of thought and decision dominated, perhaps as never before, by two forces: a deeply seated economistic reasoning and a politics of growth that cuts across the political spectrum. According to prevailing economistic thought, consumption is nothing less than the purpose of the economy. Economic activity is separated into supply and demand, and demand—that is, consumer purchasing behavior—is relegated to the black box of consumer sovereignty. The demand function is an aggregation of individual preferences, each set of which is unknowable and can only be expressed in revealed form through market purchases. Thus analytic and policy attention is directed to production—that is, to the processes of supplying consumers with what they desire. Getting production right means getting markets to clear and the economy to grow. If a problem arises in this production-based, consumer-oriented economy, corrections are naturally aimed at production, not consumption.

Running in tandem with this reasoning is a simple but compelling political fact: expanding the stock of available resources and spreading the wealth throughout the population carry a much lower political price tag than trying to redistribute resources from the haves to the have-nots. Economic growth, facilitated at every turn by public policy, becomes the lubricant for civic processes of democratic planning and compromise.[9]

The dominance of economistic reasoning and the pragmatism of growth politics conspire to insulate from policy scrutiny the individual black boxes in which consuming is understood to occur. As a result, an entire realm of questions cannot be asked. No one in public life dares—or needs—to ask why people consume, let alone to question whether people or societies are better off with their accustomed consumption patterns. People consume to meet needs; only individuals can know their needs and thus only the individual can judge how to participate in the economy. Consumption becomes sacrosanct. If water supplies are tight, one must produce more water, not consume less. If toxics accumulate, one must produce with fewer by-products—or, even better, produce a cleanup technology—rather than forgo the production itself. Goods are good and more goods are better. Wastes may be bad—but when they are, more productive efficiencies, including ecoefficiencies and recycling, are the answer. Production reigns supreme because consumption is beyond scrutiny.

One might think that environmental activism would offer a different logic, a new way of approaching problems related to throughput, growth, consumerism, or the "more-is-better" trap. But in fact many mainstream environmentalists—especially in the United States and, it seems, increasingly elsewhere—have embraced the production-oriented logic. Consumption, if addressed at all, is raised only obliquely. Because production is the problem, regulation of producers becomes the answer. Producers must internalize the cost of pollution or simply cease their abusive activities. Forests are overharvested because timber companies are shortsighted, greedy, or ignorant of proper management techniques, all warranting a change in incentive structures via laws and regulations.

Box 1.1
It's Not Population

Intuitively, people know that consumption contributes to environmental problems. We burn logs or write on paper and trees get cut. For those familiar with IPAT, the formula scientists Paul Erhlich and John Holdren developed to suggest that environmental impact (I) is a function of population (P), affluence (P), and technology (T), it is clear that consumption (affluence) is one of the big three drivers of environmental change. And yet discussion of consumption, whether in the academic, policy, or activist communities, tends to confound consumption with population growth (more people means more consumption) or to interpret overconsumption as a technology problem (more efficient use of resources means less use).

A survey of major commodities reveals that increases in resource use can only be explained in part, often only in small part, by increases in population. Three such commodities—forest products, food, and water—are indicative. The percentage increases in nearly all categories of all three commodities significantly exceed the percentage increases in population.

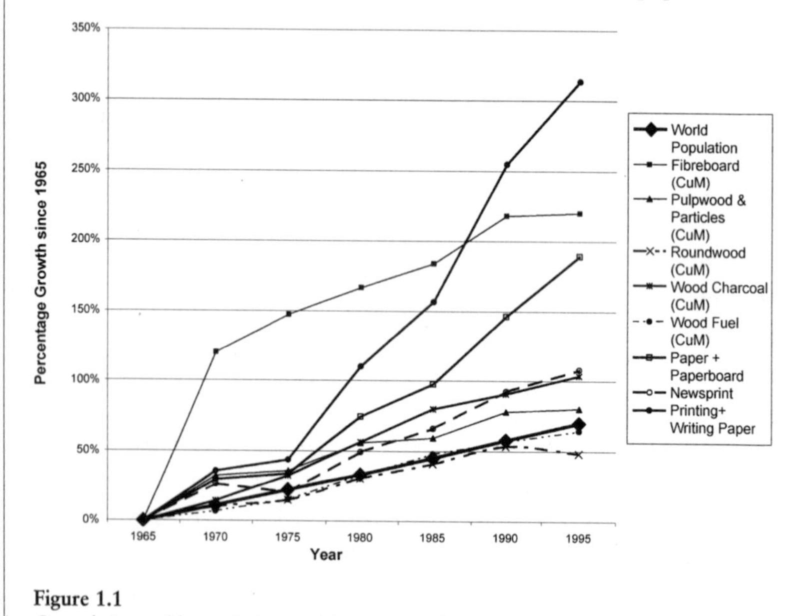

Figure 1.1
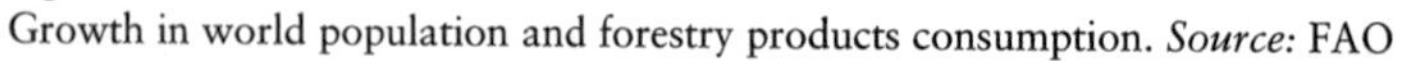
Growth in world population and forestry products consumption. *Source:* FAO

Box 1.1 ***continued***

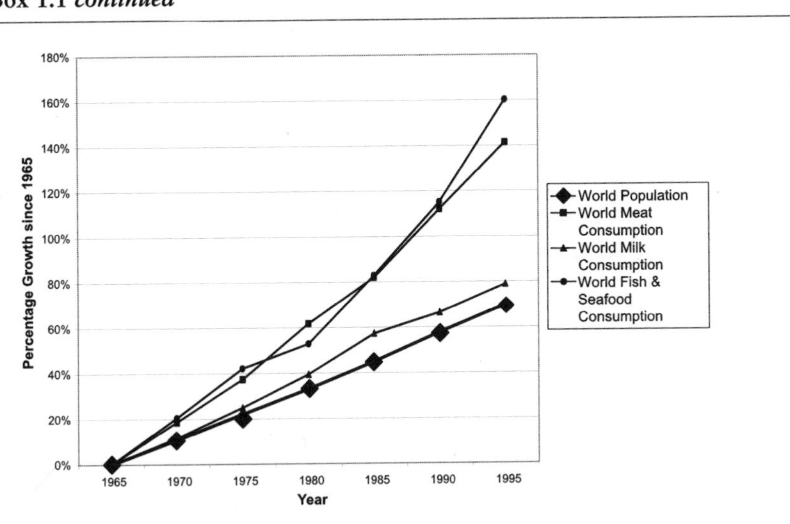

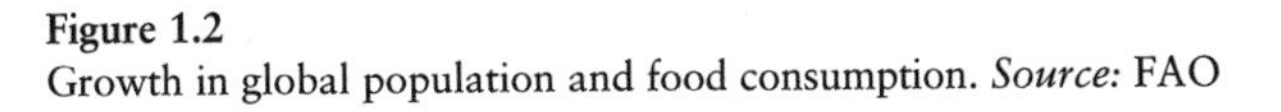

Figure 1.2
Growth in global population and food consumption. *Source:* FAO

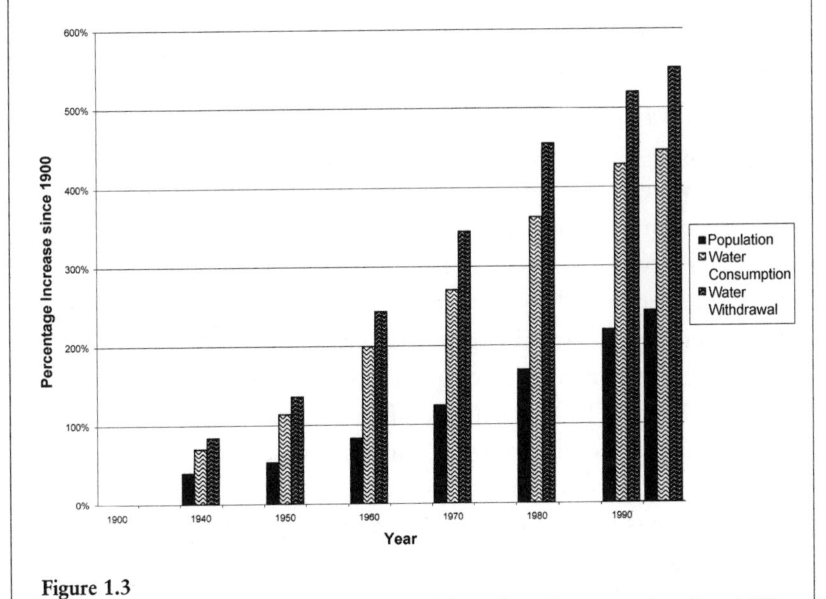

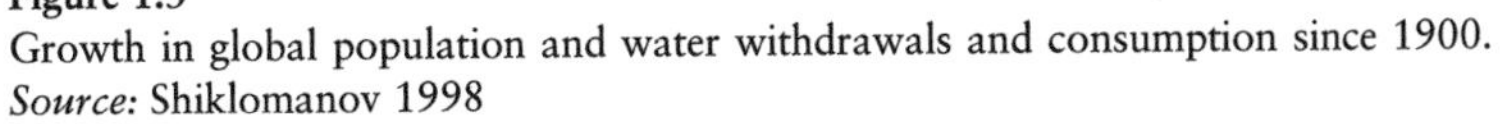

Figure 1.3
Growth in global population and water withdrawals and consumption since 1900. *Source:* Shiklomanov 1998

When these do not work, forests must be set aside from production. If such measures push production offshore, then environmentalists must go offshore, too, helping other countries to develop their regulatory apparatus or promoting international environmental law and organization. In mustering their energies for these campaigns, the largest environmental organizations have spent considerably less time and effort questioning the forces that compel those ever-larger harvests, the ever-more-intensive use of a tract of timberland, and the unending search for new forest frontiers. They tend not to challenge whether society really "needs" more paper (let alone more paper per capita) or the lowest possible prices on wood products. That, once again, would be to enter into the forbidden territory of consumer sovereignty.

An illustration comes from the 1999 annual meeting of the Governing Council of Resources for the Future (RFF), a U.S. natural-resources think tank staffed largely by economists. A member of RFF's board of directors suggested that the size of new houses and the number of miles people drive daily are, as indicators of sustainability, moving in the wrong direction. "The environmental movement is very middle class," she observed, "and its organizations do not challenge middle class values." A deputy director of Environmental Defense—an influential American environmental NGO that works with business to achieve market solutions to environmental problems—replied to the effect that "while few environmentalists were willing to dispense with, for example, air conditioning, they are receptive to producing it with the least damage to the ecology." She then observed that "everybody in China wants a car."

The statement is telling. When consumption concerns are raised in mainstream environmental circles, they are too often dismissed on their own terms, readily converted to questions of production and technology (see boxes 1.2 and 1.3), or shunted off as someone else's problem in the form of looming developments in faraway places. Perhaps for reasons of political calculation, perhaps out of fear or an inability to challenge mainstream consumer values, there is a much greater willingness to examine the way things are done, especially the way things are produced, than to question the purposes served or not served by the doing of those things.

If a citizen or student of environmental affairs finds little guidance on consumption issues in the policy and environmental realms, the academic

Box 1.2
Nor Is It Technology

Examination of technology improvements, in this case increased fuel efficiencies for private automobiles, suggests that technology change may exacerbate resource use because it spurs ever more "consuming" behavior—for example, buying bigger vehicles or more vehicles or both (however more fuel efficient) and driving further and faster. Technology may help reduce environmental impact, but only if all else remains constant. It is precisely the "all else"—the consuming behavior—that must be distinguished and addressed if overall impact is to be reduced.

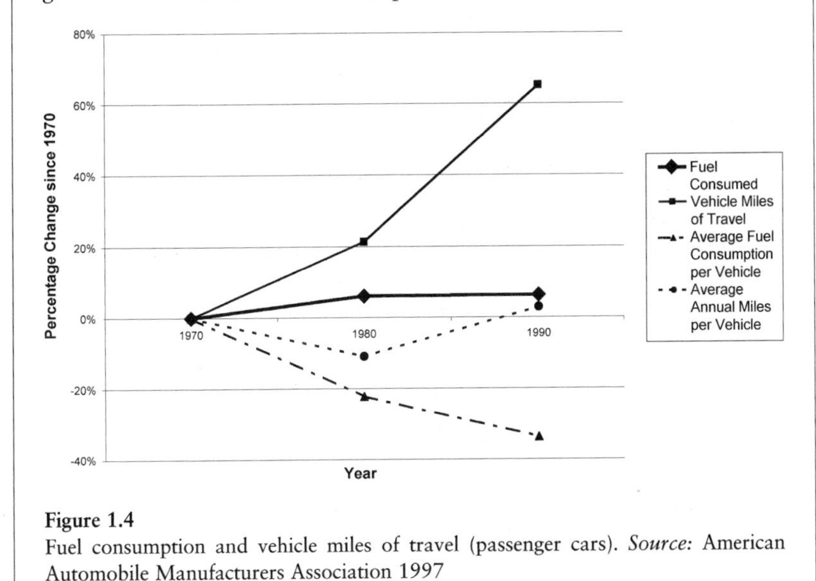

Figure 1.4
Fuel consumption and vehicle miles of travel (passenger cars). *Source:* American Automobile Manufacturers Association 1997

world is no better. As the contributors to this volume can attest, the academy has not been particularly helpful in the search for insights. Universities may exist to push the frontiers of knowledge and to experiment with alternative approaches to meeting human needs. But like the policy realm, much of the social sciences has come under the sway of economistic reasoning; the field of economics is almost universally treated as the most scientific—and, thus, most advanced—of the social sciences. A large body of economic literature exists on "consumer theory," but its analytic goal is to better estimate demand curves, not to ask whether and how consumption patterns contribute to or solve social

Box 1.3
Supply Doesn't Matter

Running short of timber, oil, or electricity? In times of shortfall, society's first response is usually to increase *supply*. Produce more and all is well. Mathematics professor Evar D. Nering of Arizona State University crunches the numbers, though, and shows that it simply will not work. Even conservation and efficiency measures (as those two terms are commonly used) will not do it, not in the long run, and not with exponentially increasing rates of consumption, a premise of modern political economies.

Imagine a 100-year supply of a fixed resource, a pool of oil for instance. At current rates of consumption, this pool will last 100 years. If, however, the rate of consumption grows by, say, 5 percent a year, the pool will last about 36 years. If the supply actually turns out to be much larger, say, 1,000 years' worth at current rates of consumption, this larger pool will be drained in 79 years. And if this pool turns out to be the mother lode of all pools, one that would last 10,000 years at current rates of consumption, the supply will only last 125 years at a rate of consumption that grows at 5 percent per year.

Estimates of known reserves vary a lot, "but the point of this analysis," says Professor Nering, "is that it really doesn't matter what the estimates are. There is no way that a supply-side attack on America's energy problem can work."

This is a disturbing conclusion, especially for those who take ever-increasing consumption as given and believe modern societies can produce their way out of their resource problems. In fact, if decision makers had a choice of doubling the supply (the current favorite option, even among many environmentalists who count on alterative sources) or halving the growth rate, the choice is clear. Doubling the size of oil reserves adds at most 14 years if society uses it at the currently increasing rate. Halving the rate—that is, cutting in half the growth in consumption—will nearly double the life expectancy of the supply, no matter what that supply is.

So this is where conservation and efficiency come in, right? Better fuel economies, for instance. Not so, says Professor Nering. "If we increase the gas mileage of our automobiles and then drive more miles, for example, that will not reduce the growth rate." It is the consuming behavior, the "driving more miles" and all the incentives and structural factors that compel such behavior, that must change.

In short, resource crises will not solve themselves, certainly not if increasing supply is the only legitimate response, as it certainly is in a production-oriented society such as the United States. These simple calculations, even without a notion of ecological or waste-sink capacity, show that consumption itself must be tamed—to the tune of a zero or even negative rate of growth.

Source: Evar D. Nering, "The Mirage of a Growing Fuel Supply," *New York Times*, op-ed page A21, June 4, 2001.

and environmental problems. Mainstream political science, preoccupied as it is with sorting out struggles over the division of spoils and burdens in a growing economy, is similarly blind to the consumption question.[10] Sociology and anthropology have taken questions of consumerism more seriously—shedding much light, for example, on the status and meaning of consumption in modern society. But the message of much of this work seems to be either that (1) we are what we consume, in the sense that consumption is a critical aspect of giving meaning, status, and identity; or (2) consumption is manufactured by structural forces that create a "consumption trap" beyond the control of the individual and the community.[11] Both of these conclusions push the act of questioning or controlling overconsumption beyond the grasp of academic insight. Psychology likewise has a tradition of examining sources of satisfaction and linking it to work and income. But much of this work tends either to further a critique of materialism or to support product marketing.[12] In all, little attention is paid to externalities, social or environmental.[13]

Political economy, a diffuse field aimed at bridging the behavioral and institutional aspects of material provisioning, may be better situated to examine consumption critically. Again, however, a production-sided logic often prevails, especially among mainstream political economists who command the most attention and prestige. At the international level, most analysis focuses on shifting modes of production, new international divisions of labor, the increasing role of finance, and the impact of all these on international security and the stability of the state. The environment enters only when specific environmental problems push onto the agendas of intergovernmental organizations or interstate negotiators.[14] Consumption enters, if at all, primarily in the "everyday life" variant of political economy. But even here the central concern has been to capture individual- and household-level expressions of these larger, often-global, forces. If one such expression is, say, debt-driven or status-driven purchasing to the detriment of self-provisioning, community integrity, and environmental services, then this purchasing and its effects are understood as the inevitable manifestation of global trends. It is almost as if the analyst says, in the end, "isn't this interesting," but does not dare ask whether such trends add up to a meaningful, let alone sustainable, society.

None of this is meant to denigrate the many important accomplishments of individuals and groups in the policy-making, activist, and academic realms struggling to put issues related to consumption, throughput, scale, and patterns of resource use on the table for discussion. Our point is not to deny the existence of these important efforts but to point to their fragmented, isolated character and the tough upstream journey they face. Small wonder, then, that individuals looking into these realms from outside see little being accomplished.

Our own professional experience reveals a similar disconnect between mainstream approaches to environmental problems and consumption. A small example: In 1996 a handful of scholars, including the three of us and several contributors to this volume, launched the first in a recurring series of panels on the political ecology of consumption at the annual meeting of the International Studies Association. Over the next few annual meetings these panels attracted several scholars working on a range of linkages among consumption and ecology. We quickly discovered that we shared a yearning for a fresh and more encompassing analytic perspective. We also found that our panels fell well outside the mainstream of "environmental" topics at the annual meetings, which emphasized interstate cooperation, sovereignty issues, links between environmental damage and violent conflict, political controversies such as NAFTA, and problems of global governance. More than once we found our panel scheduled for a small room far from the center of conference activity. One year we met in a converted hotel room several floors removed from the conference center, with the panelists squeezed between the minibar and the bathroom.

Nevertheless, we were struck by the size of the audience drawn to these panels and the shared frustration of audience and panelists alike with the failure of the social sciences to grapple seriously with problems of consumption and environmental degradation. We were also struck by the recurring tendency of the panel discussion to return to a few key themes: the unsustainability of current trends; the enormous political obstacles to placing consumption-related concerns on elite agendas in the governmental, NGO, corporate, and academic worlds; and the need to see consumption not just as an individual's choice among goods but as a stream of choices and decisions winding its way through the various stages of extraction, manufacture, and final use, embedded at every step in social relations of power and authority. The chapters in this volume, many of which first appeared as presentations in this series of panels, are an outgrowth of that collective quest for a serious and sustained discussion of these themes.

As these discussions progressed, we were each teaching courses in our respective home institutions in which consumption served as an important element or even the major focus. These courses ranged from freshman lecture courses on environment and society to graduate-level research seminars in political economy or environmental politics. The strongly favorable reactions from our students have convinced us that a large gap in the college curriculum now exists. A realm of scholarly, civic, and activist concerns is not being met by courses on environmental law, environmental regulation, and environmental policy, nor by appeals for more recycling and new partnerships with industry, nor by better biophysical understanding of processes of global change.

These experiences in the conference hall and the classroom alerted us to what many others had already discovered—that a significant portion of American society yearns for a less harried, less materialist, less time-pressed way of life, and that many know that their individual consumption and the consumption of their society as a whole are threatening environmental life-support systems.[15] Our professional experiences have exposed us as well to a parallel longing for alternatives to conventional political, economic, and ecological analysis that might more fully diagnose the challenges before us and sketch the paths to a future that works.

This volume is our collective attempt at such sketchwork. One consequence for us has been to deliberately focus our analytic attention on Northern, especially North American, patterns of consumption as well as attempts in North America to confront consumption. We have resisted the temptation to point the finger elsewhere—at growing populations and growing consumerism in China and India, for instance—on the premise that overconsumption is, at root, a problem of the North. Our approach flows from the conviction that troubling, problematic issues of consumption and consumerism can no longer be swept under the rug, and the deepening sense that ordinary people in North America and all over the world are coming to ask important questions about the means and ends of economic life. Now is the time, we believe, to follow that troubled intuition and break from the norms of polite political, economic, and environmental discourse to raise a set of tough questions:

How much is too much? How much is enough? Who decides? And who wins and who loses in the process?

A Provisional Framework

Given our dissatisfaction with prevailing, fragmentary approaches to consumption and its externalities, we seek an alternative perspective, a new angle on the consumption problem. We highlight here three critical themes as a provisional framework: emphasis on the social embeddedness of consumption; attention to the linkages along commodity chains of resource use that shape consumption decisions; and stress on the hidden forms of consuming embedded in all stages of economic activity. These themes stand in contrast to the "production angle" and its underlying assumption of an economy with ever-expanding throughput of material and energy in the human system—an assumption that exists as if ecological, psychological, and social capacity were infinitely malleable and extendable. From our "consumption angle," we assume just the opposite: that there are fundamental biophysical, psychological, and social limits that can be ignored or stretched or disguised only in the short term and only at increasing social, political, and economic cost. From the production angle, ever-increasing production is logical; displacement of costs onto others in time and space is normal competitive behavior. From the consumption angle, ever-increasing throughput and displacement of costs is ultimately destructive and self-defeating. In highlighting the dangers of exceeding social capacity and risking ecological overshoot, our intent is to question underlying assumptions, to stimulate thought, and to point to new forms of intervention.

The Social Embeddedness of Consumption

The first step in developing a more ecologically and socially attuned framework for studying and acting on overconsumption is to recognize, as many sociologists and others clearly have, that a consumer's choices are not isolated acts of rational decision making. Rather, such choices are often significant parts of an individual's attempt to find meaning, status, and identity. Moreover, far from being autonomous exercises of power by sovereign consumers, such choices in modern political economies are heavily influenced by contextual social forces (notably advertising and media images of the good life) and subject to structural features that often make it convenient, rewarding, even necessary, to increase consumption. Embedding consumption in a larger web of social relations leads us to ask about the influences on consumption choices, including the location of power in structuring those choices.

We see two particularly useful concepts emerge from this focus on the social embeddedness of consumption. Both identify important social processes that tend to be missed in conventional perspectives, perspectives that assume atomistic rationality and privilege power as consumer sovereignty. As Michael Maniates discusses in chapter 3, *individualization* is the tendency to ascribe responsibility for all consumption-related problems to freestanding individuals. In this approach, the larger social forces impinging on purchasers' decisions are ignored. A second important concept is Jack Manno's idea, in chapter 4, of *commoditization*—the tendency of commercial forces to colonize everyday action, converting more and more of life's activities to purchase decisions.

Chains of Material Provisioning and Resource Use

A second component of our framework is the commodity chain of material provisioning in which consumption consists of a linked stream of resource-use decisions. Prevailing models of economic activity typically focus on the interaction of producers' and consumers' market behavior via the forces of supply and demand. We take a different perspective, viewing economic transactions as a series of linked decisions running along a commodity chain that begins with primary resource extraction and continues through final purchase, use, and disposal. These two ends of the chain are linked by a series of nodes that include investment, manufacturing, processing, distribution, marketing, and retailing. In practice, commodities are produced via a complex, weblike network of economic relations. But the simple, linear notion of a chain is a useful first-cut approximation that directs attention both upstream and downstream from what is otherwise the conventional emphasis on individual choices of atomized consumers. The commodity-chain approach reminds us that consumption decisions are heavily influenced, shaped, and constrained by an entire string of linked choices being made, and power being exercised, as commodities are created, distributed, used, and disposed of.

Just as our emphasis on the social embeddedness of consumption leads us to highlight critical processes of individualization and commoditization, so too are critical underlying processes revealed by the emphasis on a linked commodity chain of resource-use decisions. One such process is *distancing*, a term we use to refer to the severing of ecological and social feedback as decision points along the chain are increasingly separated along the dimensions of geography, culture, agency, and power (see chapter 5, by Thomas Princen). Rather than viewing consumption as the exercise of consumer sovereignty in the context of perfect (or near-perfect) information, the concept of distancing highlights the increasingly isolated character of consumption choices as decision makers at individual nodes are cut off from a contextualized understanding of the ramifications of their choices, both upstream and downstream.

A second critical process revealed through the resource-chain perspective is the disproportionate exercise of power and authority at certain critical nodes along the chain. The commodity-chain perspective reminds us that the power to shape consumption choices resides at various nodes all along the chain. In chapter 6, Ken Conca argues that we are witnessing the *downstreaming* of economic power in a post-Fordist global economy. Across a wide swath of global economic activity, changes in the organization of production and exchange have shifted power away from control of the means of production and toward what we can think of as control of the means of consumption—enhancing the centrality of consumption in environmental regulatory strategies and raising the stakes surrounding consumption for environmental social movements.

Production as Consumption

Combining the elements of socially embedded consumers and linked chains of resource-use decisions leads to a third theme of our provisional framework: that "consuming" occurs all along the chain, not just at the downstream node of consumer demand. Nodes of raw-material extraction and manufacturing, for example, represent not just production and value *added*, but also consumption and value *subtracted*. Producers are consumers; production is consumption. An important implication of this idea is that what is being consumed at each node is not obvious. At the node of primary resource extraction it might be the tree or the fish, or it might be the ecosystem integrity of the forest or the fishery. At the node of final purchase it might be an apple, or a person's attention, or a community's social fabric.

Another implication of this view is that responsibility shifts from the individuated consumers-as-final-demanders to actors at all nodes of the chain. Producers may add value as they satisfy downstream demand, but they also risk value depletion; they *consume* value by producing. In using up resources both natural and social, they impose costs on the environment and on people—be they purchasers, workers, caregivers, neighbors, or citizens.

This *consumption angle* on resource use offers a corrective to the production-centered perspective that dominates contemporary discussions of economic affairs, including environmental protection. In that perspective, raw materials feed manufacturing and distribution to produce what people want. It follows that, because goods are good and would not be produced if people did not want them, more goods—and more production—must be better. A productive economy is, as a result, one that produces more goods for a given input (thus increasing the economy's "productivity"), yields more choices for consumers, and increases output. When production creates problems such as pollution, the productive answer is to *produce* correctives such as scrubbers, filters, and detoxifiers. So goes the logic of production, productiveness, productivity, and products—construing all things economic as producing, as adding value, as, indeed, progress. The consumption angle turns this around to self-consciously construe economic activity as consuming, as depleting value, as risking ecological overshoot, as stressing social capacity.

Plan of the Book

The chapters that follow apply these themes in varying degrees and at different levels of analysis. Some authors confront consumption in their analysis, others in the subjects they study. All work toward three applied goals of the volume: to place "cautious consuming" alongside "better producing" as an activist and policy response to environmental problems, to demonstrate that such an alternative perspective is far from wishful thinking, and to provide a preliminary map of pressure points for change.

Part I of the book (The Consumption Angle) lays important conceptual groundwork. Thomas Princen (chapter 2) critiques the prevailing focus on production and the lack of biophysical grounding, then develops a "consumption angle" with notions of overconsumption and misconsumption. Michael Maniates (chapter 3) highlights the rise of "individualization" and the narrowing of environmental imagination, while Jack Manno (chapter 4) develops the concept of commoditization, outlining the forces that drive more and more everyday activity into the commercial realm. These analytic concepts together help orient many of the remaining chapters and suggest avenues of intervention.

Part II (Chains of Consumption) draws on these ideas but shifts to a macrolevel of analysis. It begins with Princen's setup chapter (chapter 5) on the concept of distancing. Ken Conca extends this central idea of distancing with his piece on commodity chains in a globalizing economy (chapter 6). In like fashion, Jennifer Clapp analyzes preconsumer and postconsumer waste as the consuming of social and ecological capital (chapter 7), and Richard Tucker offers a historical perspective on natural-resource commodity chains in the Americas (chapter 8). All four chapters emphasize the importance of economic organization, the distribution of market and political power, and the manufacturing of demand as costs are shaded and distanced.

Moving to the realm of practice, many people are already implementing the recommendations—implied and explicit—that flow from the first two parts of this book. Part III (On the Ground) reports on and analyzes cases of individual and collective confrontation with consumption. Individuals across North America, for example, are stepping off the work-and-spend treadmill, forgoing income in favor of time spent in more rewarding, less material activities (chapter 9, by Maniates). As Marilyn Bordwell shows in chapter 10, others are organizing to take on a $170 billion/year industry with "subvertisements" and "uncommercials." Local cooperative communities are creating their own currencies to reduce the outflow of capital and jobs and to redefine linkages among work, value, and exchange (chapter 11, by Eric Helleiner), and transnational networks of individuals and organizations are building institutions for forest product certification (chapter 12, by Fred Gale). Some individuals go so far as to generate their own power, ridding themselves of dependency on centralized, fossil-fuel-based energy utilities (chapter 13, by Jesse Tatum).

Each of these case studies emphasizes the tensions among individual choices, social mobilization, and collective action within a given social and ecological context. Together, they highlight pressure points for political change as suggested by the more theoretical chapters of parts I and II. One point of potential change, clearly, is final consumption and practices such as "green consumption" and "voluntary simplicity." But as we have argued here and will argue further (see, especially, chapters 3, 6, 9, and 14), strategies that emphasize individual action alone are insufficient in the face of overconsumption. Political action—organizing for collective change—is needed.

One strategy for political change is to move "upstream," using the node of final consumption merely as the entry point for what becomes a more comprehensive critique and restructuring. This approach is perhaps best illustrated in chapter 11, on local currency movements in which local communities create new chains with radically shortened geographic and social distances. Still other responses seek to reduce distancing within existing chains, as in forest product certification (chapter 12). And there are, finally, more hard-hitting tactics. One is to shine a bright light on the asymmetric and often-abusive power arrangements inherent in dominant systems of production and consumption, as in the case of Adbusters (chapter 10) and the opponents of the globalization of waste (chapter 7). Illuminating abuses, asymmetries, and the costs of consumer culture—an important end in its own right—raises consciousness and holds powerful actors accountable. Another tactic is to explore ways of opting out entirely, exposing the impermanence and false economy of consumer culture; "home power" is a case in point (chapter 13).

We make no claim to capturing the full range of social experiments with confronting consumption. Certainly others come to mind: cities that provide infrastructure for farmers' markets to bring fresh produce to the city and reconnect town and country; citizens who campaign against urban sprawl to expose the hidden costs of subsidized private transportation and development. Nor do we suggest that participants in the movements and activities described here see consumption as the central problem or interpret that problem in the same way. Indeed, we document the rich diversity of analyses, ideologies, and norms underpinning these activities, as well as the diverse entry points innovators have found in the commodity chains and consumption streams of modern industrial society. We remain struck by the burgeoning of such activity, and wonder whether and how it might coalesce into a "sustainable consumption" movement, one that resists the temptation to point the finger of blame elsewhere, one that squarely challenges overconsumption and the disturbing trends of individualization, commoditization, and distancing (chapter 14). We see tremendous potential for learning from the different approaches and experiences captured by these responses and the movements they spawn. These approaches and experiences provide a basis for building—through further reflection and action—on the concepts and frameworks we introduce and extend in the first two-thirds of the book.

I

The Consumption Angle

2

Consumption and Its Externalities: Where Economy Meets Ecology

Thomas Princen

Analytic and policy approaches to environmental problems can be roughly grouped into two categories. There are those that take current resource-use practices as given and look for marginal improvements. And there are those that presume current practices are unsustainable, possibly catastrophic if pursued to their logical conclusions, and that look for alternative forms of social organization. Research within the economic strands of social science disciplines such as political science, sociology, and anthropology has been preponderantly in the first category, what might be termed *environmental improvement*. Pollution control, environmental movements, and environmental organizations are common topics.

At the same time that social science has focused on environmental improvement, those who chart biophysical trends say marginal change is not enough. Every time a "state of the environment" report comes out, authors across the ideological spectrum call for a fundamental shift in how humans relate to the natural world. Some call for global citizenship, others for spiritual awakening. But nearly all call for a drastic overhaul of the economic system, a system that is inherently and uncontrollably expansionist, that depends on ever-increasing throughput of material and energy, and that risks life-support systems for humans and other species. And then the best prescriptions these analysts, who are not students of human behavior for the most part, come up with are changes in taxes—classic marginal tinkering.

If the social sciences are going to make a contribution commensurate with the severity of biophysical trends, it must do better than analyze environmental improvement measures. Social scientists must develop analytic tools for the analyst (biophysical and social alike) and an effective vocabulary for the policymaker and activist that allow, indeed encourage, an escape from well-worn prescriptions that result in marginal change at best.

The difficulty in conducting such a transformative research agenda, I submit, lies in two facts. One is the reluctance or inability of social scientists to ground their theorizing in the biophysical, a problem I only touch on here.[1] A second is the fact that the economic strands of the various disciplines focus on *production*. Economic sociology concerns itself with issues of labor and management, economic history with the rise of industrialism, economic anthropology with subsistence provisioning, and political economy with the political effects of increasing trade, finance, and development. *Consumption* is nearly invisible. These strands of research adopt the position of the dominant social discipline—economics—and accept consumption as a black box, as simply what people do at the end point of material provisioning, as the reason for all the "real stuff" of economic activity, that is, production. The economy produces goods and goods are good so more goods must be better. There is little reason to investigate consumption, except to estimate demand functions. Consumers, after all, will only purchase what is good for them and producers, as a result, will only produce what consumers are willing to pay for.

When the prevailing social concern was insufficient production, shortages of food and shelter for a growing population, inadequate investment and risk taking, this stress on production was understandable.[2] It is also understandable when natural resource abundance and unending waste-sink capacity, at home or abroad, could be safely assumed. But today, such an ecologically "empty world" cannot be reasonably assumed. Humans are stressing ecosystem services and causing irreversible declines around the world, on land and water and in the atmosphere. What's more, the contemporary economic system is stressing societies at the individual, family, community, and national levels. The biophysical and social trends are unsustainable and cannot be corrected through more tinkering—that is, more environmental improvement.

Under these conditions, one must ask if the exclusive focus on production might itself contribute to abuse of resources, to the neglect of serious environmental change, especially change entailing irreversibilities and the diminution of ecosystem services, and to societal stress. One must at least ask if the predilection for *environmental improvement* might obscure, indeed, help drive, serious environmental change and do so by promoting *production*, since enhanced production, however implicit, is the overriding normative goal of the economic strands of the social sciences.

This chapter is an attempt to point in an alternative direction, what I term the *consumption angle*. The task is straightforward in the initial stages of conceptualizing: reject the production angle, adopt its polar opposite, the consumption angle, and play out its implications. The result is to show how the consumption angle raises questions outside the production angle. The first step, however, is to play out the nature of the production angle and its associated "environmental improvement" approach and show how they neglect throughput and irreversibility issues.

Before proceeding, however, it is worth noting that, although such initial conceptualization is, in many ways, straightforward, the more operational it becomes the trickier it gets, as will be evident in the hypothetical example at the end of this chapter. This trickiness, I suspect, is not due so much to the difficulties of constructing an alternative logic, one grounded in the biophysical, as it is to the hegemony of the production angle. When the idea of production as the core of economic activity is pervasive, problems in the economy (like ecosystem decline and community deterioration) are logically construed as indeed, production problems, problems to be solved with more or better production. If more, even better, production makes only marginal improvements, if it increases risk or material throughput,[3] it only postpones the day of reckoning. Contradictions mount and risks proliferate. The challenge is to push beyond the production angle, to chart an analytic perspective that at once eschews the production orientation and raises difficult questions about excess resource use.

The Production Angle

The coincidence of a production angle on economic matters and an "improvement" perspective on environmental matters is not accidental. When economic activity or, most broadly, humans' material provisioning, is preponderantly production oriented, the only logical way to deal with problems of production—for example, pollution or deforestation—is to "produce better." If automobiles are polluting, manufacturers produce catalytic converters. If they are consuming too much gasoline, manufacturers produce more efficient engines. If traffic is congested, planners produce more roadway and traffic signals. If suburban growth exceeds population growth, "smart growth" is pursued. If flooding destroys property, engineers build better levees. If aquifers are being drawn down, agriculturalists sink deeper wells. If a fish stock is being depleted, distributors develop markets for "trash species." If slash piles left after a logging operation create visual blemishes or a fire hazard, processors make particleboard out of the slash.

In all these examples, the operation is "improved," made more efficient, or the impacts are softened. But the fundamental problem is skirted or displaced in time or space. Pollutants cannot exceed absorptive capacity. Suburban growth is still growth—that is, the conversion of farmland to residential and commercial use while previously used land is left abandoned and degraded. Aquifers are still "mined" unless their extraction rate is below their regeneration rate. Aquatic systems are still disrupted, possibly irreversibly, if one species after another is fished out. And so on.

What is more, the production angle pervades all sectors of modern industrial society, not just the industrial. Consider the position of a major environmental NGO in the United States, the Natural Resources Defense Council, with respect to gas guzzling, private transportation trends, and the National Petroleum Reserve in Alaska: "It is time to ask what kind of energy policy this country really needs. Sport-utility vehicles (SUVs) are getting as little as 12 miles to the gallon. By making small improvements in the fuel economy of SUVs and other light trucks, we could save ten to forty times the estimated oil holdings of the *entire* reserve."[4]

The prevalence of the production angle on economic and environmental issues and the inadequacy of this perspective for dealing with "full-world," ecologically constrained conditions, suggest the need for an alternative perspective. The tack taken here is to develop a perspective centered on production's apparent flip side, consumption. This perspective maintains the focus on economic issues—that is, on the appropriation of resources for human benefit. To do so, I characterize two approaches. One is to retain the prevailing production-consumption, supply-demand dichotomy where consumption is largely wrapped up in the black box of consumer sovereignty. Certainly extensive study has been carried out on consumption within microeconomics (consumer theory) and marketing, and, in recent years, growing literatures have emerged in sociology, anthropology, and social history.[5] What has been missing in these lines of work, though, is explicit analysis of the *externalities of consumption.* How do decisions of consumers, individually and collectively, contribute to the displacement of costs in space and time? How do personal lives change as expression, identity, and status shift to purchasing and display? How does the polity change as democracy is increasingly defined as a vote in the marketplace?[6] In addition to the neglect of externalities, these literatures have largely ignored the role of power, whether it be the power some actors marshal over consumers or the power, potential or realized, consumers marshal to counter existing practices. Consumption all too often is treated as a passive process, indeed, merely a natural result of "real economics," namely, production and its variants of growth, investment, trade, and innovation.

The second approach to developing the consumption angle is to flip the production angle entirely around, to stand it on its head and construe all economic activity as "consuming," as using up, as degrading. This approach pushes the analytic gaze to the opposite extreme from that of the prevailing production angle where goods are good and more goods are better. As will be seen, this approach lends itself to an *ecological conception* of economic activity, where consideration of environmental impact is not just an add-on but is integral to the analysis. Goods may be good but cautious consuming is better.[7]

Consumption as Product Use

Consumption as the necessary complement to production is eating the apple, burning the log, wearing the socks. Research on consumption and its externalities must examine such decisions and influences for their biophysical impacts. A conventional starting point is the decision to purchase. From the prevailing production angle, especially that of retailing, whatever happens after purchase is of little concern unless the consumer's anticipation of subsequent decisions affects the purchase decision. But from an environmental impact perspective, the critical decision is a combination of purchase and product-use decisions, where, in some cases, major purchases drive resource use[8] and, in others, the patterns of use are most important.[9]

Disaggregating the relative impacts of purchase and use decisions is certainly critical to the consumption and environment agenda. But a more extensive approach would be to go beyond product to consider the nonpurchase decision. That is, individuals consume to meet needs. Sometimes those needs can only be met with purchased items—say, grain, electric power, and high-technology equipment. But many other needs can be obtained through productive effort, individually or collectively. Fresh produce can be purchased at a grocery store or grown at home. Personal transportation can entail driving to work or walking (or at least walking partway). Community members can raise funds to purchase playground equipment and pay to have it installed, or they can collect materials and build it themselves. If one has a need for musical experience, one can buy an album or call a few musician friends over for a jam session. In each of these examples, a priori, one cannot know for sure which activity has the least environmental impact. But an initial and plausible operating assumption is that the commercial, purchased choices are more a part of the current trends: *ever-increasing throughput.*

Little if any research has been done on peoples' choices *not* to purchase or to seek less consumptive, less material-intensive means of satisfying a need. The reason may be obvious: it is very hard to get an analytic or empirical handle on an act that entails not doing something. But my hunch (and it can only be a hunch, given the state of knowledge on this kind of question) is that this gap exists in large part because the question is out of, or contrary to, the dominant belief system where value is presumed to inhere in market transactions. A consumption perspective that is more expansive, that recognizes that individuals actually meet their needs with noncommercial or relatively nonmaterial means, makes *the nonpurchase decision a critical focus of inquiry.*

Research that retains the consumption-production dichotomy, then, must address product use, not just purchase. What is more, both post-purchase decisions and nonpurchase decisions must be included in the analysis. At least two empirical questions arise. One, under what conditions do individuals switch from purchasing a high-environmental-impact item to a relatively low-impact item, when impact is evaluated not just in production but in the use of the product itself? This question might fit existing research programs, including those of energy use,[10] household metabolism,[11] industrial ecology,[12] and market research.[13] Two, and this may well be the most difficult yet most important question, under what conditions do individuals opt for a *noncommercial or relatively nonmaterial response to meet a need*? Research does exist on intrinsic satisfaction as it relates to conservation behavior,[14] subjective well-being,[15] and work and leisure.[16] Much of this research could be extended to consumption patterns and their environmental impacts.

Conducting such research within the framework of the supply-demand, producer-consumer dichotomy is important, as noted, because production has been the dominant focus not only in economics but in the economic strands of other disciplines. It may also be the safest research tact, given the hegemony of the economistic belief system. Unpacking the demand function for environmental impacts can enrich existing research traditions and inform policymaking and do so without challenging their underlying assumptions. But for those seeking a more transformative approach to environmental problems, an approach that goes beyond "environmental improvement," the prevailing dichotomy is probably more of a hindrance than an aid. It tends to constrain the analysis to market functioning (and malfunctioning) where "the environment" is merely an externality.

A more radical approach, one that challenges this dichotomy and its propensity to relegate consumption to a black box or to the marginal status of emotion or personal values, is to treat all resource use as consuming and ask what risks are entailed in patterns of resource acquisition, processing, and distribution. This approach is more consistent with the ecological economics perspective where human economic activity is seen as an open subset of a finite and closed biophysical system.[17] Consuming is that part of human activity that "uses up" material, energy, and other valued things.

Consumption as "Using Up"

A definition of consumption that transcends the supply-demand dichotomy would start with biophysical conditions and their intersection with human behavior. That intersection, following from systems theory, has many attributes but key is ecological feedback: signals from the biophysical system that are picked up and reacted to by individuals and groups in the social system.[18] At its most basic and general level, the human behavior that intersects with the biophysical realm can be termed *material provisioning*—that is, the appropriation of material and energy for survival and reproduction.

Material Provisioning

All human activity can be divided among overlapping sets of behavior that include reproduction, defense, social interaction, identity formation, and material provisioning. Three broad categories of material provisioning are hunting/gathering, cultivating, and manufacturing. The question then is, what aspects of each category of material provisioning are best construed as consumption? Alternatively, if the activities encompassed by hunting/gathering, cultivating, and manufacturing are each construed as consumption rather than as production, what insights are gained? Answering this first requires a general definition of consumption itself.

Consumption, according to *The American Heritage Dictionary of the English Language*, is to expend or use up, to degrade or destroy. Thermodynamically, it is to increase entropy. Biologically, it is capturing usable material and energy to enhance survival and reproduction and, ultimately, to pass on one's genes. Socially, it is using up material and energy to enhance personal standing, group identity, and autonomy.

Hunting/Gathering A defining characteristic of consuming behavior, therefore, is that it is the feature of material provisioning that permanently degrades material and energy and serves some purpose to the individual or to the group. Within hunting/gathering, consumption begins when the deer is shot or the apple is picked and ends when the user has fully expended the material and energy in that deer or apple. It is important to stress that, in hunting/gathering, the consumption act is only the appropriation *of the item* and its ingestion. The one deer and the one apple is permanently degraded, not the deer herd or species and not the apple tree or species. This level of consumption is the most fundamental biologically and, indeed, is integral to all life. When some argue that consumption is "natural," they are right—at this level.

Cultivation In cultivation, one begins to see both the extension of consumption beyond single items and the external effects of consumption. Consumption in cultivation begins when a forest is cleared or a grassland plowed. It ends when the crop is harvested and the wood burned or the bread eaten. What is expended—used up or degraded—is not just the wood fiber or seed of individual plants. Rather, it is, first and foremost, the ecosystem that preceded the cultivation and, second, the cultivated plants that no longer function within integrated ecosystems. Cultivation may be conventionally thought of as production—that is, as adding value. But from the consumption angle, a perspective grounded in the biophysical, cultivation is a set of degrading behaviors—clearing, breeding, harvesting, and ingesting.[19]

Manufacturing Whereas cultivation involves rearranging extant plants and animals, manufacturing, quite literally, is making things by hand. It is applying human labor and ingenuity to create wholly new substances. Ecologically, it draws on more than the available soil and water and associated ecosystems. In particular, manufacturing extends consumption beyond the direct use of individual organisms and ecosystems to the use of energy sources and waste sinks. Converting a log into lumber and then furniture entails an expenditure of low-entropy fuel and the disposal of waste material and heat. From the production angle, this is value added. But from the consumption angle, it is using up secondary resources (energy and waste-sink capacities) to amplify and accelerate the use of primary resources (forests, grasslands, fisheries, and so on). Consuming here may entail permanent and unavoidable depletion as with fossil fuels or a temporary drawdown with the possibility of regeneration as with soil buffering.

Both cultivation and manufacturing risk permanent degeneration in ecosystem functioning. But manufacturing is generally more risky due to the separation of activity from primary resources. High technology and global finance are extreme examples where so-called "wealth creation" is far removed, some would argue completely removed, from a natural-resource base. The consumption angle directs attention to the heightened risks of such distanced material provisioning.[20]

In sum, an ecologically grounded definition of consumption takes as a starting point human material provisioning and the draw on ecosystem services. It is distinguished from definitions that begin with market behavior and ask what purchasers do in the aggregate, and from definitions that start with social stratification and ask how consumption patterns establish hierarchy or identity. The potential of such an ecological definition is to escape the confines of both limits-to-growth and economistic frameworks that tend to prescribe top-down, centralized correctives for errant (i.e., overconsuming) human behavior. An ecological approach to consumption directs attention to ecological risk and the myriad ways clever humans have of displacing the true costs of their material provisioning. The next step in conceptualizing the consumption-environment nexus is to specify what excessive or maladaptive consumption is. In particular, it is to ask how a given act of consumption (e.g., eating the apple, converting the forest, manufacturing the chair) can be interpreted or judged. I start with the broad biophysical context in which consuming behavior can be interpreted as "natural" or "background" and then consider both ecological and social definitions of degradative consumption, what I call *overconsumption* and *misconsumption.*

Excess Consumption: Three Interpretive Layers

A strictly ecological interpretation takes consumption as perfectly "natural." To survive, all organisms must consume—that is, degrade resources. This interpretation of a given consumption act is *background consumption.* It refers to the normal, biological functioning of all organisms, humans included. Every act of background consumption by an individual alters the environment, the total environmental impact being a function of aggregate consumption of the population. Individuals consume to meet a variety of needs, physical and psychological, both of which contribute to the ability of the individual to survive and reproduce.

From this limited, *a*social, *non*ethical interpretation of consumption, all consumption patterns and consequences are natural, including population explosions and crashes and irreversibilities caused by the expansion of one species at the expense of other species. If, however, the interpretation is modified to include *human concern* for population crashes, species extinctions, permanent diminution of ecosystem functioning, diminished reproductive and developmental potential of individuals, and other irreversible effects, then "problematic consumption" becomes relevant. Two interpretive layers are *overconsumption* and *misconsumption.*

Overconsumption is the level or quality of consumption that undermines a species' own life-support system and for which individuals and collectivities have choices in their consuming patterns. Overconsumption is an aggregate-level concept. With instances of overconsumption, individual behavior may be perfectly sensible, conforming either to the evolutionary dictates of fitness or to the economically productive dictates of rational decision making. Collective, social behavior may appear sensible, too, as when increased consumption is needed in an advanced industrial economy to stimulate productive capacity and compete in international markets. But eventually the collective outcome from overconsuming is catastrophe for the population or the species. From a thermodynamic and ecological perspective, this is the problem of *excessive throughput.*[21] The population or species has commanded more of the regenerative capacity of natural resources and more of the assimilative capacity of waste sinks than the relevant ecosystems can support. And it is an *ethical* problem because it inheres only in populations or species that can reflect on their collective existence. What is more, for humans it becomes a *political* problem when the trends are toward collapse, power differences influence impacts, and those impacts generate conflict.

The second interpretive layer within problematic consumption is *misconsumption*, which concerns individual behavior. The problem here is that the individual consumes in a way that undermines his or her own well-being even if there are no aggregate effects on the population or species. Put differently, the long-term effect of an individual's consumption pattern is either suboptimal or a net loss to that individual. It may or may not, however, undermine collective survival. Such consumption can occur along several dimensions.

Physiologically, humans misconsume when they eat too much in a sitting or over a lifetime or when they become addicted to a drug. The long-term burden overwhelms the immediate gratification. Psychologically, humans misconsume when, for example, they fall into the advertiser's trap of "perpetual dissatisfaction." They purchase an item that provides fleeting satisfaction, resulting in yet another purchase. Economically, humans misconsume when they overwork—that is, engage in onerous work beyond what can be compensated with additional income. With more income and less time, they attempt to compensate by using the additional income, which is to say, by consuming.

Ecologically, humans misconsume when an increment of increased resource use harms that resource or related resources and humans who depend on the resource. In the short term, if one builds a house on a steep, erosion-prone slope, the construction itself increases the likelihood of massive erosion and the destruction of one's consumption item, the house. In the longer term, if one uses leaded housepaint, one's children or grandchildren are more likely to have developmental problems.

Misconsumption, then, refers to individual resource-using acts that result in net losses for the individual. They are not "rational" or sensible in any of several senses—psychological, economic, or healthwise. And, once again, they may or may not add up to aggregate, ecological decline. The question that critically defines the consumption and environment research agenda at this, the individual level, is, what forms of individual misconsumption lead to collective overconsumption? Put differently, when is overconsumption not simply a problem of excessive throughput—that is, a problem of too many people or too much economic activity—and when is it a question of the inability of individuals to meet their needs in a given social context? When, in other words, do individuals simultaneously wreak harm on themselves *and* on the environment through their consumption patterns?

These questions are important because they point toward potential interventions that make sense at both levels and without requiring evolutionarily novel human behavior such as global citizenship[22] or authoritarian command structures.[23] These questions point toward win-win, "no-regrets" policies that simultaneously produce improved human welfare and reduced ecological risk to humans' life-support system. A critical area of research, therefore, is the intersection of misconsumption and overconsumption where individuals and society together can potentially benefit from improved consumption patterns. This may offer the greatest, and certainly the easiest, opportunities for interventions. But a second area is at least as important yet more vexing. This is consumption patterns that involve individually satisfying behavior with net benefits to the individual and, say, to that individual's kin, yet net harms to others. This is unavoidably a distributional question and, hence, a moral and political issue. Below I explore part of this moral and political dimension by considering how producers must exercise restraint and resistance when demand is overwhelming.

A Simple Application

To grasp how a critique of the production angle and a preliminary ecological conceptualization of a consumption angle may be operationalized in a research-and-action agenda, imagine that a resource is stressed, say, more timber is being harvested in a watershed than the forest ecosystem can regenerate. What is more, the primary reason is demand. Consumers want more of the timber than the ecosystem can bear and they can pay a sufficiently high price or marshal enough coercion to compel high production.[24] Imagine further that the forest users—that is, those who decide harvest rates and management techniques—are responding entirely to this demand, managing the forest and choosing harvest rates and practices that best fit that demand. As demand increases, they increase the harvest rate in the short term and, for the longer term, plant, say, fast-growing species. Their entire enterprise is production oriented and demand driven. More demand, expressed either through higher prices or increased orders or both, compels more production. The only question forest users and others ask is, how can demand be anticipated and then met? Consequently, the timber owner tries to improve harvesting methods to get more usable logs off each acre and the mill owner tries to improve milling to get more sellable board feet out of each log. Builders and retailers develop new construction methods to get more square feet of building or more pieces of furniture out of a load of lumber. And all step up production with more labor and more hours of operation. To deal with the loss of the forest, government officials and activists might require replanting or mandate a set-aside to preserve a portion of the forest. They might require buffer strips along streams to reduce siltation and beauty strips along roadways to reduce aesthetic loss.

All these production-oriented measures fall within the realm of "environmental improvement." For a given level of harvest they generate more usable product or less environmental damage. But the harvest is, indeed, *given*—that is, given by demand, by some combination of human need and desire and agents to supply (or stoke) that demand. And all via a supply of money that exists completely outside of ecological carrying capacity. Such production-oriented measures may be able to accommodate more of the demand or ameliorate the environmental effects. But when demand continues to increase and then exceed supply (in an ecological sense), the real issue regarding overharvesting is, indeed, the demand, not the supply. Better forest management practices, less wood waste, more efficient milling, lower transportation costs, rehabilitation, and set-asides will have little effect on the excessiveness of the demand.[25] Use of the forest may appear to be a production issue, but when overharvesting is the concern, it is really a consumption issue. For both analytic and behavior-change reasons, it should be investigated from the consumption angle.

Before doing so, it is important to point out that the production angle starts with a set of conditions that, in contemporary industrial society, are taken as the baseline, the starting point from which all else progresses. What is more, this baseline is unecological. If policymakers want to increase employment, the central banker stimulates demand through the money supply and interest rates. Financial signals start in the capital city, work their way through planners, designers, and builders to retailers and processors and, eventually, in this case, to the timber owner, who hires more workers and develops new technologies to cut more trees. The financial stimulus occurs as if ecological constraints are irrelevant. Indeed, the financial signal exists completely independent of signals from the ecosystems that must adjust.

Signals from elsewhere in the commodity chain operate similarly. If members of the wood-products industry want to capture market share in rot-resistant timber, say, they convince municipalities to mandate pressure-treated lumber in outdoor applications. What were once trash species become highly marketable and demand rises. Production again increases, and all as if there were no ecological constraint, as if ecosystems were mere *inputs* to the economy, not a *foundation* of the economy.

The production angle is, thus, inherently unecological. If countervailing biophysical signals happen to work their way from the forest to the timber owner to processors, distributors, and retailers (let alone to money-supply managers), they are overwhelmed by the presumption of net benefits from more production: producers produce goods, goods are good, more goods are better. Consumers benefit as revealed by their willingness to pay. (Note how the notion of consumer sovereignty is integral to the production angle.) But as many have argued, economic growth, conventionally defined and measured, can be "uneconomic," even on its own terms, let alone on ecological terms. It can lead to net harm, especially when ecosystem services, family and community integrity, and future generations are taken into account.[26]

If the production angle is inherently unecological, if it naturally overwhelms feedback that would otherwise reveal long-term net harm, then the consumption angle, if it has analytic and policy utility, ought to do just the opposite. It should direct analytic attention to what is lost, to what risks are incurred when, in this example, the harvest rate exceeds the regenerative rate of the forest ecosystem. Following the framework outlined above, I begin the consumption angle within the production-consumption, supply-demand dichotomy, then shift to material activity up and down the chain of resource-use decisions.

Within the production-consumption dichotomy, the first observation is that the production-oriented measures do not solve the problem of overharvesting. They mitigate the damage or extend the time until the forest is completely cut over. When resource-use decisions are largely governed by agents and managers and consumers highly removed from the forest itself, the problem is not inefficiencies or lack of political will or greed. The problem is not inefficient use of logs and lumber or the political difficulties of creating parks and buffer strips. Indeed, the problem is not the intricacies of meeting ever-increasing demand, satisfying customers, stockholders, and workers. Rather, the problem is the demand itself, the array of signals and incentives coming from actors highly distanced from ecological constraint.[27]

A consumption perspective therefore asks about the *nature of the demand*, what is otherwise forbidden territory in the production angle where consumers are sovereign. The consumption angle asks whether the increasing demand is simply a matter of a growing population in need, say, of shelter; whether the price paid by buyers reflect full costs, social and ecological (however measured); whether consumption is facilitated, maybe subsidized, by low-cost transportation infrastructure or easy credit;[28] whether the benefits of new products are highlighted while the drawbacks are shaded; whether retailers launched a new line of luxury furniture or builders doubled the average house size (yet again). It asks whether consumers have the option of choosing less wood-intensive means of meeting the same needs. What is more, the consumption angle raises questions of nonpurchase. Is some segment of the population forgoing purchase of the product and, if so, is it because of income, availability, information, or alternative means of meeting the same needs? In other words, is nonpurchase a meaningful option and, if so, does the increasing demand truly reflect a net social-welfare gain, the implicit assumption in the production angle?

Shifting to consumption as "using up," a comprehensive analysis would examine each decision node from initial extraction to ultimate use and disposal. Here I focus on processing and initial extraction. From the production angle, a mill owner "produces" lumber from logs. Equally logically, yet from the consumption angle, the mill owner *consumes* logs from the timber owner's forest. That is, what is used up is a log and its alternative uses. The log is irretrievably converted to one item—lumber—never to be used for veneer or paper or larger-dimension lumber.

Applying the same logic to the timber owner and the forest, what is used up, it would appear at first glance, is trees. Each tree cut is irreversibly removed, its ability to photosynthesize and provide habitat for other plants and animals, completely eliminated. But at some rate and extent of harvest, more than single trees' capacity to put on cellulose is consumed. Each tree constitutes a node in a complex system. As some trees fall to the ax (or to wind or insects), the system, being adaptive and resilient, adjusts. But as more trees fall, a threshold is passed and the integrity of the system is compromised. At this level of harvest, the system itself and all the species it entails and ecosystem services it provides are, indeed, consumed—that is, used up. Efficient production, erosion control, and preserves (unless on a system scale) do nothing to alter this ultimate effect of ever-increasing harvest. They may postpone the threshold or soften its impact, but the forest ecosystem is still consumed.

From this analytic perspective, all decision makers along the chain from extraction to end use are "consumers." As such, they and policy makers and citizens who condition their choices logically must ask what is "used up," what services are put at risk, what features of the primary resource—the forest ecosystem (not the trees)—are being eliminated. Moreover, to ask these questions is unavoidably to ask about long-term effects. If the ecosystem is degraded, will there always be another? Do the benefits accrued by each actor along the material chain accumulate so as to unambiguously override the risk or loss of the ecosystem (especially if such a judgment is made in an open, well-informed forum rather than in the market)? Are future generations likely to recognize and accept the value of the trade-off: more human capital (lumber and furniture and their associated technologies) for less natural capital?

These are questions of long-term sustainable resource use, not environmental improvement. They take high-integrity ecosystems as given, indeed necessary, not the social organization for resource use and economic expansion that happens to be hegemonic at this historical juncture. But I should stress that these questions derive logically from the consumption angle, not from the arbitrary or ideologically derived positions of those who "value nature" or who profess concern for future generations. These questions come from turning the dominant perspective on its head and asking not what goods are produced (and presuming that all such goods consumed are "good"), but what services, *what forms of social and natural capital, are consumed.* Just as the production angle presumes goods are good and more goods are better, in the consumption angle, *life-support systems are presumed fragile and critical irreversibilities possible*. "Cautious consuming" is not only prudent but rational. In addition, not only do these questions derive logically from the consumption angle, whether emphasizing the demand side or viewing all economic activity as consuming, the perspective itself is at least as logical as the production angle. In fact, although the production angle may be the most logical in an "empty-world," frontier economy, in a "full world," in an ecologically constrained economy, the consumption angle can be judged more logical. It does, after all, draw attention to ecosystem functioning as an integral part of the analysis, not as an add-on, not as an "externality," the prevailing approach in the production angle.

Finally, I should add that instances do exist where production-oriented measures are eschewed by resource users. Under some conditions timber owners, fishers, and others will deliberately limit output.[29] Such measures entail behaviors by producers that tend to be ignored, if not suppressed, when a production perspective is dominant. Two such behaviors that derive logically from the consumption angle are *restraint* and *resistance*.

If forest users have a strong interest in their own long-term economic security and if they depend on the forest, it is sensible to avoid or modulate demand-driven harvest decisions. When demand is low, these forest users would harvest little and invest little. But they would also shift to different forest uses, from timber harvesting only, say, to hunting and fishing and tourism, as well as to nonforest means of making a living.[30] Multiple use, therefore, makes sense as a survival strategy for the user, as a means of ensuring long-term economic security, not just as public policy. If demand is high, they would increase the harvest rates only to a point. Beyond that point, a point determined not by short-term economic opportunity but by a sense of ecological limits, by a risk-averse approach to complex natural systems and to the users' own economic security, they would logically *restrain* their harvests so as not to jeopardize future use and those other uses. What is more, if demand is so intrusive, so overwhelming via temptingly high prices or coercion (force or law), then a second behavior would logically be *resistance*. Users would develop organizational, legal, or, if necessary, coercive means of their own to resist the intrusion, limit their harvest, and thus maintain the resource over the long term.[31]

Again, the problem here may appear to be one of production—that is, harvest rates. But it is really one of *consumption* vis-à-vis production. The foci of conventional production analysis—issues of investment, management, extraction, pricing, processing, distributing—tend not to ask questions about restraint and resistance among producers. Quite the contrary, the productive enterprise is precisely one of opening markets, lowering prices, gaining efficiencies, and capturing market share—in short, increasing production. It is a process that sees the *addition* of value, not the *subtraction* of value, not the *risks* to multiple uses or to residents' economic security or to the long-term viability of the supporting ecosystem.

In sum, this simple application suggests that a consumption angle on resource-use problems—especially problems of ecological overuse—compels examination of decisions among extractors and processors that tend not to get asked from a conventional production perspective. Among these decisions are those associated with the general behaviors of restraint, that is, self-limiting behavior, and of resistance to destructive intrusions. Comparing cases where restraint and resistance are prominent with those in which they are not and applying indices of sustainable practice would be a useful research direction.

Conclusion

By making consumption more visible analytically, certain activities become more prominent. From a production angle, the simple-living movement, home power, and local currencies (part III) are trivial instances of protest; they are of little political or economic consequence. From a consumption angle, however, they are concrete expressions of concern and resistance. They represent a sense that too much of what is important in day-to-day life is lost through the lens of ever-more production meeting the (presumably) insatiable desires of people as consumers. These cases not only give meaning to consumption, but they give meaning to economic activity as being more than that which ascribes value only to what is produced and sold in the open market and that assigns people the role of consumer, not producer and certainly not citizen. If simple living, home power, and local currencies are trivial by conventional (read, production) measures, they are not trivial representations of the widespread discontent with consumerist society.

In short, the consumption angle is a means of "rethinking how humans relate to nature." It is a way to, in effect, wipe the slate clean with respect to how analysts, policymakers, and citizens understand social organization for resource use. It puts aside, or goes back to the origins of, the neoclassical economic model and asks what model would have been most useful *given* ecological constraint, *given* the lack of unending frontiers and infinite waste sinks, and *given* the inability to find a technical substitute for everything from petroleum to the ozone layer. The consumption angle not only allows for consideration of "full-world," ecologically constrained conditions, but places ecosystem functioning up front and central. It does so by generating questions that ask what is consumed, what is put at risk, what is lost. And it does so without restricting the questions to consumer products or even industrial inputs but by going all the way back up the decision chain to organisms and ecosystems and biogeochemical processes. It also does so by drawing attention to behaviors and movements that otherwise tend to escape those who hold the production angle sacrosanct: restraint and resistance with respect to ever-increasing demand, simple living, home power, and local currencies with respect to lifestyle and economic life. Finally, the consumption angle lends itself to explicit assignment of responsibility for excess throughput. This stands in marked contrast to the production angle, where actors routinely escape responsibility via distanced commerce and the black box of consumer sovereignty.

3

Individualization: Plant a Tree, Buy a Bike, Save the World?

Michael Maniates

"But now," says the Once-ler, "now that you're here, the word of the Lorax seems perfectly clear. UNLESS someone like you cares a whole awful lot, nothing is going to get better. It's not. SO . . . catch!" calls the Once-ler. He lets something fall. "It's a Truffula seed. It's the last one of all! You're in charge of the last of the Truffula seeds. And Truffula Trees are what everyone needs. Plant a new Truffula. Treat it with care. Give it clean water. And feed it fresh air. Grow a forest. Protect it from axes that hack. Then the Lorax and all of his friends may come back."
—Dr. Seuss[1]

Most people are eagerly groping for some medium, some way in which they can bridge the gap between their morals and their practices.
—Saul Alinsky[2]

One of the most successful modern-day children's stories is *The Lorax*, Dr. Seuss's tale of a shortsighted and voracious industrialist who clear-cuts vast tracks of Truffula trees to produce "Thneeds" for unquenchable consumer markets. The Lorax, who "speaks for the trees" and the many animals who make the Truffula forest their home, politely but persistently challenges the industrialist, a Mr. Once-ler, by pointing out again and again the terrible toll his business practices are taking on the natural landscape. The Once-ler remains largely deaf to the Lorax's protestations. "I'm just meeting consumer demand," says the Once-ler; "if I didn't, someone else would." When, finally, the last Truffula tree is cut and the landscape is reduced to rubble, the Once-ler—now out of business and apparently penniless—realizes the error of his ways. Years later, holed up in the ruins of his factory amidst a desolate landscape, he recounts his foolishness to a passing boy and charges him with replanting the forest.

The Lorax is fabulously popular. Most of the college students with whom I work—and not just the ones who consider themselves environmentalists—know it well and speak of it fondly. My children read it in school. The 30-minute animated version of the book often finds its way onto television. The tale has become a beloved organizing touchstone for environmentalists. In years past, for example, the EcoHouse on my campus has aired it as part of its Earth Day observations, as did the local television station. A casual search through the standard library databases reveals over 80 essays or articles in the past decade that bear on or draw from the book. A more determined search of popular newspapers and magazines would undoubtedly reveal additional examples of shared affection for the story.

All this for a tale that is, well, both dismal and depressing. The Once-ler is a stereotypical rapacious businessman. He succeeds in enriching himself by laying ruin to the landscape. The Lorax fails miserably in his efforts to challenge the interlocking processes of industrial capitalism and consumerism that turn his Eden into a wasteland. The animals of the story are forced to flee to uncertain futures. At the end of the day the Lorax's only satisfaction is the privilege of being able to say "I told you so," but this—and the Once-ler's slide into poverty—has to be small consolation. The conclusion sees a small boy with no evident training in forestry or community organizing entrusted with the last seed of a critical species. He's told to "plant a new Truffula. Treat it with care. Give it clean water. And feed it fresh air. Grow a forest. Protect it from axes that hack. Then the Lorax and all of his friends may come back." His chances of success are by no means high.

So why the amazing popularity of *The Lorax*? Why do so many deem it "the environmental book for children"—and, seemingly, for grown-ups too—"by which all others must be judged?"[3] One reason is its overarching message of environmental stewardship and faith in the restorative powers of the young. The book recounts a foolish tragedy that can be reversed only by a new and, one hopes, more enlightened generation. Surely another reason is the comfortable way the book (easily trivialized by adults as *children's* literature) permits us to look squarely at a set of profoundly uncomfortable dynamics we know to be operating but find difficult to confront: consumerism, the concentration of economic power, the mindless degradation of the environment, the seeming inability of science (represented by the fact-spouting Lorax himself) and objective fact to slow the damage. The systematic undermining of environmental systems fundamental to human well-being is scary stuff, though no more so than one's own sense of personal impotence in the face of such destruction. Seuss's clever rhyming schemes and engaging illustrations, wrapped around the twentieth-century tale of economic expansion and environmental degradation, provide safe passage through a topic we know is out there but would rather avoid.

There is another reason, though, why the book is so loved. By ending with the charge to plant a tree, *The Lorax* echoes and amplifies an increasingly dominant, largely American response to the contemporary environmental crisis. This response half-consciously understands environmental degradation as the product of individual shortcomings (the Once-ler's greed, for example), best countered by action that is staunchly individual and typically consumer based (buy a tree and plant it!). It embraces the notion that knotty issues of consumption, consumerism, power, and responsibility can be resolved neatly and cleanly through enlightened, uncoordinated consumer choice. Education is a critical ingredient in this view—smart consumers will make choices, it is thought, with the larger public good in mind. Accordingly, this dominant response emphasizes (like the Lorax himself) the need to speak politely, and individually, armed only with facts.

For the lack of a better term, call this response the *individualization of responsibility*. When responsibility for environmental problems is individualized, there is little room to ponder institutions, the nature and exercise of political power, or ways of collectively changing the distribution of power and influence in society—to, in other words, "think institutionally," as UC Berkeley sociologist Robert Bellah says.[4] Instead, the serious work of confronting the threatening socioenvironmental processes that *The Lorax* so ably illuminates falls to individuals, acting alone, usually as consumers. We are individualizing responsibility when we agonize over the "paper-or-plastic" choice at the checkout counter, knowing somehow that neither is right given larger institutions and social structures. We think aloud with the neighbor over the back fence about whether we should buy the new Honda or Toyota hybrid-engine automobile now or wait a few years until they work the kinks out. What we really wish for, though, is clean, efficient, and effective public transportation of the sort we read about in science fiction novels when we were young—but we cannot vote for it with our consumer dollars since, for reasons rooted in power and politics, it is not for sale. So we ponder the "energy stickers" on the ultraefficient appliances at Sears, we diligently compost our kitchen waste, we try to ignore the high initial cost and buy a few compact-fluorescent lightbulbs. We read spirited reports in the *New York Times Magazine* on the pros and cons of recycling while sipping our coffee,[5] carefully study the merits of this and that environmental group so as to properly decide on the destination of our small annual donation, and meticulously sort our recyclables. And now an increasing number of us are confronted by opportunistic green-power providers who urge us to "save the planet" by buying their "green electricity"—while doing little to actually increase the quantity of electricity generated from renewable resources.[6]

The Lorax is not why the individualization of responsibility dominates the contours of contemporary American environmentalism. Several forces, described later in this chapter, are to blame. They include the historical baggage of mainstream environmentalism, the core tenets of liberalism, the dynamic ability of capitalism to commodify dissent, and the relatively recent rise of *global* environmental threats to human prosperity. Seuss's book simply has been swept up and adopted by these forces. Were he alive, Seuss would probably be surprised by the near deification of his little book. And his central character, a Lorax who politely sought to hold a corporate CEO accountable, surely would be appalled that his story is being used to justify individual acts of planting trees as the primary response to the threat of global climate change.[7]

Mark Dowie, a journalist and sometimes historian of the American environmental movement, writes about our "environmental imagination," by which he means our collective ability to imagine and pursue a variety of productive responses (from individual action to community organization to whole-scale institutional change) to the environmental problems before us.[8] My claim in this chapter is that an accelerating individualization of responsibility in the United States is narrowing, in dangerous ways, our "environmental imagination" and undermining our capacity to react effectively to environmental threats to human well-being. Those troubled by overconsumption, consumerism, and commodification should not and cannot ignore this narrowing. Confronting the consumption problem demands, after all, the sort of institutional thinking that the individualization of responsibility patently undermines. It calls too for individuals to understand themselves as citizens in a participatory democracy first, working together to change broader policy and larger social institutions, and as consumers second. By contrast, the individualization of responsibility, because it characterizes environmental problems as the consequence of destructive consumer choice, asks that individuals imagine themselves as consumers first and citizens second. Grappling with the consumption problem, moreover, means engaging in conversation both broad and deep about consumerism and frugality and ways of fostering the capacity for restraint. But when responsibility for environmental ills is individualized, space for such conversation becomes constricted. The individually responsible consumer is encouraged to purchase a vast array of "green" or "ecofriendly" products on the premise that the more such products are purchased and consumed, the healthier the planet's ecological processes will become. "Living lightly on the planet" and "reducing your environmental impact" becomes, paradoxically, a consumer-product growth industry.[9]

Skeptics may reasonably question if the individualization of responsibility is so omnipresent as to warrant such concern. As the next section of this chapter shows, it is: the depoliticization of environmental degradation is in full swing across a variety of fronts and shows little sign of abating. The chapter continues with a review of the forces driving this individualization; in particular, it implicates the rise of global environmental problems and the construction of an individualized politics around them. How might these forces be countered? How can the politics of individualization be transcended? How might our environmental imagination be expanded? I wrestle with these questions in the final section of this chapter by focusing on the IPAT formula—a dominant conceptual lens within the field of environmental policy and politics, which argues that environmental impact = population × affluence × technology.

A Dangerous Narrowing?

A few years back Peter Montague, editor of the Internet-distributed *Rachel's Environmental and Health Weekly*, took the Environmental Defense Fund (EDF)[10] to task for its annual calendar, which this powerful and effective organization widely distributes to its more than 300,000 members and many nonmembers too. What drew Montague's ire was the final page of EDF's 1996 calendar, which details a ten-point program to "save the Earth" (EDF's phrase):

1. Visit and help support our national parks
2. Recycle newspapers, glass, plastic, and aluminum
3. Conserve energy and use energy-efficient lighting
4. Keep tires properly inflated to improve gas mileage and extend tire life
5. Plant trees
6. Organize a Christmas-tree recycling program in your community
7. Find an alternative to chemical pesticides for your lawn
8. Purchase only brands of tuna marked "dolphin-safe"
9. Organize a community group to clean up a local stream, highway, park, or beach
10. Become a member of EDF.

Montague's reaction was terse and pointed:

> What I notice here is the complete absence of any ideas commensurate with the size and nature of the problems faced by the world's environment. I'm not against recycling Christmas trees—if you MUST have one—but who can believe that recycling Christmas trees—or supporting EDF as it works overtime to amend and re-amend the Clean Air Act—is part of any serious effort to "save the Earth?" I am forced to conclude once again that the mainstream environmental movement in the U.S. has run out of ideas and has no worthy vision.[11]

Shortly after reading Montague's disturbing and, for me, surprising rejection of ten sensible measures to protect the environment (many of which I myself practice), I walked into an introductory course on environmental problems that I often team-teach with colleagues in the environmental science department. The course, like many taught at the undergraduate level, strives to integrate the natural and social sciences, challenging students to consider not only the physical cause-and-effect relationships that manifest themselves as environmental degradation, but also to think critically about the struggles for power and influence that underlie most environmental problems. That day, near the end of a very productive semester, my colleague divided the class of about 45 students into smaller "issue groups" (energy, water, agriculture, and so on) and asked each group to develop a rank-order list of "responses" or "solutions" to environmental threats specific to that issue. He then brought the class back together, had each group report in, and tabulated their varied "solutions." From this group of 45, the fourth most recommended solution to mounting environmental degradation was to ride a bike rather than drive a car. Number 3 on the list was to recycle. The second most preferred action was "plant a tree," and the top response was, again, "plant a tree" (the mechanics of tabulating student preference across the issue groups permitted a singularly strong preference to occupy two slots).

When we asked our students—who were among the brightest and best prepared of the many we had worked with over the years—why, after 13 weeks of intensive study of environmental problems, they were so reluctant to consider as "solutions" broader changes in policy and institutions, they shrugged. Sure, we remember studying these kinds of approaches in class, they said, but such measures were, well, fuzzy, mysterious, messy, and "idealistic."[12]

The end of the day came soon enough and I began my walk home, a pleasant half-mile stroll. The next day was "garbage day" and my neighbors were dutifully placing their recyclables, carefully washed and sorted, on the curbside. I waved hello and we chatted about the weird weather and all that talk about "global climate change." I made my way through my own front door to find my daughters camped out in front of the television, absorbed in a rare predinner video. The evening's selection, a gift from a doting aunt, was from the popular "Wee Sing" series. Entitled *Under the Sea*, the production chronicles the adventures of a small boy and his grandmother as they interact with a variety of sea creatures on the ocean floor. Dramatic tension is provided by the mysterious sickness of Ottie, a baby otter meant to tug at the heartstrings of all but the most jaded viewers. The story's climax comes when the entire cast discovers a large pile of garbage on the coral reef, a favorite playground of Ottie, and then engages in a group clean-up of the site while singing a song extolling the virtues of recycling and condemning the lazy, shortsighted tendencies of "those humans."[13] My daughters were enthralled by the video: its message about the need to take personal responsibility for the environment resonated clearly with all that they were then learning about the environment, in preschool and kindergarten respectively.

As I reflect now on these past events, I wonder if they are getting the wrong message, ubiquitous as it has become. Consider the following:

• Despite repeated and often highly public criticism of the "10 simple things to save the planet" focus of its calendars, the EDF pushes forward undaunted. Its 2000–2001 calendar again offers "10 tips to help our planet," which again revolve around individual consumer action: recycle, use energy-efficient lighting, avoid the purchase of products that come from endangered species.

• A colleague recently received a small box in the mail with an attached sticker that read "Environmental Solutions—Not Just Problems." Inside was a peat pot filled with soil in which was growing a pine-tree seedling, together with a piece of paper about 2 inches square that said "Rather than sensationalize the problems in our world, *Environmental Science* provides your students with the tools to develop their own opinions and focus on solutions. Keeping with this theme, you and your students can decide where to best plant the enclosed seedling and watch it grow throughout the year." The package was a promotion for one of the most widely used undergraduate environmental-science textbooks.

• These days, my students argue that the best way to reverse environmental degradation is to educate children now in school. When pressed, they explain that only a sea change in the choices individual consumers are making will staunch the ecological bleeding we are now facing—and it is too late to make much of dent in the consumer preferences of young adults like themselves.

• The biggest environmental issue to hit our community in the last decade has been the threatened demise, for lack of funding, of "drop-off" centers for recycled products. Primary-school students have distributed their artwork around the theme of "Save the Planet—Recycle" (presumably with their teachers' encouragement). Letters to the editor speak gravely of myriad assaults on the planet and the importance of "buying green" and recycling if we are to stop the destruction. And this is not a phenomenon limited to small-town America; a friend visiting Harvard University recently forwarded a flyer, posted over one of the student copy machines, with a line drawing of planet earth and the slogan "Recycle and Do Your Part to Save the Planet." Recycling is a prime example of the individualization of responsibility.

• Despite the criticism by some academics of the mega-hit *50 Simple Things You Can Do to Save the Earth* (a small book outlining 50 "easy" lifestyle changes in service of sustainability), publications sounding the same theme proliferate.[14]

• My daughters (now in first and fourth grade), like so many children their age, remain alert to environmental issues. A favorite book of the younger one is *The Berenstein Bears Don't Pollute*, which speaks to the need to recycle and consume environmentally friendly products. The older one has been drawn to computer games, books, and movies (e.g., *Free Willy*) that pin the blame for degraded habitat, the loss of biodiversity, and the spread of environmental toxins on Once-ler-like failings (shortsightedness, greed, materialism) of humans in general.

In our struggle to bridge the gap between our morals and our practices, we stay busy—but busy doing what we are most familiar and comfortable with: consuming our way (we hope) to a better America and a better world. When confronted by environmental ills—ills many confess to caring deeply about—Americans seem capable of understanding themselves almost solely as consumers who must buy "environmentally sound" products (and then recycle them), rather than as citizens who might come together and develop political clout sufficient to alter institutional arrangements that drive a pervasive consumerism.[15] The relentless ability of contemporary capitalism to commodify dissent and sell it back to dissenters is surely one explanation for the elevation of consumer over citizen.[16] But another factor, no doubt, is the growing suspicion of and unfamiliarity with processes of citizen-based political action among masses of North Americans. The interplay of state and market after World War II has whittled the obligations of citizenship down to the singular and highly individualized act of voting in important elections. The increasing fragmentation and mobility of everyday life undermines our sense of neighborhood and community, separating us from the small arenas in which we might practice and refine our abilities as citizens. We build shopping malls but let community playgrounds deteriorate and migrate to sales but ignore school-board meetings. Modern-day advances in entertainment and communication increasingly find us sitting alone in front of a screen, making it all seem fine. We do our political bit in the election booth, then get back to "normal."[17]

Given our deepening alienation from traditional understandings of active citizenship, together with the growing allure of consumption-as-social-action, it is little wonder that at a time when our capacity to imagine an array of ways to build a just and ecologically resilient future must expand, it is in fact narrowing. At a moment when we should be vigorously exploring multiple paths to sustainability, we are obsessing over the cobblestones of but one path. This collective obsessing over an array of "green consumption" choices and opportunities to recycle is noisy and vigorous, and thus comes to resemble the foundations of meaningful social action. But it is not, not in any real and lasting way that might alter institutional arrangements and make possible radically new ways of living that seem required.

Environmentalism and the Flight from Politics

The individualization of responsibility for environmental ills and the piecemeal, counterproductive actions it produces have not gone unnoticed by analysts of contemporary environmental politics. Over a decade ago, for example, social ecologist Murray Bookchin vigorously argued that:

> It is inaccurate and unfair to coerce people into believing that they are *personally* responsible for present-day ecological disasters because they consume too much or proliferate too readily. This privatization of the environmental crisis, like the New Age cults that focus on personal problems rather than on social dislocations, has reduced many environmental movements to utter ineffectiveness and threatens to diminish their credibility with the public. If "simple living" and militant recycling are the main solutions to the environmental crisis, the crisis will certainly continue and intensify.[18]

More recently, Paul Hawken, the cofounder of the environmentally conscious Smith and Hawken garden-supply company and widely published analyst of "eco-commerce," confessed that

> it [is] clear to me ... that there [is] no way to "there" from here, that all companies are essentially proscribed from becoming ecologically sound, and that awards to institutions that had ventured to the environmental margins only underlined the fact that commerce and sustainability were antithetical by design, not by intention. Management is being told that if it wakes up and genuflects, pronouncing its amendes honorable, substituting paper for polystyrene, we will be on the path to an environmentally sound world. *Nothing could be farther from the truth. The problem isn't with half measures, but the illusion they foster that subtle course corrections can guide us to the good life that will include a "conserved nature" and cozy shopping malls.*[19]

Bookchin and Hawken are reacting, in large measure, to a 1980s transformation in how Americans understand and attack environmental problems. The 1980s was a decade in which reenergized, politically conservative forces in the United States promoted the rhetoric of returning power and responsibility to the individual, while simultaneously curtailing the role of government in an economy that was increasingly characterized as innately self-regulating and efficient. Within this context, responsibility for creating and fixing environmental problems was radically reassigned, from government, corporations, and the environmentally shortsighted policies they were thought to have together fostered, to individual consumers and their decisions in the marketplace.

This shift was altogether consistent with then-President Reagan's doctrine of personal responsibility, corporate initiative, and limited government.[20] The new conventional wisdom rejected environmental regulations that would coerce the powerful to behave responsibly toward the environment and slap them hard if they did not. Instead, an alternative environmental politics of "win-win," zero-coercion scenarios flourished, in which a technological innovation here or an innocuous change in policy there would, it was argued, produce real reductions in environmental degradation *and* higher corporate profits. This "win-win" approach continues to dominate American environmental politics, and a vast range of environmentally friendly, economically attractive technologies, from compact fluorescent lights to ultra fuel-efficient automobiles, are showcased as political-economic means toward a conflict-free transition to a future that works. These kinds of technologies make environmental sense, to be sure, and they typically make economic sense as well, once one accounts for the full range of costs and benefits involved. However, they often fail to make "political sense," insofar as their wide diffusion would drive a redistribution of political or economic power.

As cleaner and leaner (i.e., more efficient) technologies surfaced in the 1980s as *the* solution to pressing environmental ills, responsibility for the environmental crisis necessarily became increasingly individualized. The new technologies, it was thought, would take root and flourish only if consumers purchased them directly or sought out products produced by them. A theory of social change that embraced the image of consumers voting with their pocketbook soon took root. Almost overnight, the burden for fundamental change in American patterns of consumption and production shifted from government (which was to be trimmed) and corporations (which were cast as victims of government meddling and willing servants to consumer sovereignty—see chapter 14), onto the backs of individual consumers.

Scholars of environmentalism, however, caution us against too enthusiastically fixing blame for the individualization of responsibility on the Reagan years. Tendencies toward individualization run deep in

American environmentalism; Ronald Reagan merely was adept at tapping into them. Some analysts, for instance, note that mainstream environmentalism has technocratic, managerial roots and thus has always been a polite movement more interested in fine-tuning industrial society than in challenging its core tenets.[21] Environmentalism's essential brand of social change—that which can be had by tinkering at the margins and not hurting anyone's feelings—makes it a movement that tends naturally toward easy, personalized "solutions."

Others pin the blame for the individualization of responsibility on the bureaucratic calcification of mainstream, "inside-the-Beltway" environmental groups.[22] Buffeted by backlash in the 1980s, laboring hard to fend off challenges to existing environmental regulations in the 1990s, and unsure about how to react to widespread voter apathy in the 2000s, mainstream environmental groups in the United States have consolidated and "hunkered down." To survive as nonprofit organizations without government financing (as is common in other countries), these U.S. NGOs have had to avoid any costly confrontation with real power while simultaneously appearing to the public as if they are vigorously attacking environmental ills. The result: 10 easy steps to save the planet of the sort proffered each year by the Environmental Defense Fund.

Other scholars draw attention to the classical liberal underpinnings of environmentalism that bias environmentalism toward timid calls for personal responsibility and green consumerism.[23] As Paul Wapner, a professor at American University, notes:

> Liberal environmentalism is *so* compatible with contemporary material and cultural currents that it implicitly supports the very things that it should be criticizing. Its technocratic, scientistic, and even economistic character gives credence to a society that measures the quality of life fundamentally in terms of economic growth, control over nature, and the maximization of sheer efficiency in everything we do. By working to show that environmental protection need not compromise these maxims, liberal environmentalism fails to raise deeper issues that more fundamentally engage the dynamics of environmental degradation.[24]

And yet mainstream environmentalism has not always advanced an individualized consumerist strategy for redressing environmental ills. Even during the turn of the last century, a time of zealous rediscovery of the wonders of efficiency and scientific management, "the dynamics of conservation," observes famed environmental historian Samuel P. Hays, "with its tension between the centralizing tendencies of system and expertise on the one hand and the decentralization of localism on the other," fueled healthy debate over the causes of and cures for environmental ills.[25] Throughout the twentieth century, in fact, mainstream environmentalism has demonstrated an ability to foster multiple and simultaneous interpretations of where we are and where we should be heading.

But that ability has, today, clearly become impaired. Although public support for things environmental has never been greater, it is so because the public increasingly understands environmentalism as an individual, rational, cleanly apolitical process that can deliver a future that works without raising voices or mobilizing constituencies. As individual consumers and recyclers we are supplied with ample and easy means of "doing our bit"—green consumerism and militant recycling becomes the order of the day. The result, though, is often dissonant and sometimes bizarre: consumers wearing "save the earth" T-shirts, for example, speak passionately against recent rises in gasoline prices when approached by television news crews; shoppers drive all over town in their gasoline-guzzling SUVs in search of organic lettuce or shade-grown coffee; and diligent recyclers expend far more fossil-fuel energy on the hot water spent to meticulously clean a tin can than is saved by its recycling.

Despite these jarring contradictions, the technocratic, sanitary, and individualized framing of environmentalism prevails, largely because it is continually reinforced. Consider, for example, recent millennial issues[26] of *Time* and *Newsweek* that look to life in the future. They paint a picture of smart appliances, computer-guided automobiles, clean neighborhoods, ecofriendly energy systems, and happy citizens. How do we get to this future? Not through bold political leadership or citizen-based debate within enabling democratic institutions—but rather via consumer choice: informed, decentralized, apolitical, individualized. Corporations will build a better mousetrap, consumers will buy it, and society will be transformed for the better. A struggle-free ecorevolution awaits, one made possible by the combination of technological innovation and consumer choice with a conscience.

The "better mousetrap theory of social change" so prevalent in these popular news magazines was coined by Langdon Winner, a political science professor and expert on technological politics, who first introduced the term in an essay on the demise of the appropriate technology movement of the 1970s:

A person would build a solar house or put up a windmill, not only because he or she found it personally agreeable, but because the thing was to serve as a beacon to the world, a demonstration model to inspire emulation. If enough folks built for renewable energy, so it was assumed, there would be no need for the nation to construct a system of nuclear power plants. People would, in effect, vote on the shape of the future through their consumer/builder choices. This notion of social change provided the underlying rationale for the amazing emphasis on do-it-yourself manuals, catalogues, demonstration sites, information sharing, and "networking," that characterized appropriate technology during its heyday. Once people discovered what was available to them, they would send away for the blueprints and build the better mousetrap themselves. As successful grass-roots efforts spread, those involved in similar projects were expected to stay in touch with each other and begin forming little communities, slowly reshaping society through a growing aggregation of small-scale social and technical transformations. Radical social change would catch on like disposable diapers, Cuisinarts, or some other popular consumer item.

Like the militant recyclers and dead-serious green consumers of today, appropriate technologists of the 1970s were the standard-bearers for the individualization of responsibility. The difference between then and now is that appropriate technology lurked at the fringes of a 1970s American environmental politics more worried about corporate accountability than consumer choice. Today, green consumption, recycling, and Cuisinart–social change occupy the heart of U.S. ecopolitics. Both then and now, such individualization is alarming, for as Winner notes:

> The inadequacies of such ideas are obvious. Appropriate technologists were unwilling to face squarely the facts of organized social and political power. Fascinated by dreams of a spontaneous, grass-roots revolution, they avoided any deep-seeking analysis of the institutions that control the direction of technological and economic development. In this happy self-confidence they did not bother to devise strategies that might have helped them overcome obvious sources of resistance. The same judgment that Marx and Engels passed on the utopians of the nineteenth century apply just as well to the appropriate technologists of the 1970s: they were lovely visionaries, naive about the forces that confronted them.[27]

Though the inadequacies of these ideas is clear to Winner, they remain obscure to the millions of American environmentalists who would plant a tree, ride a bike, or recycle a jar in the hope of saving the world. The newfound public awareness of global environmental problems may be largely to blame. Shocking images of a "hole" in the ozone layer in the late 1980s, ubiquitous videos on rainforest destruction, media coverage of global climate change and the warming of the poles: all this and more have brought the public to a new state of awareness and concern about the "health of the planet." What, though, is the public to do with this concern? Academic discussion and debate about global environmental threats focuses on distant international negotiations, complicated science fraught with uncertainty that seems to bedevil even the scientists, and nasty global politics. This in no place for the "normal" citizen. Environmental groups often encourage people to act, but recommended action on global environmental ills is limited to making a donation, writing a letter, or—yes—buying an environmentally friendly product. The message on all fronts seems to be "Act ... but don't get in the way." Confronted by a set of global problems that clearly matter and seeing no clear way to attack them, it is easy to imagine the lay public gravitating to individualistic, consumer-oriented measures. And it is easy to understand how environmental groups would promote such measures; these measures do, after all, meet the public's need for some way to feel as if it is making a difference, and they sell.

Ironically, those laboring to highlight global environmental ills, in the hope that an aroused public would organize and embark on collective, political action, aided and abetted this process of individualization. They paved the way for the likes of Rainforest Crunch ice cream ("buy it and a portion of the proceeds will go to save the rainforests") because they were insufficiently attentive to a fundamental social arithmetic: heightened concern about any social ill, erupting at a time of erosion of public confidence in political institutions and citizen capacities to effect change, will prompt masses of people to act, but in that one arena of their lives where they command the most power and feel the most competent—the sphere of consumption.

Of course, the public has had some help working through this particular arithmetic. A privatization and individualization of responsibility for environmental problems shifts blame from state elites and powerful producer groups to more amorphous culprits like "human nature" or "all of us." State elites and the core corporations on which they depend to drive economic growth stand to benefit from spreading the blame and cranking the rotary of consumption.[28] And crank they will. One example of this dynamic, though not one rooted per se in global ecology, is found in a reading of the history of efforts in the United States in the 1970s to implement a nationwide system of beverage- and food-container reuse, a policy that would have assigned the responsibility for resolving the "solid waste crisis" to the container industry. The container industry spent tens of millions of dollars to defeat key "bottle bill" referendums in California and Colorado, and then vigorously advanced recycling—*not reuse*—as a more practical alternative. Recycling, by stressing the individual's act of disposal, not the producer's acts of packaging, processing, and distributing, fixes primary responsibility on individuals and local governments. It gives life to a "Wee Sing" diagnosis of environmental ills (see note 13) that places human laziness and ignorance center stage. The bottling industry was successful in holding out its "solution" as the most practical and realistic, and the state went along.[29]

The same dynamic now permeates mainstream discussions of global environmental ills. Pratap Chatterjee and Matthias Finger, seasoned observers of global environmental politics, highlight the rise of a "New Age Environmentalism" that fixes responsibility on all of us equally and, in the process, cloaks important dimensions of power and culpability.[30] They point, for example, to international meetings like the 1992 Earth Summit that cultivate a power-obscuring language of "all of us needing to work to together to solve global problems." In the same vein, academics like Gustavo Esteva and Suri Prakash lament how the slogan "think globally, act locally" has been shaped by global environmentalism to support a consumer-driven, privatized response to transboundary environmental ills. In practice, thinking globally and acting locally means feeling bad and guilty about far-off and megaenvironmental destruction, and then traveling down to the corner store to find a "green" product whose purchase will somehow empower somebody, somewhere, to do good.[31] Mainstream conversations about global sustainability advance the "international conference" as the most meaningful venue for global environmental problem solving. It is here that those interests best able to organize at the international level—states and transnational corporations—hold the advantage in the battle to shape the conversation of sustainability and craft the rules of the game. And it is precisely these actors who benefit by moving mass publics toward private, individual, well-intentioned consumer choice as *the* vehicle for achieving "sustainability."

It is more than coincidental that as our collective perception of environmental problems has become more global, our prevailing way of framing environmental problem solving has become more individualized. In the end, individualizing responsibility does not work—you cannot plant a tree to save the world—and as citizens and consumers slowly come to discover this fact their cynicism about social change will only grow: "You mean after 15 years of washing out these crummy jars and recycling them, environmental problems are still getting worse—geesh, what's the use?" Individualization, by implying that any action beyond the private and the consumptive is irrelevant, insulates people from the empowering experiences and political lessons of collective struggle for social change and reinforces corrosive myths about the difficulties of public life.[32] By legitimating notions of consumer sovereignty and a self-balancing and autonomous market (with a well-informed "hidden hand"), it also diverts attention from political arenas that matter. In this way, individualization is both a symptom and a source of waning citizen capacities to participate meaningfully in processes of social change. If consumption, in all its complexity, is to be confronted, the forces that systematically individualize responsibility for environmental degradation must be challenged.

IPAT, and Beyond

But how? One approach would focus on undermining the dominant frameworks of thinking and talking that make the individualization of responsibility appear so natural and "commonsense." Among other things, this means taking on "IPAT."

At first glance it would seem that advocates of a consumption angle on environmental degradation should naturally embrace IPAT (impact = population × affluence × technology). The "formula" argues, after all, that one cannot make sense of, much less tackle, environmental problems unless one takes into account all three of the proximate causes of environmental degradation. Population growth, resource-intensive and highly polluting technologies, and affluence (that is, levels of consumption) together conspire to undermine critical ecological processes on which human well-being depends. Focusing on one or two of these three factors, IPAT tells us, will ultimately disappoint.

IPAT is a powerful conceptual framework, and those who would argue the importance of including consumption in the environmental-degradation equation have not been reluctant to invoke it. They note, correctly so, that the "A" in IPAT has for too long been neglected in environmental debates and policy action.[33] However, although IPAT provides intellectual justification for positioning consumption center stage, it also comes with an underlying set of assumptions—assumptions that reinforce an ineffectual Loraxian flight from politics.

A closer look at IPAT shows that the formula distributes widely all culpability for the environmental crisis (akin to the earlier-mentioned "New Age Environmentalism"). Population size, consumption levels, and technology choice are all to blame. Responsibility for environmental degradation nicely splits, moreover, between the so-called developed and developing world: if only the developing world could get its population under control and the developed world could tame its overconsumption and each could adopt green technologies, all would be well. Such a formulation is, on its face, eminently reasonable, which explains why IPAT stands as such a tempting platform from which advocates of a consumption perspective might press their case.

In practice, however, IPAT amplifies and privileges an "everything is connected to everything else" biophysical, ecosystem-management understanding of environmental problems, one that obscures the exercise of power while systematically disempowering citizen actors. When everything is connected to everything else, knowing how or when or even why to intervene becomes difficult; such "system complexity" seems to overwhelm any possibility of planned, coordinated, effective intervention.[34] Additionally, there is little room in IPAT's calculus for questions of agency, institutions, political power, or collective action. Donella Meadows, a systems analyst and coauthor of *The Limits to Growth*, the 1972 study that drew the world's attention to the social and environmental threats posed by exponential growth, had long advocated IPAT. But the more her work incorporated the human dimension, including issues of domination and distribution, the more she questioned the formulation. After a 1995 conference on global environmental policy, she had a revelation:

> I didn't realize how politically correct [IPAT] had become, until a few months ago when I watched a panel of five women challenge it and enrage an auditorium full of environmentalists, including me. IPAT is a bloodless, misleading, cop-out explanation for the world's ills, they said. It points the finger of blame at all the wrong places. It leads one to hold poor women responsible for population

growth without asking who is putting what pressures on those women to cause them to have so many babies. It lays a guilt trip on Western consumers, while ignoring the forces that whip up their desire for ever more consumption. It implies that the people of the East, who were oppressed by totalitarian leaders for generations, now somehow have to clean up those leaders' messes.

And then, in ways that echo Langdon Winner's assessment of the better-mousetrap theory of social change, Meadows concludes that

IPAT is just what one would expect from physical scientists, said one of its critics. It counts what's countable. It makes rational sense. But it ignores the manipulation, the oppression, the profits. It ignores a factor that [natural] scientists have a hard time quantifying and therefore don't like to talk about: economic and political power. IPAT may be physically indisputable. But it is politically naive.[35]

One need go no further that the 1998 *Human Development Report*[36] to witness the corrosive effect of such political naïveté, especially with respect to the consumption problem. The report marks the first time a major institutional actor in the struggle for global environmental sustainability has made consumption a top policy priority. A glance at the summary language on the report's back cover is encouraging: "These consumption trends," it reads, "are undermining the prospects for human development. *Human Development Report 1998* reviews the challenges that all people and countries face—to forge consumption patterns that are more environmentally friendly, more socially equitable, that meet basic needs for all and that protect consumer health and safety." The report begins promisingly enough, with a stirring foreword by Gus Speth, a former director of the U.S. Council on Environmental Quality and, later, of the World Resources Institute, on the need to look consumption squarely in the face:

When consumption erodes renewable resources, pollutes the local and global environment, panders to manufactured needs for conspicuous display and detracts from the legitimate needs of life in modern society, there is justifiable cause for concern. [Yet] those who call for changes in consumption, for environmental or other reasons, are often seen as hair-shirt ascetics wishing to impose an austere way of life on billions who must pay for the waste of generations of big consumers. Advocates of strict consumption limits are also confronted with the dilemma that for more than one billion of the world's poor people increased consumption is a vital necessity and a vital right—a right to freedom from poverty and want. And there is the ethical issue of choice: how can consumption choices be made on behalf of others and not be seen as a restriction on their freedom to choose?

But then the tone changes. Having introduced ideas of "consumption limits" and "manufactured needs," Speth dispenses with them. It is better to reflect on the *patterns* of consumption, he says—that is, the mix of products made in environmentally destructive ways compared to those manufactured in environmentally "sustainable" ways—than on *absolute levels* of consumption itself. For those troubled by consumption, he argues, the best mix of policies are those that expand the economic production of the poor and maintain it for the rich while reducing overall environmental impact through the dissemination of environmentally benign technologies. One solves the consumption problem, in other words, by getting rich consumers and poor alike to demand ecotechnologies.

Remarkably, after promising to help forge "consumption patterns that are more environmentally friendly," it takes the *Human Development Report* just five paragraphs to steer clear of any discussion of overall limits to consumption, of paths to more fulfilling, lower-consuming lifestyles, or of the insidious dynamics of consumerism and manufactured needs. Indeed, the need to "challenge consumerism," which Speth alludes to in his foreword, is never again broached in the remaining 228 pages of the document.

The *Human Development Report* is nevertheless a fine resource for those wrestling with the complexities of international economic development. I criticize it to show how inquiry into consumption quickly bumps up against tough issues: consumerism, "manufactured needs," limits, global inequity, the specter of coercion, competing and sometimes conflicting understandings of human happiness. Dealing with these topics, as the editors of this book argue in chapter 1, demands a practiced capacity to talk about power, privilege, prosperity, and larger possibilities. IPAT, despite it usefulness, at best fails to foster this ability; at worst, it actively undermines it. When accomplished anthropologist Clifford Geertz remarked that we are still "far more comfortable talking about technology than talking about power,"[37] he surely had conceptual frameworks like IPAT in mind.

Proponents of a consumption angle on environmental degradation must cultivate alternatives to IPAT and conventional development models that focus on, rather than divert attention from, politically charged elements of commercial relations. Formulas like IPAT are handy in that they focus attention on key elements of a problem. In that spirit, then, I propose a variation: "IWAC," which is environmental *I*mpact = quality of *W*ork × meaningful consumption *A*lternatives × political *C*reativity. If ideas have power, and if acronyms package ideas, then alternative formulations like IWAC could prove useful in shaking the environmentally inclined out of their slumber of individualization. And this could only be good for those who worry about consumption.

Take "work," for example. IPAT systematically ignores work while IWAC embraces it. As *Atlantic Monthly* senior editor Jack Beatty notes, "radical talk" about work—questions about job security, worker satisfaction, downsizing, overtime, and corporate responsibility—is coming back into public discourse.[38] People who might otherwise imagine themselves as apolitical care about the state of work, and they do talk about it. IWAC taps into this concern, linking it to larger concerns about environmental degradation by suggesting that consumerist impulses are linked to the routinization of work and, more generally, to the degree of worker powerlessness within the workplace. The more powerless one feels at work, the more one is inclined to assert power as a consumer.[39] The "W" in IWAC provides a conceptual space for asking difficult questions about consumption and affluence. It holds out the possibility of going beyond a critique of the "cultivation of needs" by advertisers to ask about social forces (like the deadening quality of the workplace) that make citizens so susceptible to this "cultivation."[40] Tying together two issues that matter to mass publics—the nature of work and the quality of the environment—via something like IWAC could help revitalize public debate and challenge the political timidity of mainstream environmentalism.

Likewise, the "A" in IWAC, "alternatives," expands IPAT's "T" in new directions by suggesting that the public's failure to embrace sustainable technologies has more to do with institutional structures that restrict the aggressive development and wide dissemination of sustainable technologies than with errant consumer choice. The marketplace, for instance, presents us with red cars and blue ones, and calls this consumer choice, when what sustainability truly demands is a choice between automobiles and mass transit systems that enjoy a level of government support and subsidy that is presently showered on the automotive industry.[41] With "alternatives," spirited conversation can coalesce around questions like: Do consumers confront real or merely cosmetic choice?[42] Is absence of choice the consequence of an autonomous and distant set of market mechanisms? Or is the self-interested exercise of political and economic power at work? And how would one begin to find out? In raising these uncomfortable questions, IWAC focuses attention on claims that the direction and pace of technological development is far from autonomous and is almost always political.[43] Breaking down the widely held belief (which is reinforced by IPAT) that technical choice is "neutral" and "autonomous" could open the floodgates to full and vigorous debate over the nature and design of technological choice. Once the veil of neutrality is lifted, rich local discourse can, and sometimes does, follow.[44]

And then there is the issue of public imagination and collective creativity, represented by the "C" in IWAC. *Imagination* is not a word one often sees in reflections on environmental politics; it lies among such terms as *love*, *caring*, *kindness*, and *meaning* that raise eyebrows when introduced into political discourse and policy analysis.[45] This despite the work of scholars like political scientist Karen Litfin that readily shows how ideas, images, categories, phrases and examples structure our collective imagination about what is proper and what is possible. Ideas and images, in other words, and those who package and broker them, wield considerable power.[46] Susan Griffin, an environmental philosopher, argues the same point from a different disciplinary vantage point when she writes that:

> Like artistic and literary movements, social movements are driven by imagination.... Every important social movement reconfigures the world in the imagination. What was obscure comes forward, lies are revealed, memory shaken, new delineations drawn over the old maps: it is from this new way of seeing the present that hope emerges for the future.... Let us begin to imagine the worlds we would like to inhabit, the long lives we will share, and the many futures in our hands.[47]

Griffin is no new-age spiritualist. She is closer to rough-and-tumble neighborhood activist Saul Alinsky than ecopsychologists like Roszak, Gomes, and Kanner.[48] Alarmed by the political implications of our collective sense of limited possibility and daunting complexity, she is quick to dispense with claims so prevalent in the environmental movement that a "healed mind" and "individual ecological living" will spawn an ecological revolution. Her argument, like Litfin's, bears restating: ideas and the images that convey them have power; and though subtle, such power can and is exercised to channel ideas into separate tracks labeled "realistic" and "idealistic." Once labeled, what is taken to be impossible or impractical—"idealistic," in other words—can no longer serve as a staging ground for struggle.

Conclusion

IWAC is more illustrative than prescriptive. It highlights how prevailing conceptualizations of the "environmental crisis" drive us toward an individualization of responsibility that legitimizes existing dynamics of consumption and production. The globalization of environmental problems—dominated by natural-science diagnoses of global environmental threats that ignore critical elements of power and institutions—accelerates this individualization, which has deep roots in American political culture. To the extent that commonplace language and handy conceptual frameworks have power, in that they shape our view of the world and tag some policy measures as proper and others as far-fetched, IWAC stands as an example of how one might go about propagating an alternative understanding of why we have environmental ills, and what we ought to be doing about them.

A proverbial fork in the road looms large for those who would seek to cement consumption into the environmental agenda. One path of easy walking leads to a future where "consumption" in its environmentally undesirable forms—"overconsumption," "commodification," and "consumerism"—has found a place in environmental debates. Environmental groups will work hard to "educate" the citizenry about the need to buy green and consume less and, by accident or design, the pronounced asymmetry of responsibility for and power over environmental problems will remain obscure. Consumption, ironically, could continue to expand as the privatization of the environmental crisis encourages upwardly spiraling consumption, so long as this consumption is "green."[49] This is the path of business as usual.

The other road, a rocky one, winds toward a future where environmentally concerned citizens come to understand, by virtue of spirited debate and animated conversation, the "consumption problem." They would see that their individual consumption choices are environmentally important, but that their control over these choices is constrained, shaped, and framed by institutions and political forces that can be remade only through collective citizen action, as opposed to individual consumer behavior. This future world will not be easy to reach. Getting there means challenging the dominant view—the production, technological, efficiency-oriented perspective that infuses contemporary definitions of progress—and requires linking explorations of consumption to politically charged issues that challenge the political imagination. Walking this path means becoming attentive to the underlying forces that narrow our understanding of the possible.

To many, an environmentalism of "plant a tree, save the world" appears to be apolitical and nonconfrontational, and thus ripe for success. Such an approach is anything but, insofar as it works to constrain our imagination about what is possible and what is worth working toward. It is time for those who hope for renewed and rich discussion about "the consumption problem" to come to grips with this narrowing of the collective imagination and the growing individualization of responsibility that drives it, and to grapple intently with ways of reversing the tide.

4

Commoditization: Consumption Efficiency and an Economy of Care and Connection

Jack Manno

Opening the black box of consumption, as the editors of this book argue in chapter 1, makes it possible to ask questions about the fundamental purposes of economic activity and to imagine a society that prospers with much less *stuff*. These questions force one to consider why people organize economic systems in the first place. Is it merely to produce more things, or is it to create the material prerequisites for a good life, one that includes many of the things that money cannot buy? Since major environmental stresses are directly related to the side effects and by-products of economic activity, the question is particularly pressing. The big challenge is how to live well without undermining the natural systems on which we fundamentally depend. The solution lies in getting more with less, not more stuff but more satisfaction, not quantity but quality. This is what is meant by the concept of *consumption efficiency*: the level of social welfare and personal satisfaction obtained per unit of energy and materials consumed. Policies directed toward improving consumption efficiency could begin to disassociate individual and social welfare from increasing levels of material and energy consumption and accompanying waste products. Improved consumption efficiency means increasing satisfaction of rational human needs and wants with decreasing amount of consumption. This chapter will not attempt to directly quantify consumption efficiency[1] but rather will describe the difficulties in achieving such efficiencies under current economic conditions. It will suggest political and economic reforms that can lead to improved consumption efficiency.

Typically, the focus of efficiency analyses is on the key factors of production, labor and capital, measured in terms of input per unit of output. In practice, inputs are generally measured in terms of monetary costs.

Investments in innovation are directed toward decreasing the costs of inputs and/or increasing the value of outputs, thereby maximizing productivity. The result is that investments keep spurring production without considering the overall efficiency of how the product is ultimately consumed. In fact, for the producer, the less consumption efficiency the better, the result being ever-increasing consumption and ever-increasing dependency on the producer.

It is important to understand that technical progress that leads to improved production efficiency of capital and labor is a necessary but not sufficient condition for improvements in consumption efficiency. In general, improved efficiency of production simply lowers the costs of producing *stuff* and transfers the resulting savings toward additional consumption. Gains made in improving the fuel efficiency of the U.S. motor fleet, for example, have been more than offset by trends toward larger vehicles, more cars per household, and more miles per car.[2] A study by Peter Freund and George Martin demonstrated that even though the automobile fuel efficiency in the United States improved considerably (34 percent) between 1970 and 1990, total fuel consumption during the same period increased by 7 percent. The number of multicar families had increased and the family drove more miles.[3] This paradox is sometimes referred to as "Jevons's paradox" after economist Stanley Jevons, who pointed out in 1864 that efforts to conserve English coal by increasing the coal-use efficiency of British steam engines ended up making steam power cheaper compared to human and animal power, in the end stimulating increased coal consumption.[4] Likewise, production efficiencies unaccompanied by brakes on consumption tend to bring the consumption of energy and materials to levels greater than what existed before the production efficiencies were introduced. Energy-efficiency gains will thus only be successful in uncoupling improved quality of life from increased energy use if they are accompanied by comprehensive political and economic strategies to reduce consumption. Without such a strategy, discussed in detail later in this chapter, improved efficiency leads to lower costs and increased consumption.

Modern industrial economies provide a way of satisfying almost all human needs and wants through the individual consumer purchase of some form of commodity. There are, however, two possible additional approaches to address any particular set of needs, such as the need for transportation, for entertainment, for food and clothing, and so on. These are:

- *Need-reduction and need-prevention approach* One can reduce the need for health care through public health, hygiene, and health promotion. One can reduce the need for transportation through urban planning that clusters housing near workplaces and services. Farmers can reduce the need for commercial fertilizers through recycling soil nutrients. Homeowners can reduce the need for lawn watering by planting drought-tolerant species. The number of approaches to need reduction is limited only by creativity. If directed to need reduction, technological and social development could greatly reduce many excessive forms of consumption. Although basic needs such as the need for food, clothing, shelter, and love are difficult to reduce, perceived needs in these areas may often be in excess of basic needs and can be effectively reduced. Obviously, one key to gains in consumption efficiency lies in such need-reduction strategies.
- *Cooperative and collective approaches* There are many ways to meet needs collectively rather than individually. These approaches also hold great promise for improving consumption efficiency. Governments can reduce the need for waste-management services by preventing pollution. Public transportation can move far more people-miles per unit of energy than the same number of people in individual cars. Well-organized neighborhoods and communities create the possibilities for sharing resources through tool and appliance libraries, central worksheds, entertainment centers, cooperative arrangements for shared child and elder care, and all sorts of mutual-aid arrangements.

A healthy, balanced economy would be able to steadily improve and develop all three approaches: personal consumption, need reduction, and cooperation, directing its research and creativity toward all three areas relatively equally. The current economy is primarily an economy of commodities focused on their production and consumption. Need-reduction and collective approaches replace consumption with increased quantity and quality of relationships between individuals in community and between people and their environment. This can be considered the economy of care and connection. As we will see, however, in modern industrial societies economic forces come into play that distort economic and social development increasingly in the direction of increased consumption. Technical progress is overwhelmingly directed toward increasing the amount, variety, and availability of goods and services for purchase. One key economic force that I call *commoditization*[5] must be understood and countered if progress is to be made toward improving consumption efficiency and increasing our chances of building communities that prosper while living more lightly on the earth—the key component of environmentally sustainable development.

This chapter analyzes commoditization and suggests that social distortions wrought by it are the major systemic obstacle to improving consumption efficiency. *Commoditization* is the tendency to preferentially develop things most suited to functioning as commodities—things with qualities that facilitate buying and selling—as the answer to each and every type of human want and need. In many ways, commoditization is the same process as industrialization, but looked at through the lens of consumption.

Commoditization

One can refer to any given economic activity or sector—say, food production or health-care provision—as being more or less commoditized, by which is meant that the continuous improvement and development of those activities is more or less focused on the most marketable end products rather than on the social systems necessary for the delivery of nutritious food or the maintenance of human health. One can also refer to a given economy as more or less commoditized, to the extent that the commoditization process dominates in the allocation of resources in that economy.

Under the current industrial capitalist system of incentives and disincentives, what we consider "progress" is invariably directed toward increasing levels of consumption. There are, however, other possible paths of progress and development. As Cogoy[6] and others have pointed out, technical progress that improves the efficiency of labor could be directed toward either increasing production or reducing labor time. If directed toward freeing up more and more time, this time could be directed toward improvements in community and cultural life. This time would become the raw material for an economy of care and connection and the main source of improvements in consumption efficiency. But under present economic conditions, investment is almost always directed preferentially toward the production of commodities to be consumed rather than toward the freeing of more time.

Commoditization makes the attainment of consumption efficiency difficult at best. The fundamental thrust of consumption efficiency is increased human satisfaction with decreased consumption. The general thrust of commoditization is ever-increasing consumption. Therefore, to explore the possibilities for improving consumption efficiency, one first needs to understand commoditization.

Another concept must be understood, though, before we can fully define commoditization—*commodity potential.* If a commodity is something endowed with the qualities that facilitate exchange, then commodity potential is the potential of a thing to carry those qualities. Commodity potential is a measure of the degree to which a good or service carries the qualities that are associated with and that define something as a commodity. Goods and services can be described as having high commodity potential (HCP) or low commodity potential (LCP), or to be HCP or LCP goods and services.

Everything has some commodity potential, even if, as in the example of interpersonal relationships, it is small. Commoditization operates to increase commodity potential to its maximum, no matter how small or large that maximum may be. If we want to understand how this works, we must first identify the qualities that are associated with and define commodities. Let's first compare the difference between goods and services with low and high commodity potential. This way we can begin to understand the qualities commoditization develops and why.

The primary qualities that define a commodity are:

- Alienable, the ease with which ownership can be asserted, assigned, and transferred
- Standardizable, independence from the particularity of geography or culture
- Autonomous, the ability to be used independently, outside the constraints of social relationships
- Convenient, the ease with which it can be used
- Mobile, the ease with which something can be packaged and transported

These characteristics, along with several related qualities and their implications for consumption efficiency, are considered in more detail later in tables 4.2a and 4.2b.

The distinction between HCP and LCP goods is not absolute but rather one of degree. Goods and services more or less have the characteristics of a commodity. They are more or less alienable, standardizable, autonomous, convenient, mobile. Noncommodities are less alienable (more communal), less standardizable (attached to local ecosystems or local culture), less autonomous (goods and services that rely on a web of relationships), less convenient (involving a complex set of relationships), and less mobile. Most things are some of each. Even noncommodities like "friendship" have their commoditized service version in "psychic friends networks," personal ads, and so on. In highly industrialized societies where commoditization operates most strongly, few aspects of human life have not been commoditized to some extent. The selection pressures that favor commodities over noncommodities involve a gradual "survival of the fittest" where what is fit is by definition what is marketable.

Consider, for example, children's need for play. At one end of the scale of commodity potential are such mass-marketed toys as Barbie dolls, superhero action figures, and the packaged entertainment that accompany them. These products are inexpensive, marketed worldwide, and involve immense sums invested in product research and development as well as in packaging and marketing. Their production is energy intensive and fossil-fuel dependent and involves the highly publicized exploitation of cheap labor and mountains of industrial and postconsumer waste. In the middle of the commodity scale lie locally produced, handcrafted dolls, toys, and games usually made from renewable materials and with local or culturally idiosyncratic designs. These are the goods of the crafts market and bazaar. Also in the midscale are all the services for sale: child care, playgroups, clowns for hire, and so on. At the far end of the commodity-potential range are things not for sale such as making angels in the snow, play with found objects, group play, sing-a-longs, and all the goods of interpersonal contact.

The point here is *not* that goods with lower commodity potential are morally preferable or even always more benign. It is that LCP goods and services have the potential to satisfy human needs with less material and energy; they form the basis of need-reduction and cooperation strategies and an economy of care and connection that facilitates consumption efficiency. It is most likely that a sustainable society would support the development of both HCP and LCP goods and services. Given the selection pressures of commoditization, however, unless public policy deliberately intervenes, HCP goods and services inevitably outcompete LCP goods and services for investment and resulting allocations of time, attention, and the means of material survival. Commoditization pressures act over time to gradually and inexorably expand the number of commodities available, the geographic spread of their availability, and the range of needs for which commoditized satisfactions exist.

Let us consider several economic sectors to see the range between goods with high, medium, and low commodity potential (table 4.1). Then we will consider the qualities associated with different degrees of commodity potential (tables 4.2a and 4.2b).

If the qualities associated with commodities are privileged in an economy, and if economic rather than social or democratic forces dominate a society, then over time, more and more of that society's attention, resources, creativity, enthusiasm, and so on will be directed toward the production and reproduction of those qualities. At the same time, qualities associated with lower commodity potential will become increasingly underdeveloped in comparison.

Improved consumption efficiency—a rise in social and individual welfare with lower energy and material consumption—is increasingly difficult to achieve to the extent that commoditization drives the evolution of an economy. Like all systems, economies evolve over time. This evolution occurs through a process of natural and "unnatural" selection in which certain things survive and others do not. Those that survive become the goods and services that citizens use or consume to meet their needs. In a modern industrial economy, the key ingredients for survival are investment capital, time, attention, skill, technology, and creativity. They are the nutrients and raw materials of economic life. The allocation of these "nutrients" is affected by subtle economic pressures that "select" the options for satisfying wants and needs that are most commoditizable.

Several factors determine which goods and services are available and practically obtainable for the satisfaction of human needs and wants. These factors can be illustrated in the design of a car. Physical laws determine the range of design options, and as a result most cars are fundamentally alike: height, width, wheel span, steering mechanisms, and so on. Designs that veer too far from the physically optimal pay the price in higher fuel demands as well as in safety risks and associated liability concerns, and are selected out. Consumer preference determines most of

Table 4.1
Commodity potential

Sector	High commodity potential (products involving distant or abstract relations between producer and consumer)	Medium commodity potential (products involving direct relations between producer and consumer)	Low commodity potential (processes involving direct or cooperative social and ecological relations)
Children's play	Barbie dolls, action figures, packaged entertainment	Handicrafts, child-care, live entertainment	Direct child-led interaction with natural surroundings, group play, interpersonal goods
Food production	Commercial fertilizers, pesticides, engineered seeds, mechanization tools, genetic material	Commercial manure, stored seeds, farm animals, tools for small-farm, agricultural extension, and research services	Knowledge of soil, locally co-evolved skills and techniques
Health care	Mass-marketed drugs, diagnostic equipment, hospital supplies, insurance	Doctor-provided services, hands-on therapies and treatments	Knowledge of healing, personal health maintenance and illness prevention, life-style adaptations, sense of well-being
Energy	Grid-dispersed electricity, power-plant equipment, fossil and nuclear fuels	Renewable energy sources, energy-conservation services, wage labor	Personal energy conservation strategies, passive solar design, cooperative sharing activities
Transportation	Personal transport vehicles and the infrastructure of roads and so on that supports it	Public transportation	Transportation-reduction strategies (such as cluster housing near workplaces), walking

Environmental protection	Pollution-control equipment, waste-to-energy incinerators and equipment	Recycling, pollution-reduction/prevention services	Pollution-prevention redesign, materials and energy-use reduction strategies
Mental health	Mood-altering drugs	Therapists, fitness clubs	Peer counseling and mutual help, friendship, exercise
Finance/credit	Options, junk bonds, credit cards	Neighborhood banking, credit unions	Personal loans, gifts

Note: Commoditization is the process that favors goods that have the quality of a commodity (see tables 4.2a and 4.2b). Goods and services with low commodity potential (LCP) tend to be processes that involve direct and cooperative relationships between people or between people and the natural world. Goods and services with medium commodity potential (MCP) involve a direct exchange relationship between the purveyor of the goods and the end user. High-commodity-potential (HCP) goods and services involve highly abstract and usually distant relationships between producers and consumers.

Table 4.2a
A comparison of the key attributes associated with the high and low commodity potential and their effects on the process of development

Attributes of goods and services with high commodity potential	Attributes of goods and services with low commodity potential	Negative effects of commoditization on development	Positive effects of commoditization on development
Alienable, excludable, enclosable, assignable, patentable: simpler to establish right of ownership, easier to establish price.	Openly accessible, inalienable: difficult to establish rights, widely available, difficult to accurately price.	Privatization accelerates decline of sense of community and the common good and increases commoditization of all aspects of life. Skills and capacity for managing common property and promoting common good are underdeveloped.	Releases individual and corporate entrepreneurial energy. Ability to manage individual property and promote personal gain is highly developed.
Standardized, universal, centralized, and uniform: adaptable to many contexts.	Particular, customized, decentralized, and diverse: each culture potentially derives the best practices for its particular environmental context, leading to diverse customized goods and practices.	Reduces cultural and geographic diversity, standardized methods may not be suited to particular ecosystems, as a result efficiency potential is reduced. Locally appropriate development options remain underdeveloped.	Allows rationalization of production, economies of scale and transfer of skills. Greatly increases production efficiency.

Autonomous, depersonalized, use or practice occurs largely independent of social relationships. Primary relationship is between consumer and product.	Embedded, use or practice occurs in a web of social and ecological relationships.	Promotion of individual consumption reduces the efficiency gains made possible by sharing, increases flow of energy and materials. Excessive autonomy undermines social relationships, system redundancy and resilience.	Minimizes the complications of relationships. Advances freedom of the individual.
Embedded knowledge or skills, convenient, use simplified and inherent in design and material.	Dispersed knowledge and skills, convenience is not goal, use requires relevant knowledge and skills.	Impoverishes knowledge base, particularly at the personal, local, and regional levels.	Convenience frees human attention for other activities.
Mobile, transferable, easy to package and transport.	Rooted in local ecosystem and community.	Propensity for mobility increases flows and export of energy and materials. Local knowledge and connectivity underdeveloped.	Makes trade possible with accompanying spread of benefits. Trade and markets become highly developed.

Table 4.2b
Secondary attributes related to key attributes in table 4.2a

Attributes of goods and services with high commodity potential	Attributes of goods and services with low commodity potential	Negative effects of commoditization on development	Positive effects of commoditization on development
Contributes to production efficiency; more is produced per unit of currency expended.	Contributes to consumption efficiency; more satisfaction per unit of material and energy expended.	Neglects the potential for achieving sustainability through increased satisfaction with less material and energy.	Increased production efficiency creates more wealth and greater availability of material goods and services.
Product-oriented: ends are accomplished through products.	Process-oriented: ends are accomplished through processes and interactions involving people in relationship with their environment.	Discourages systems thinking, keeps attention on parts rather than wholes. Undermines capacity for ecosystem approaches to decision making. Overdevelops competitive skills and underdevelops collaborative skills.	Produces cornucopia of products.
Responsive: services that respond to needs and fix problems.	Anticipatory: services designed to avoid needs and prevent problems.	Focus on need satisfaction rather than prevention encourages the expansion of neediness and associated chronic dissatisfaction. Anticipatory and preventive capacities underdeveloped in society.	The capacities associated with responsiveness are highly developed; wide range of needs are addressed. Problem-solving capacities and tools are developed.

Table 4.2b (continued)

Attributes of goods and services with high commodity potential	Attributes of goods and services with low commodity potential	Negative effects of commoditization on development	Positive effects of commoditization on development
Embedded energy: production is energy intensive. Packaging, transportation, and promotion add to energy embedded in the product.	Dispersed energy: energy is used and dissipated at the site of the activity or point of exchange or consumption.	Concentration of energy causes ecological disruption at the point of its release. Commoditization of fuel facilitates dramatic increase in energy availability and use. Decentralized energy strategies underdeveloped.	Energy sector highly developed. Commercial energy increasingly replaces human labor and leads to expansion of leisure.
High capital intensity, low energy productivity, low labor intensity, high labor productivity.	Low capital intensity, high energy productivity, high labor intensity, low labor productivity.	Eliminates jobs, encourages replacement of workers with fossil-fuel energy.	Increased productivity frees capital to invest in new productive activities, creating new jobs.
More stable, predictable, reliable.	More variable, unpredictable, unreliable.	Predictability tends toward simplification, including loss of diversity and redundancy in ecosystems.	Increased predictability and reliability benefit all human activities.

Table 4.2b (continued)

Attributes of goods and services with high commodity potential	Attributes of goods and services with low commodity potential	Negative effects of commoditization on development	Positive effects of commoditization on development
Design resists and/or alters natural flows and cycles.	Design follows and mimics natural flows and cycles.	Failure to promote and use ecological designs leads to increased energy use and more waste.	Overcoming ecological constraints opens more possibility for economic growth.
Abstract, distanced, less direct ties to physical base of reality.	Concrete, tied to physical and biological constraints.	Reduces knowledge and awareness of physical basis of human life and culture.	Overcoming physical constraints opens more possibility for economic growth.
Path-breaking, break from bonds of place and tradition. Relationships structured by contract.	Co-evolutionary and traditional, evolves in the context of specific ecosystem and culture, relationships often structured by custom.	Loss of traditions and traditional knowledge.	Overcoming cultural constraints opens more possibility for economic growth.
Short-term, large return on investment.	Long-term, stable returns.	Increases speculation, reduces investments in sustainable opportunities.	Increases wealth and its accompanying benefits.

Table 4.2b (continued)

Attributes of goods and services with high commodity potential	Attributes of goods and services with low commodity potential	Negative effects of commoditization on development	Positive effects of commoditization on development
Efficient, the most exchange value for a given investment.	Sufficient, optimal service for minimal expenditure of material and energy.	Reduces capacity to develop low-impact living, accelerates commoditization.	Efficiency frees reserves for other wealth-producing activities.
Contributes to GNP, GNP growth measures commoditization.	Contributes little to GNP, the less commoditized a good or service, the less it contributes to GNP.	Public policy goals become tied to growth in size of economy rather than improvements in quality of life.	GNP represents accurate measure of economic activity and is closely related to improved quality of life.

the variability within the range of what physical laws determine is practical. These choices are the result of the interplay of options and motivations, including disposable income, status seeking, comfort, and practical considerations such as size of family and the purposes for which the vehicle would be used. The result is a range of available vehicles and features that represent the optimal balance of possible customer satisfactions. Neither these physical limits nor consumer choices, however, account for the selection of automobile transport over other LCP options for moving people and goods, such as well-designed public transport, urban designs that minimize transport needs, and other approaches to minimizing the need for the personal automobile. A transport system based on the personal automobile represents the transportation option with the greatest commodity potential. Cars are individually owned, their operation is nearly globally standardized, they allow tremendous individual autonomy, they are always available and simple to use, and they greatly expand individual mobility.

This phenomenon, the selection of HCP over LCP solutions to personal and social needs, can be observed in sector after sector. It is a self-reinforcing, positive-feedback mechanism. People grow increasingly dependent on HCP goods to meet their needs, much as people have become dependent on cars as suburbs grow and public transportation becomes comparatively underdeveloped. Society increasingly invests in the infrastructure to support a highly commoditized lifestyle. Since HCP goods receive by far the greater amount of research and development (R&D), they invariably appear to be more advanced and competitive. As a result, it feels modern or more advanced to purchase and use the more commoditized products. It is then logically compelling to assume that the proliferation of HCP goods is the very meaning of development; to suggest alternative, currently less developed pathways to social development is to be unrealistic, even reactionary. Ways of meeting human needs that rely on LCP goods and services appear on the surface to be less developed, more backward, and less capable of meeting broad human needs despite the fact that they are the key to improving consumption efficiency. Their proponents and serious proponents of environmentally sustainable economic development face this problem of unfair comparisons. Approaches or products with high commodity potential receive far greater R&D investment. Over time they become more "developed" and appear more practical. R&D investment is not the sole determinant of success nor the sole determinant of outcome in this evolutionary selection process, but it is a good indicator of what the economy will identify as potentially successful in the context of commoditization. The remainder of this chapter will describe the implications of this with examples from agriculture and pollution control.

Agriculture

No economic sector has been as highly affected by the pressures of commoditization as agriculture. Commoditization operates on both the inputs and outputs of the production process, preferentially investing in commercial chemical fertilizers, pesticides, machinery, and standardized crops suited for long shelf life, transport, and branding while underinvesting in the development of local site-specific knowledge and skills of soil management, site-specific agronomy, and diverse crops with a high mix of nutritional qualities. As a result of this pattern of preferential investment, more progress occurs in highly commoditized industrial agriculture than in alternative agricultural methods that depend less on commercial inputs and more on highly evolved, site-specific skills and methods. Hence, when comparing the future prospects for feeding the world's people with different agricultural methods, small-scale, low-input, LCP methods appear less capable. But this comparative disadvantage is the result of past distortions. If as much development attention and investment had been going into improving low-input alternatives, these alternatives could be as highly developed and productive as more industrial agriculture. Its comparative underdevelopment then becomes the argument to justify the disproportionate share of investment in research and development that industrial agriculture receives. It becomes a vicious circle in which underdevelopment becomes the excuse for further underdevelopment. This is exactly the way commoditization affects and distorts development practice.

Agriculture is also one of the most environmentally damaging sectors, and the increasing use of energy and other inputs is highly implicated in this damage. In the United States, agriculture is responsible for 57 percent of the pollution in freshwater lakes. Significant improvements in consumption efficiency will be required to reduce this damage while providing adequate nutrition for people. Improvements in consumption efficiency mean better nutrition for more people with lower consumption of energy and materials in producing, processing, packaging, and delivering food. These improvements can also mean, at the individual level, more nutrition per calories consumed. Consider the distinction between HCP *products* and LCP *processes* in the agriculture industry:

Agricultural-system products (high commodity potential)
Proprietary hybrid and patented seeds
Insecticides, herbicides, and fungicides
Commercial fertilizers
Farm machinery
Fuel
Farm-management books and magazines

Agricultural-system processes and skills (low commodity potential)
Soil protection and management
Water conservation and management
Knowledge of soil, climate, local pests
Energy conservation and management
Nutrient cycling and enhancement
Crop rotation and placement
Rural networks of mutual aid
Pest control and management

The selection pressures of commoditization help determine which farming methods are widely adopted. These pressures favor approaches that rely on HCP inputs and produce HCP crops best suited for broad marketing, rejecting LCP crop varieties that rely on LCP agricultural methods. Given the tendency toward standardization that characterizes commoditization, one result is to gradually select out crop varieties uniquely adapted to particular growing conditions. The result is greatly reduced agricultural and genetic diversity.[7] Preindustrial traditional agronomy bred many different crop varieties, each uniquely suited for a specific set of growing conditions. Each farming community and culture bred its own wide range of crop varieties. Under the pressures of commoditization, the tendency is to engineer the soil through HCP inputs to recreate the conditions favorable to standard high-yielding varieties rather than creating unique LCP crops for unique soil and climate conditions. This soil engineering is accomplished through the application of a variety of agricultural chemicals and fertilizers and the use of massive farm machinery. In this way, the same crop varieties can be grown in many different locations.

With respect to genetic engineering in the absence of commoditization pressures, this new technology might be used to create economical crop varieties, each suited to particular growing environments. This could conceivably be a great boon for local food production and rural self-sufficiency, especially if marginal land is brought into production. Most genetic engineering research and development, however, is directed toward promoting qualities that either maximize attributes associated with high commodity potential, such as long shelf life, portability, and standardization, or maximize qualities that enhance and promote the adoption of a particular brand of product, such as product-specific herbicide resistance.

Society's resources, represented by the amount of time, attention, and money directed toward R&D, is understandably oriented toward whatever will yield the greatest return on investment. Low-input, indigenous, and environmentally sustainable agricultural methods may rely less on purchased fertilizers, pesticides, and other inputs but they produce lower crop yields. Or so it is widely assumed. According to this reckoning, the cost of a large-scale transition to sustainable methods would simply be too costly and risky to be worthwhile. For example, former Secretary of the U.S. Department of Agriculture Earl Butz was reported to have responded to those arguing for greater support for organic agriculture, "Show me the first 10,000 Americans who are prepared to starve to death and then I'll do something."[8] But the assumption that sustainable agriculture is necessarily less productive is misleading.

Conventional industrial agriculture is indeed more productive in producing HCP products but not necessarily in growing nutritious food. There are indeed economies of scale that can be achieved by industrialized farms, but improved labor and capital productivity—not necessarily land productivity—realize these savings. In terms of land productivity, small farms, which rely heavily on skilled labor, can yield more per acre than large farms. The most productive farming turns out to be small labor-intensive, gardenlike cultivation systems with mixed crops, shifting cultivation, and a high degree of nutrient recycling. Such systems are capable of producing three times as much per unit area as highly mechanized, capital- and energy-intensive agricultural production systems utilizing minimal human labor.[9] This productivity consists of the total of multiple yields in a mixed-crop farm. Industrial farms are far more productive in producing a single crop for market in terms of yield per worker-hour. But low-input agriculture is more productive in terms of yield per area, and far more productive in terms of yield per unit of inputs such as chemical fertilizer, pesticides, fuel, and irrigation water. In addition, studies have shown that the nutritional content of organically grown produce can be significantly higher than in conventionally grown equivalents,[10] suggesting that consumption efficiency is higher at the same time that environmental impact is lower.

For farms larger than intensively cultivated gardens, small farmers produce about twice as much per hectare as do large, industrial-scale farmers, while using only one-fourth or one-fifth as many purchased inputs.[11] Studies in modern Mexico have shown the advantages of traditional companion planting methods. These methods rely on the mutually beneficial characteristics of corn, beans, and squash. The corn stalk provides structure on which the bean plant can climb, while the broad squash leaves shade out weed growth. Insect-attracting plants are deliberately planted at the edge of the fields to draw pests away from the crop, while other plants are cultivated in the field to deliberately attract insects that feed on the most damaging of crop insect pests. Fields planted in this diversified pattern have been able to increase yields by as much as 50 percent over methods of planting a single highly marketable commercial variety.[12]

Most traditional farming methods rely heavily on skilled labor. But labor in a commoditized economy is increasingly expensive relative to inputs with higher commodity potential: machinery and raw materials. Since some of the highest costs of any operation are associated with labor, particularly skilled labor, investments naturally focus on reducing costs through mechanization and standardization. Eliminating skilled labor removes the very resource most essential to sustainable agriculture—people with intimate, detailed knowledge of particular lands and soils. A single technician giving instructions to untrained field hands on an industrial farm can manage several thousand acres using standard recipes for determining the amounts and timing for the application of pesticides and fertilizers. Organic and other forms of sustainable agriculture, on the other hand, require highly trained and experienced labor and diverse management skills. Because resources for agricultural training are distorted by commoditization, almost no formal training options are available for individuals interested in obtaining the skills for sustainable farming. Internships are few, college training programs even fewer.

One example of how this lack of skills and training affects agricultural practice has been the disappointing progress in implementing integrated pest management (IPM) programs as an alternative method of pest control. In the 50 years since farmers started using DDT for protecting crops, use of pesticides has grown thirty-three-fold. Worldwide, 50 million kg of pesticides were applied to crops in 1945; by the mid-1990s annual worldwide pesticide use had jumped to 2.5 billion kg.[13] Pests continue to develop resistance and become even more harmful as a result, leading to increasing pesticide use and the continuing need for new and more targeted pest-management products. All the while, pesticides continue to spread into the water and air, endangering other plants and animals as well as humans. The problems associated with increased reliance on chemical pesticides have been known for decades. Excellent alternatives have been developed, mostly under the heading of IPM. These systems are designed to support and promote natural pest resistance through companion planting, the planting of hardier breeds, the introduction of beneficial insects and pathogens, and the controlled use of narrowly targeted and nonpersistent pesticides. Despite considerable success with these methods, they have not been widely adopted, except on a few crops, and have not significantly affected the amount of pesticides used worldwide. The failure to adopt IPM stems from two facts: the full cost of pesticides in terms of environmental health effects is not reflected in their price, and the implementation of IPM requires careful management, experimentation, and observation—all LCP services.[14] IPM, like other efforts designed to decrease the environmental impacts of farming, requires just the kind of site-specific knowledge and understanding that has been selected against by the force of commoditization and is in decline worldwide.[15]

Commoditization and associated industrialization consistently distort development toward capital-intensive and less productive large industrialized farms that are devoted to the production of a single or just a few crops. This, as David Barkin and others have noted, encapsulates the experience of rural development worldwide. Emphasis is invariably placed on producing a single commodity without understanding the role of agriculture in peasant society, culture, and economy.[16] Cropland devoted to single crops (usually wheat, corn, or rice) has expanded worldwide with heavy consumption of pesticides and fertilizers. Some land is simply rotated between corn and soybeans. Farm animals are increasingly raised apart from croplands, further reducing the opportunities to recycle animal waste as soil nutrients and greatly adding to the pollution load on streams and rivers near chicken houses and feedlots.[17] The number of hog and dairy operations declined by 70 percent from 1969 to 1992 while production, now highly concentrated, remained stable.[18] At present, 90 percent of the world's food supply comes from only 15 species of crop plants and 8 species of livestock among the estimated 10 million species of plants and animals in the world.[19]

The process of commoditization in agriculture is reflected in the ongoing replacement of human labor, animal power, and renewable energy with fossil fuel. Each calorie of food we eat from high-input agriculture embodies several calories of fossil-fuel energy. From petroleum come the synthetic chemicals in pesticides. Oil powers the production of chemical fertilizers, moves agricultural commodities around the world, drives the farm machinery, raises the irrigation water, and on and on. High-input agriculture requires about 3 kcal of energy derived from fossil fuels for every 1 kcal of human food produced. Food production accounts for about a third of all energy used. The United States uses three times as much energy per capita for food production as developing countries use for all energy-consuming activities combined, including food production.[20]

This measure, the amount of food energy produced per unit of energy expended in food production, is a measure of energy efficiency—a form of consumption efficiency. The energy efficiency of modern conventional agriculture declined in the United States throughout the period 1920–1973.[21] Industrial agriculture is tremendously *inefficient* in terms of energy. This inefficiency is directly related to the economic force of commoditization. Fossil fuel has much greater commodity potential (see tables 4.1 and 4.2) than other forms of energy do. The fertilizers and pesticides produced from fossil fuels have much more commodity potential than do alternative methods of soil enhancement and pest control. The preferential development of commodities through the process described here as commoditization greatly distorts agricultural development toward massive energy inefficiencies. These inefficiencies will likely continue until public policy intervenes to direct development toward low-input and less environmentally disruptive agriculture.

Consider rice production. Modern rice farmers get a negative 1-to-10 energy return. In other words, they consume up to ten times as much energy to produce the food than the resulting food yields in calories.[22] Compare this, for example, to the traditional rice farmers of Bali, who are reported to produce 15 calories of food energy for every 1 calorie of energy used, and even higher yields are sometimes obtained. Many of the negative environmental consequences of conventional agriculture result from this massive dependence on nonrenewable fossil-fuel energy.

Because of commoditization, organic agriculture in general cannot compete for private investment capital. Up-front public funding of research, development, and agricultural extension is essential for organic farming to succeed. To achieve the full potential of organic farming requires time and the knowledge and skills of the farmers to build up the humus content of the soil. Without public financial support, it is extremely difficult for a farmer to stick out the lag time between start-up or changeover from standard farming practices to organic practices. In addition, organic farming is very labor intensive, and the costs of sufficient labor can be prohibitive.[23]

When farmers switch to organic methods, they experience a short-term decline in productivity for some crops in terms of yield per unit area and a dramatic rise in productivity in terms of yield per unit of energy input. A 1980 USDA study, however, not only found increased energy efficiency on organic farms but roughly equivalent yields per acre of organic versus standard methods. For wheat, there was no significant difference; for soybeans, organic methods produced 14 percent higher yields. Most studies show a 20 to 30 percent decline in productivity in the short term, with productivity increasing slowly but steadily in organic farms over time.[24]

Agricultural research dollars go overwhelmingly to benefit the dominant agricultural paradigm promoting HCP products and processes. The Organic Farming Research Foundation analyzed the U.S. Department of

Agriculture (USDA) Current Research Information System (CRIS) to assess the pertinence to organic farming of the currently funded research.[25] Of the 4,500 projects the authors reviewed, only 301 led to results that were relevant to organic practice. Funding for these projects amounted to about one-tenth of 1 percent of the USDA annual research and education budget. In another study, the CRIS database was searched for projects that included the term *sustainable* or *low-input* in their title or abstract. These 122 projects were then analyzed for content. According to Molly Anderson, "Few of the 122 projects showed the broad scope that writing about alternative agricultural research emphasizes. Only 22% dealt with entire farms, 25% looked at both crops and livestock and 19% studied general processes from which basic agroecological principles could be learned."[26] Both these studies analyzed the CRIS database that includes only publicly funded research and probably greatly *over*estimates the actual share of total agricultural research directed to sustainable or low-input agriculture. Privately funded research tends to be even less relevant to organic farmers and other forms of alternative agriculture.

Instead of comparing averages, the performance of the *best*, most productive low-input farmers should be compared with the best, most productive high-input farmers. In general the best, most experienced organic farmers can achieve yields equal to or better than conventional farmers while decreasing energy use, building up humus in the soil, dramatically reducing soil erosion, reducing nitrate concentrations in groundwater, and enhancing plant resistance to disease.[27] Studies undertaken by the Rodale Institute concluded that experimental plots using standard organic-farming techniques had similar yields to plots using conventional methods over a 15-year period. While the experimental crops produced as well as the conventional plots, they also enriched the soil. Carbon levels rose dramatically, and nitrogen losses were half of what they were in the conventional high-input plots.[28] Well-maintained soil acts as a carbon sink, preventing large amounts of carbon from entering the atmosphere, not inconsequential when industrialized societies are straining to figure out ways to reduce the increase in atmospheric carbon dioxide to stem global warming.

The stakes continue to grow. It will be increasingly difficult to continue to raise yields enough to meet the needs of growing populations when the amount of cropland is declining on a per-capita basis. Farmland is disappearing under human settlement and environmental degradation. Although new areas continue to be opened, the most fertile lands were long ago put under the plow and much was later buried under cities and suburbs. Such land was, in short, consumed. Since the 1950s, total land area in grain cultivation has increased by around 19 percent while global population has increased 132 percent, resulting in a decline of 50 percent in the amount of grain area per person.[29] Heavy uses of fertilizers and pesticides have more than doubled average yields, so that per-capita production has continued to increase. We have begun to see potential saturation points beyond which increasing fertilizer use will not likely increase yields to the same degree as initial applications once did.[30]

Agriculture is in a bind. Industrialized methods have led to reduced natural fertility in farmlands worldwide. This reduced fertility makes it more difficult to begin the transition to lower-input methods. At the same time, the amount of available good cropland declines; cities and suburbs are spreading out into surrounding farmlands as the land becomes more economically valuable to grow houses and strip malls than to produce food. Calls for the continued intensification of agriculture are likely to increase along with charges that proponents of alternative, low-input agriculture are irresponsible. The results of the "unfair comparison" are only likely to worsen. We approach what looks like a choice between feeding growing populations of hungry people and protecting the environment. This is a false choice.

All agriculture disrupts natural ecosystems. That is what agriculture is meant to do, replace natural ecosystems with ones deliberately manipulated to produce more of the plants and other organisms people find useful and desirable. However, disruptions can be limited, first by limiting the number and amount of agricultural inputs such as fertilizers and pesticides. While it may not be possible to maximize both environmental protection *and* crop yield, each can be optimized in relation to the other. The goal of agricultural policy and practice should be to optimize harvest *and* ecological integrity, which will maximize neither. It should be to increase consumption efficiency. Such measures will only work within an economic context that balances commercial and noncommercial values. This brings us to the most important problem created by commoditization, a problem which constantly distorts human development by giving overwhelming preference to commercial values. This makes it extremely difficult to find the optimal balance between environmental protection and production—the goal of sustainable development in agriculture and in every other sector of the economy.

The world is in a bind, too. World per-capita food production has increased by 15 percent from the 1960s to the 1990s. But per-capita figures are misleading. Modern commoditized agriculture has been enormously successful in producing a wide variety of food for the tables of those who can afford it. For those left out of that market for one reason or another, the result has been a devastating loss of capacity to achieve self-sufficiency in food. Thus there has been an enormous increase in the number of people on the planet currently considered malnourished, more than half of the present world population of around 6 billion people.[31] Current agricultural practices contribute significantly to all the major environmental problems facing the world: global climate change, loss of biological diversity, polluted and overdrawn water resources, spread of toxic chemicals, and air pollution. The combination of trends necessitates a shift in agriculture toward more benign and sustainable practices. In addition, agriculture in general suffers from problems of diminishing returns. The first crop planted in a cleared field yields more than subsequent crops. The initial application of fertilizers can boost production substantially, but later applications are less effective. The same is true for irrigation water. The first applications of pesticides are more effective than later ones, after pests begin to adapt and evolve pesticide-specific resistance.

The good news is that because of the underdevelopment of organic methods, we are a long way from reaching the limits of the potential gains to be had there. We have considerable knowledge of how to farm sustainably and an ancient history with many successes. The bad news is that commoditization pressures continue to skew investment toward research and development in the opposite direction of sustainability. In addition, ecological disruption changes the ecosystems within which traditional farming methods have evolved. Once these changes occur, traditional practices based on knowledge of local ecosystems are no longer viable. We are losing our inheritance by consuming our natural and social capital, as the editors of this book put it. Global climate change makes these phenomena worse still. All agriculture depends on relatively stable climate. A shift into a new climate regime would severely complicate our task. Commoditization is preventing us from achieving sustainable development, and in the process grossly limiting development of our full human potential. The more extensive the commoditization, the more ecologically disruptive it is. The longer it lasts, the more underdeveloped alternatives become and the more difficult the road to sustainability becomes.

Environmental Pollution Control and the 4Rs

The field of environmental protection is subject to the same commoditization forces. The distinction between products with high commodity potential and processes with low commodity potential are the following:

Environmental products (high commodity potential)
Cleanup equipment and tools
Energy-efficient appliances
Waste-management equipment and services
Environmentally friendly products
Photovoltaic cells
Biomass fuels
Parks and zoos

Environmental system processes (low commodity potential)
Energy- and materials-conservation programs
Ecological design
Watershed management
Voluntary simplicity
Community building and resource sharing
Environmental education
Waste-reduction programs
Extended producer responsibility
Habitat protection and conservation

Many analysts look to nature for models of effective resource conservation. In natural systems there is no such thing as waste, there are only by-products that become resources for another organism or process. By mimicking nature in the design of our production and waste-management systems, energy and materials efficiency can be enormously improved and wastes thereby dramatically reduced. In the process, consumption efficiency can be improved as well. Every step, from obtaining and processing raw materials to delivering final goods to repair and maintenance to final disposal, could be organized with the intention of minimizing the waste of energy and materials.[32]

There is a widely accepted formula for waste minimization, the *4Rs:* Reduce, Reuse, Recycle, and Recover. These four approaches are typically depicted in order of effectiveness and priority:

- *REDUCE the amount of waste produced.* Reduce unnecessary packaging, improve energy and materials efficiency in both production and consumption, and reduce the amount of material and energy required to provide any given service.
- *REUSE materials.* The waste materials and energy that cannot be eliminated should be captured and reused in another part of the production process. Waste can also be greatly reduced by repair and remanufacture of existing consumer products rather the manufacture and sale of new ones. For example, reconditioning a car to make it last for another ten years requires 42 percent less energy and significantly less material than manufacturing a new car.[33]
- *RECYCLE used materials.* Materials that cannot be recovered for reuse, repair, or remanufacture should be recycled for use as raw materials somewhere else in the economy.
- *RECOVER materials.* Finally, some components of the waste stream that are not suited for reuse as raw materials in other processes can sometimes be recovered as fuel for energy production or steam generation.

Each of these Rs is preferred over simple disposal. A formal policy on waste reduction was established in the United States by the 1976 Resource Conservation and Recovery Act (RCRA), which originally focused on requiring industry to modify its production processes to encourage and facilitate recycling of raw materials. This approach, however, was considered too expensive by manufacturers, who successfully convinced the administration and Congress that RCRA should instead support landfill improvements and the construction of a new generation of solid-waste incinerators.[34] Waste-handling services and incineration have far greater commodity potential than does waste reduction, which in effect has negative commodity potential.

The priority ranking for the 4Rs[35] is the exact opposite of the order of commodity potential. Energy recovery yields electricity, usable energy at its most portable and marketable. Recycling produces some products, and the things that have been most recycled in the United States—paper and aluminum—are the most marketable. Reuse tends to reduce the consumption of new goods. And as already noted, use reduction has virtually negative commodity potential. Although the 4Rs approach and prioritization are regularly advocated by environmental agencies in the United States and elsewhere, actual practice reflects another set of priorities.

A closer look at recycling as it is presently organized in the United States provides a good example of how commoditization distorts what is ostensibly an environmentally beneficial activity. The biggest problem with recycling is that it has never been fully integrated into a system of waste reduction that includes the entire life cycle of a product, from raw-materials extraction to production and consumption. In such a system, recyclability would be designed and manufactured into products up front and manufacturers would be responsible for seeing that their products are recycled at the end of their lives. Consumer products would contain materials meant to last beyond their first use, materials designed to be recycled. A variation of this is becoming increasingly the case for automobiles, where a used-parts market has always thrived, and somewhat less so for major appliances. But for the majority of small appliances and the vast majority of consumer goods, this is far from the case.

Recycling as a stand-alone service rather than as part of an overall waste-reduction strategy is invariably organized to produce raw materials for sale. This is consistent with the pattern we have been describing of privileging products over processes. In today's U.S. economy, it is considerably less expensive to throw away a damaged radio, for example, than it is to pay to have it repaired. This is a direct result of commoditization policies, which lower prices for energy and raw materials while raising the cost of labor by taxing wages and not energy. As commoditization drives innovation, goods that were once repairable no longer are. For example, in recent years the sports-shoe industry has replaced rubber and leather with synthetic materials. As a result, the athletic shoes are now unrepairable.

Products are not designed for reuse, and therefore recycling programs must first transform the material they collect into usable materials again, a process that uses additional raw materials and energy and produces considerable pollution.[36] Even though science can produce materials that are easily recyclable, investments in this research and development are stymied because of the ambiguity of property rights over recycled materials in subsequent uses and liability for the environmental effects of postconsumer waste. As I have argued, the ease with which property rights can be established is the most significant characteristic of goods and services with high commodity potential.

One proposal for improving recycling and materials recovery is to require manufacturers to take back products, especially large consumer goods and appliances, once they have completed their service life. This would create significant incentives for manufacturers to design longevity and recyclability into their products. Companies facing such requirements or liability for the environmental effects of their waste may shift to leases rather than sales. This is already happening in some instances. For example, some large carpet manufacturers lease carpets to major institutional customers with the intention of recovering the worn carpets and using the materials in the production of new carpeting. Leasing is possible for a wide range of consumer goods.

Since the 1970s, the generation of waste products by industry in the United States has declined significantly. This is often pointed to as evidence that environmental problems can be solved by continued economic growth and that any efforts to slow growth to achieve environmental ends are counterproductive. However, taking a consumption perspective as described in chapters 1 and 2, it becomes apparent that only industrial waste has declined, while *consumer waste has risen precipitously*. This is possible because a large percentage of the material thrown away by Americans was produced in other lands. For example, annual production of metals in the United States is more than 1.5 tons per capita, down from a maximum of close to 2 tons in the early 1970s. However, this decline merely reflects the fact that the United States is increasingly dependent on imported ores and metals. U.S. metal *consumption* is now more than 2.5 tons per capita.[37] In effect, U.S. environmental gains have been purchased by exporting industrial production and associated pollution elsewhere. Declines in industrial waste have not been matched by a decline in domestic refuse. *Per-capita* municipal solid waste has increased from 2.7 pounds per day in 1960 to 4.5 pounds per person per day in 1990, meaning that not only are we producing more garbage because our population has grown, but each of us actually consumes more products and discards more waste.[38] What is more, the presence of plastics, the most highly commoditizable material in the waste stream, has grown by a factor of 40, while metal and glass, materials with lower commodity potential, have increased only slightly. As a percentage of total waste, plastics have increased from less than half a percent to nearly 10 percent, a twentyfold increase.

In many ways, plastics are the ultimate commodity materials. Almost infinitely malleable, they have greatly increased our capacity for standardization and packaging. Mostly synthesized from petroleum, plastics greatly expand the amount of fuel energy embodied in the products of everyday life. Their malleability makes them particularly well suited for molding operations, which greatly increase the labor savings. Plastics are difficult to recycle and plastic products almost impossible to repair. They have become a huge portion of the waste stream and degrade extremely slowly in the environment, if at all. The by-products of their oxidation, particularly when burned, produce a considerable portion of the dioxins and other toxic pollutants in the air, water, and soil.[39]

Recommendations and Conclusions

Western-style economic development has flourished largely because societies invented legal and institutional mechanisms that favored commoditization and expansion. If noncommercial values such as human rights and ecological integrity are to be serious goals of public policy, legal and political instruments designed to favor the noncommodity satisfaction of human wants must be adopted to counterbalance the force of commoditization. Commercial law is now evolving into a global legal framework designed to unleash commercial energies worldwide by minimizing the capacity of states to restrict access to markets. As a result, commoditization pressures are expanding worldwide. Since the legal and political actors unleashing these forces operate at the global level, countervailing pressures must also operate globally. But since noncommodity solutions to human needs and wants are inherently local, the effects of these countervailing forces must be felt at the local level. New legal and political capacity to stimulate investment in community-based, less commoditized satisfactions for human needs and wants must devolve to the level nearest to the people with those needs and wants. There have been several efforts to describe the emergence of global civil society as a precursor to a governance capacity that can act with some effect and authority at both the global and local levels.[40] None of this work addresses consumption efficiency, however, let alone commoditization and its potential for stimulating material and energy throughput. At the same time, nation-states must invent new legal frameworks that allow localities to innovate economically and that protect them from the colonizing impulses of global forces and actors.

The forces that drive the increasing commoditization of all aspects of life derive naturally from the structure of the economy. Resources (wealth, income, and associated power) are allocated to those who are successful at selling. Intuition is enough for almost everyone to understand that not everything in the good life can be packaged and sold. Yet the full development of those noncommodities also requires time, attention, and resources. Public policy can go a long way toward correcting the distortions of development wrought by commoditization by making certain that noncommercial values and consideration of the common good play a role in the allocation of resources. Policy tools that counterbalance commoditization pressures fall into the following main categories:

- Increased public investment in research and development and additional subsidization of beneficial goods and services with low commodity potential.
- Taxes and fees that help make the prices of environmentally and socially damaging commodities reflect their true costs. Examples include the 1990 U.S. tax on ozone-depleting chemicals, taxes and fees linked to discharges, and carbon dioxide taxes in Sweden and Norway.
- Investing in "natural capital." This includes various methods to encourage or require investments in protecting and restoring the natural environment to an equivalent degree to which profit-making activities deplete the natural environment.
- Protecting ecological integrity. This would entail mandating limits to human disruption of key ecological regions and global ecosystemic processes, thereby encouraging consumption efficiency.
- Investing in human development. Public investment would be directed toward beneficial qualities associated with low commodity potential.
- Tax reforms that decrease taxes on income from labor (the least commoditizable of production inputs), thereby decreasing the cost of labor in relation to energy and materials.

- Protecting the rights of workers. Human labor has far lower commodity potential than machinery and energy. Policies that subsidize job creation, particularly in community building, environmental restoration, and other caretaking tasks directly counteract the effects of commoditization.

The sum of these efforts would begin to create a parallel economy of care and connection that can counter the negative effects of the domination by the economy of commoditization. An economy of care and connection, because it is directed toward development of goods and services with lower commodity potential and higher consumption efficiency, will require considerable involvement of a democratically accountable public sector at the local, national, and global levels. At the global level, what is clearly needed are democratically accountable institutions with the authority and capacity to act on the same international scale as the multinational corporations. To promote effective policies at the global level we will need an international labor movement, an international consumer movement, an international democracy movement, and so on. At the same time, noncommodity alternatives to human-need satisfaction are inherently local, and any governance reform for sustainable development must include increased civic capacity at the local and neighborhood levels.

II

Chains of Consumption

5

Distancing: Consumption and the Severing of Feedback

Thomas Princen

Power and influence typically enter discussions of environmental degradation as a lack of political will among policy makers, greed among corporations, or ignorance among consumers. An analytical approach, by contrast, what might be termed a political economy of degradation, focuses on the day-to-day decisionmaking of key actors, especially producers and consumers, their interactions, and the environmental impacts of their decisions. This approach relies for its explanation of environmental change less on such nebulous notions as political will or greed or even lack of information, and more on actors' incentives to realize immediate gains in a given institutional setting. It also presumes that individual choices can be perfectly rational while the collective outcome is suboptimal, even destructive, for all. It assumes that even where collective action problems do not arise, strategic interaction can result in displaced or deferred depletion of natural capital. In short, a political economy of degradation posits that, where better attitudes and values or more data are not enough, analysis of the distribution of short-term costs, along with many long-term impacts, will generate a better explanation and, eventually, better prescription.

A political economy of degradation, of production and consumption, and, as a corollary, of sustainability, must account not only for the full range of costs but the sources of those costs. It must consider how the pursuit of wealth can, deliberately or not, lead to uncounted costs and unaccountable actors. Rather than taking these costs as external and treating them as a production failure which can be internalized or a consumption choice that can be corrected with more information or as a contentious item that can be negotiated or as a necessary side effect that must be tolerated, it needs to treat them as part of competitive business strategy. Furthermore, such cost generation should be assessed as a product of both production and consumption decisions, whether or not the costs are deliberately or knowingly externalized. Finally, it must consider how differences in power (political, financial, informational) contribute to such costs.

In this chapter, then, I explore the conditions of cost generation and externalization, especially those that are largely unintended and inadvertent. I first argue that business strategy and state policy creates a never-ending search for frontiers, defined in political, economic, and ecological terms. Then, from the production angle, I develop the concepts of *shading*, that is, obscuring of costs, and, from the consumption angle, *distancing*, the separation of production and consumption decisions, both of which impede ecological and social feedback. I conclude with four specific propositions and two general implications.

The Quest for Frontiers

For a business firm, the ideal economy is a frontier economy. Politically, a frontier economy lacks jurisdictional authority. Resource users are those who claim rights but do not need to accept responsibility for the resource. Resistance by downstream recipients of externalized costs is insignificant. Economically, a frontier economy provides free resources and waste sinks. Equivalently, there is always another frontier to move to when the resource is exhausted. Ecologically, resources and sinks in a frontier economy either have infinite regenerative capacity or they can always be replaced or substituted via mobility and technology. Negative feedback only takes the form of financial returns on investment, never full ecological costs returning to the firm. Business strategy with respect to resource use can be viewed as a search for approximations of this ideal. Being strategically competitive means entering early to extract the resource and exiting promptly when political resistance mounts or costs return.

Although there are few, if any, true frontier economies today, three factors lead policy makers and business people to continually try to construct them. Whether or not the resulting simulations are truly frontiers in an ecologically meaningful sense, they do foster the same kind of short-term, cut-and-run, mine-and-depart behavior. Put differently, because firms always have an incentive to reduce costs, they continually face a binary choice between efficiency seeking and cost externalizing. In a frontier economy, they tip toward cost externalizing.

The first factor is governmental promotion of industry where "internal subsidies" are created for firms, external costs are passed to some or all of the government's own citizens, and boom-and-bust economies are the norm.[1] Often these are manufacturing or "brown" industries that are lured to enhance the tax base or provide jobs.

The second factor is the so-called technology- or information-based economies which have the appearance of creating unlimited growth via infinitely expandable, "clean" pools of data and knowledge. Evidence typically given to support this view is that natural resource production only accounts for some 5% of advanced economies. There is, however, no evidence that such economies actually limit, let alone, reduce the throughput of materials and energy, nor that such economies could survive without a net inflow of resources from abroad. "Advanced" production may actually entail considerable cost externalization, what some have termed ecological "shadow"[2] or "footprint."[3] Such production may simply extend the cost return time by obscuring the true consequences of such production and by separating economic actors' decisions so as to render those consequences unintelligible, what I describe below as shading and distancing.

The third factor is jurisdictional discontinuity. From the firm's perspective, the more its transactions cross jurisdictional boundaries, the more it is operating, de facto and often de jure, in a frontier economy. The effect can be seen by comparing local and international economies. Local economies, especially natural resource–based economies, are embedded in a mosaic of institutional arrangements, some governmental and legally enforceable, some cultural and enforced by societal norms.[4] If a recipient of externalized costs cannot negotiate satisfactorily with or find redress from a producer, that party can generally appeal to a higher authority. Such appeals stop abruptly at a border, however, especially an international border. Thus, in the classic case of externality, downstream recipients of a factory's wastes need merely travel upstream to register a complaint or negotiate compensation. By contrast, Mexican recipients of North American waste (or of surface water depletion) will encounter myriad legal, cultural, and political obstacles crossing the border to seek redress—and vice versa.[5]

The jurisdictional discontinuity derives largely from a defining characteristic of the state, that is, its ability to protect its citizens from claims by foreigners. At the same time that states exercise their prerogative to protect their citizens, they promote those practices that separate and distribute the costs of production to their benefit. If the costs of production can be externalized beyond state boundaries and the benefits internalized, both the state and the firm benefit. The state benefits indirectly via enhanced tax revenues and the avoidance of costly regulation. International borders are probably best suited to make this separation, especially when the gains from trade can be invoked and such gains are readily measured. The maquiladoras in Latin America typify such activity by the state and selected firms. The benefits of foreign assembly go primarily to Northern firms and consumers via lower prices, whereas the costs, especially those that do not enter the gains-from-trade calculation—e.g., worker injuries, local pollution—go to the workers and their communities.

In short, incentives arise for producer and state alike to export costs and to do so as if other jurisdictions are mere frontiers. For the firm, these incentives derive in part from the need to be competitive. For the state, they derive in part from the need to assist domestic industries in international markets which generates revenues and relieves unemployment. But incentives for firms and states also derive from the jurisdictional quandary created by constantly changing technologies and markets. Such economic changes generate institutional demands that few governments are equipped to handle.[6] Pollutants have long time lags between cause and effect and endangered species pit workers against environmentalists. "Going abroad," especially for short-term gains, becomes a tempting way for business and state alike to escape such demands. Even where this incentive is minor, promoters of innovation and market expansion who seek mobility, open markets, and a cheap and compliant labor force, have little to say about the requisite institutional climate and what Polanyi long ago recognized as the crucial need to match institutional rates of change with economic rates.[7] Instead, these dominant economic actors cross boundaries, opting for simulations of frontiers where resources and sinks are abundant (at least from the perspective of the home firm and country) and recipients of external costs are few or have little clout (again, at least with respect to the home country). The jurisdictional divide is a convenient means of taking a hands-off approach to these "messy political" issues. If ecological or social decay should follow an investment or market opening, as is happening in southern Mexico and parts of China, the firms and states merely move on to the next frontier. What appear to be efficiencies to some (namely, the beneficiaries of such expansion) is cost externalization to others (namely, to those who incur the costs, immediately and in the long run). Whether net benefits result is an open question, one I will take up further under shading. What can be said, however, is that to the extent expanding markets are indeed characterized by simulated frontiers, businesses that appear to be creating wealth often are actually converting renewables—resources and sinks—into non-renewables. Resource extraction becomes mining, not long-term management or stewardship. Under these conditions, the net benefits over time are, at best, questionable. For prescriptive purposes, this suggests that, on a global basis, it is not necessary to prove sustainability by calculating with precision net benefits when, in the aggregate, resources, especially those that are potentially renewable, are being mined. Frontier economies are, by themselves, not sustainable and certainly perpetual simulations of frontiers are not.

If firms and states find incentives to tip their cost reduction strategies from efficiency seeking to cost externalizing by simulating frontiers, they can be successful in part because certain costs are, or can be made, less visible. I use the term "shading" with its several denotations and connotations to suggest that, in a firm's relations with buyers, suppliers, consumers, governments, and the public at large, immediate benefits are emphasized while full costs of production are depreciated. What is more, from a strategic point of view, the firm tends to select technologies and production patterns that engender the least resistance from recipients of external costs and from those who represent them (e.g., environmental NGOs) and that minimize the likelihood of costs returning to the firm in a decision-relevant time period.

Shading Costs

Firms can control some of their costs by varying the factor mix, negotiating prices for labor and supplies, and investing in new technologies.

But just as all human activity has some environmental impact, all production generates some costs beyond the purview or control of the firm. As a result, some costs are not just externalized by the firm, that is, knowingly sent downstream. Rather, they are rendered invisible to the firm and to others. This I term "shading." Shading occurs when the known cost of what firms send downstream is only a part of the full ecological and social costs. Environmental policies, especially government measures that depend on the firm's knowledge of impacts, cannot internalize all such costs. In part, this is a cognitive issue. But in large part it is a structural and strategic issue, a result of firms operating in market structures where competition is not the classic case of perfect competition but the oligopolistic case of competitive strategy.[8] In this setting, firms' opportunistic behavior vis-à-vis each other and governments confounds internalization efforts. In a non-strategic, non-opportunistic world of competitive markets, most externalized costs can be addressed by monitoring, measuring, and regulating, in short, by internalizing costs. But in the strategic world of business, arguably the one that prevails domestically in capitalist societies and certainly internationally, internalization may not be enough. Four usages of shading suggest how internalization approaches are likely to miss the unknown, the misperceived, or the deliberately manipulated costs of production.

The first usage of shading is to cast a shadow or obscure an image. As firms highlight the benefits of production, they simultaneously shade the costs. In this sense, shading, as when a tree shades out competitors, is passive, it is without malicious intent or even neglect. Just as the tree, by maximizing leaf exposure, is just being a tree, the firm, by being competitive, by highlighting benefits and shading costs, is just doing business. But in just doing business, it may cast a shadow on more than its own competitors. It may, however inadvertently, cast a shadow on the health of nearby citizens and ecosystems. If the firm's benefits are realized in the short term (say, the time from initial development to positive net returns) and if the shadow entails long-term and, especially, irreversible effects—whether for individuals, communities, or ecosystems—the costs are neither trivial nor "mere externalities."

Such passive shading occurs when that which can be predicted—i.e., the benefits of production—coincides with the interests of those who promote new technologies and markets. That is, because the benefits of new technologies are predictable (since this is why they were developed) and many costs are necessarily unpredictable or highly uncertain, benefits inevitably receive more initial attention. Moreover, because developers of the technology or market have the incentive to highlight benefits, the technology is, in effect, presumed innocent of significant deleterious effects. The burden of proof that the technology produces net harms, falls on the recipients of those harms. But there is always a lag time between the application of the technology and the realization of its full costs. Due to scientific, cognitive, and cultural lapses, that time may be days, decades, or, in the case of fossil fuels and global warming, even centuries. During this time, the costs are real but invisible. And the time of their invisibility may be extended if a new technology supplants the old or if a firm develops a new application or relocates its production facilities. Such shading of costs can be quite unintentional or inadvertent, yet exacerbated by accelerating economic activity. The long-term consequences, however, especially if irreversibilities ensue, may be profound for the environment and for the society, not to mention for the firm itself.

To illustrate, DDT was widely heralded as a miracle insecticide during and immediately after WWII.[9] Only a couple decades later did its true costs become evident, resulting in a ban on use in the US and many other countries. Similarly, CFCs were considered for decades to be "inert" wonder chemicals that became essential to refrigeration, cleaning, and other industrial purposes. It was almost by accident that its stratospheric ozone depleting properties were discovered.

In the second usage, shading occurs when a firm deliberately outcompetes others by, among other things, weakening their ability to operate. Thus, the botanical metaphor would be a parasite such as mistletoe that extracts resources from its host tree, debilitates the tree in the process, and then propagates and spreads to other trees to continue the process. Shading occurs over time because the extractive agent and its beneficiaries enjoy the benefits during extraction but leave the costs behind. The mistletoe, for most of its life in the tree, is inconspicuous and apparently harmless. Similarly, a mining company enjoys (and to some extent distributes) the benefits during mining and then leaves the environmental and social costs to those remaining. The mistletoe does not suffer because its host is debilitated nor does the mining company because there are either always more sites or the company can reorganize.

As with passive shading, the mistletoe is just being a parasite and the firm that extracts and moves on is just doing business. But it is active in the sense of being parasitical. The temporal separation of costs and benefits allows the benefits to be highlighted ex ante, while the costs are born ex post.

It is important to emphasize that, at this point in the argument, no malicious intent nor deliberate attempt to export costs is necessary to demonstrate (at least logically) that firms may indeed externalize costs and that such externalization may be a function of everyday competitive strategy as well as governmental policy. To the extent that empirical testing confirms this component of cost externalization, the argument constitutes the strong case. That is, if firms, by just engaging in standard business practice, including the proliferation of new technologies and the expansion of markets, generate significant externalities, then ethical or legal arguments or even scientific verification are not necessary to justify strict regulation of such practices, if not wholesale restructuring of industrial production. The structure of business, especially the lags between benefits and costs and the institutional lags in responding, are enough to justify such changes when irreversibilities ensue and life-support systems are threatened.

The third usage of shading is associated with shading prices. Here, marketing techniques range from a simple "sale" or "come-on" to tactics aimed at gaining brand name loyalty to dumping and predatory pricing. Prices are not shaded to be competitive in the economic sense of reducing costs through greater productive and allocative efficiencies. Rather, they are shaded to be competitive in the marketing sense of capturing share.[10] Such price reduction is necessarily short term since the entire point is to gain a temporary market advantage, an advantage that, under economically competitive conditions, could not be maintained. Thus, after the competition is driven out, a firm employing, say, predatory pricing can increase prices or produce an inferior product or pass some of the costs of the predatory strategy on to others or to the environment.

To illustrate, in the North American west before railroads, meat production was a highly localized operation. Farmers raised pigs and local butchers prepared and sold them. Feed was grown locally and the by-products disposed locally. With the railroad, however, and especially refrigerated cars, packing shifted to metropolitan centers such as Chicago. Production and distribution became nationwide, even global in the late 19th century. Packers with capital and political connections drove out local butchers by employing least-cost technologies and engaging in predatory pricing. The industry was soon dominated by a handful of meat packers. The rapid expansion of the market accelerated the conversion of prairie and bison to cornfields, fenced pastures, and feedlots. With large, capital-intensive plants and their associated chemical research laboratories, the entire operation was extremely efficient, from feedlot fattening and slaughtering to the use of virtually every body part and the distribution of the products nationwide and abroad. At the same time, packers disguised and sold bone and offal and spoiled meat and sent all remaining wastes into nearby streams. Historian William Cronon concludes that the packers

> ... sold what they should have thrown away—and yet did little to prevent pollution from the wastes that finally washed down their sewers.... The packers drove honest butchers out of business with deceitful products, so that in the end there would be nothing left but the Big Four [packers] and their foul meats. The Chicago packers had wasted honesty and community alike in their single-minded drive to extract every last penny from the wretched animals that walked through their doors.[11]

The benefits of such modern meat production were, of course, considerable. But the costs to natural systems, especially to the prairie and aquatic ecosystems, and to public health were also considerable, yet invisible to most. Shading was in part passive and occurred through the diffusion of new technologies—railroads and refrigeration. The prairie disappeared incrementally as local, national, and international demand for feed and meat rose. But a significant, and otherwise largely neglected, part of the cost shading was direct and active and, where the only visible losers were "inefficient" local butchers, legal.

Thus, the notion of shading suggests that normal business practices, from the passive, inadvertent efforts to reduce costs to the active, parasitical, or predatory techniques, can be perfectly legal and acceptable to nearly all concerned. When costs are born by a few or dispersed to the many, and when costs are displaced over time and space, such business activity appears not only normal but desirable as net benefits can be presumed. Yet net benefits are rarely demonstrated, especially with respect to long-term effects. Moreover, the calculations are extremely difficult if not impossible to make. But from a strategic perspective, a possibly overriding reason why net benefits are assumed is that those who benefit from a new technology or a market expansion are also those who make many of the key decisions. They are the ones who gain the railroad right-of-ways and who manipulate finances to gain economic power to drive out the small producers. With benefits highlighted and costs shaded—at least in the near term—such business activity, even that bordering on outright exploitation and destruction, is presumed to have net benefits.

Many of the techniques for highlighting benefits and shading costs are standard business practices and constitute much of what is taught in business schools under the rubric of "competitive strategy." From the perspective of the firm, competition is not becoming a firm with "normal economic returns" as economists would characterize competitive industries. Rather, competition is indeed outdoing competitors by weakening them or by cornering a market. The endgame for the strategically competitive firm is not to be one firm among many in a competitive market; it is becoming a monopolist to appropriate economic rents in the form of above normal (i.e., above competitive) prices. Aside from power utilities, it is only a temporary monopoly the strategically competitive firm seeks. Once competitors enter, the dominant firm changes the product or develops a different product. Thus, what is competitive strategy to a firm is a shifting monopolist with shaded costs to others. Although many of these costs are economic and are borne by losing firms, many are passed on to consumers and downstream recipients. Because this form of shading applies to a short-term business strategy, it can readily generate long-term or environmental costs. The important question, then, is, to what extent does shifting monopoly lead to environmental degradation? Two mechanisms appear to translate shifting monopoly as competitive strategy to environmental degradation as predation. One is to shorten time horizons (or increase discount rates) and the other is, once again, to simulate the cut-and-run practices of a frontier economy.

Shortened time horizons can be seen by comparing the shifting monopoly to the firm that succeeds by grounding its operations firmly in a community. The shifting monopolist is, by design, flexible and adaptable. The most successful (and, it seems, the most admired in the business community) are those who can move freely from one market niche to another, those who can easily integrate vertically or outsource as market conditions for suppliers change, those who can move capital and production readily capturing favorable returns and inputs as markets, labor forces, and consumers develop. By contrast, the locally owned firm succeeds by developing a stable base in a community, relying upon the local labor force and on many local suppliers and buyers, all of whom in turn rely on the company. This firm must still be adaptable in the face of changing external factors such as market prices and migration. But successful adaptation for the local firm to, say, a market downturn means, among other things, cutting back on production and on employee hours. It does not mean eliminating production altogether and cutting back on employees, at least not to the degree and with the facility of mobile firms. It does not mean constantly finding the best return on assets with its capital, plant, and equipment. It does not mean pulling up stakes whenever economic times are tough or opportunities suddenly emerge elsewhere.[12]

The difference between the shifting monopolist and the community-based firm is not just big versus small, nor primarily mobile versus immobile. Rather, the crucial difference with respect to resource use is time horizon and the institutional factors that influence that horizon.[13] To illustrate, Norwegians have been fishing for cod in the Lofoten Islands of northern Norway for centuries. Their privately owned enterprises range from one-man operations with tiny boats to 100 foot ships with crews of a dozen or more. But virtually all fishermen have a long, intergenerational history of fishing and, especially in the last century, a history of self-regulation. They deliberately limit technologies and use rights that threaten the resource. Factory trawlers and seine ships from outside the region, those with technologies and capital to make their operations highly mobile—and highly short-term and unregulated in behavior—have always been prohibited.[14] The result is, by virtually any definition, sustainable use.

Similarly, in the timber industry in North America and probably in many other regions, virtually the only firms that can be described as sustainable—i.e., conducting business over generations of trees and people—are small, family-owned operations that have weathered, along with their communities, the vicissitudes of changing technologies and markets. Unlike their highly capitalized competitors, they do not pull up stakes when a forest is depleted. With respect to that resource, they are in business for the long term.

The second mechanism translating shifting monopoly to environmental degradation is the simulation of cut-and-run operations in a frontier economy. With changes in technology and marketing, a firm finds new markets and new, albeit temporary, monopolies. Just as the cut-and-run, mine-and-depart operators leave behind degraded landscapes with depleted resources, the shifting monopolist creates one product after another, each of which is distinguished not by its contribution to a sustainable society but by its own temporary ability to sell. And when the shifting monopolist moves on, the products, its waste stream, and an entire line of workers from extraction to sales, are left behind. The costs are all "legitimate": during production the benefits are presumed to exceed the costs. But those costs—both direct and indirect—were incompletely assessed in the first place, they were shaded, and left for others to absorb. To illustrate, some transnational agribusinesses incorporate the anticipation of insurmountable pest problems and rising discontent among local farmers in their investment strategies. In one case in Mexico, investors in export melons planned for only a seven-year production cycle in a given site with continuous investments in "new regions of Mexico in order to have production sites available as the anticipated ecological disruption, economic crisis, and social unrest emerged. The vacating exporters left diminished economic opportunities and increased social inequity in their wake."[15]

The fourth usage of shading is that of shady dealings and thus connotes the truly disreputable side of doing business. It includes familiar examples of illegal waste dumping and safety violations as at Bhopal. This usage involves more than illegal business activity, however. It includes those activities that are technically legal but cost society in ways not yet accounted for by existing legal institutions. For this form of shading, the term "cheating," which implies neither illegal nor predatory activity, may be most appropriate. Cheating is making individual, short-term gain at the expense of the larger business community and, ultimately, the larger social and ecological community. It exploits gaps or lapses in institutional structures due to either technological changes (and their unforeseen consequences) or to power disparities among producers or consumers. To illustrate cheating on institutional lapses, the conversion of prairie to cornfield in the western United States was greatly facilitated by both the railroad and the grain elevator. The new technology of the grain elevator was readily exploited by operators in the mixing and pricing of grades of grain and by speculators in the cornering of markets. Farmers were often the losers, not to mention prairie ecosystems. When the entire market was threatened by such cheating, the Chicago Board of Trade and the Illinois legislature instituted strict regulations.[16]

Cheating on weaker actors can be illustrated by activities of the transnational pesticide industry. The dangers of pesticide use to workers is well established in the United States and elsewhere to the point where many chemicals are outright banned for domestic use. But producers use the same chemicals in many Southern countries where safety measures are notoriously lax. What is more, as pesticide residues have hurt the exports of crops due to agricultural restrictions on imports by Northern countries, pesticide manufacturers have shifted to pesticides that are less persistent but more toxic to workers.[17]

In sum, from the production angle on commerce, costs are easily rendered invisible in an expansive, rapidly changing, strategically oriented business climate. Some technologies and markets—e.g., pesticides and lead additives—may be more prone than others to such shading. But it is important to stress that, in general, it is the very nature of new technologies and new markets that creates the impression of ever new frontiers, however simulated. This impression also provides the rationale for discovering yet more frontiers when the current one is exhausted. Ultimately, it provides the justification and the logic for mining resources, as opposed to managing them over the long term. In part, then, market expansion is frontier expansion. Not only does expansion relegate more and more resources to market forces, a pervasive critique of the market from many environmentalists. But the very dynamism of such expansion, the continuous creation of new products and processes and markets, leads to cognitive and institutional lags between initial realization of benefits and eventual discovery of full costs.

Shading as competitive strategy is short term but, because it can be reproduced, because one can, in principle, always create a new product and find a new market, it has the appearance of being long term, what some would term "sustainable." In fact, much of the discussion in the fast-growing "business and environment" literature implicitly uses sustainability in terms of sustaining current business practice.[18] But such usage has neither ecological nor social meaning when the rates of technological and market changes exceed the institutional capacities to detect and deal with long-term effects.

To this point, I have addressed cost externalization as a function of producers' strategies. A fuller set of propositions, a "theory" of the political economy of environmental degradation, must also account for *consumers'* propensities to externalize costs. To bring in this side of the economic equation, the consumption angle, I assume that consumer decisions based on incomplete information about the conditions of production are not likely to account for their long-term and environmental impacts, including both the use of the resource and the disposal of the endproducts. That is, commercial patterns that separate consumers from the consequences of their behavior are likely to weight consumption decisions toward narrowly self-interested consumption and away from long-term, intergenerational, and non-human concerns. This is not to say that consumers would otherwise be altruistic. It is reasonable, however, to assume that, when feedback is effective and costs cannot be displaced, consumers are capable of adjusting their consumption patterns and restraining their use of resources.[19] That the contemporary political economy breaks feedback and externalizes costs is the puzzle in need of explication. Consumers in a dynamic, expansive economy are more likely to be insulated from the consequences of their choices. They are left with little basis for their decisions beyond price. This insulation occurs in part through the separating of production and consumption decisions along a chain of resource decisions, what I term "distancing."

Distancing

"Distance" is the separation between primary resource extraction decisions and ultimate consumption decisions occurring along four dimensions—geography, culture, bargaining power, and agency. At one extreme, zero-distance is production and consumption by one household or individual; at the other extreme, it is global, cross-cultural, and among agents of disparate abilities and alternatives. As with shading, the environmental impacts of these consumption and production decisions are not always obvious, intended, or even known to decision makers. In the following, I compare a local farmer's market with the international food trade to illustrate the dimensions of distance and to generate propositions that relate consumption and production to ecological impact.

Geographic Dimension

In a farmers market, the physical distance farmers are likely to travel is small. It is constrained by the cost of transportation and by competing farmers' markets. Consumers are similarly constrained. In contrast, farmers selling in international markets have virtually no constraints on physical distance because transportation is readily available and intermediaries can facilitate transactions.

From a resource use perspective, that is, from the perspective of the soil and water supporting agricultural production, the physical distance itself may have only a minor impact, except for questions of spoilage, packaging, transmission of pests, and the like. Aggregate demand for the product is not affected by the physical distance per se. As discussed below, bargaining and agency are more critical dimensions of distancing and are often, but not always, associated with physical distance. These dimensions bear directly on critical resource management issues of ecological feedback, monitoring, and exclusion. Nevertheless, because geographic distance often interacts with these other dimensions, it can be a useful proxy.

Cultural Dimension

In the local market, as a consumer I know something about local farmers. I have a pretty good idea of who they are, what they value, and where they fit in my society. I certainly can see them face-to-face. I can see how they set up their stall, and how they display their produce. When I ask about the produce, how it was grown, how to cook it, and so forth, I am likely to get a straight answer. At a minimum, in this setting deception is more difficult with those who are perceived to be part of a repeated game of buying and selling.[20]

In addition, I know, and the farmer knows, and we each know that the other knows that, with the geographic distance so short, I could travel to the farm and check things out. Of course, it is very unlikely that I or any other customer will actually do so. But from a strategic perspective, the important calculation is whether there is a significant probability of just one visit—from a customer, from a member farmer, or from, say, a reporter. If direct monitoring is likely, or if farmers so perceive it, and if their sales depend on their reputations as growers of quality produce, then production short-cuts will be risky.

These conditions do not apply to the international grower. When I buy grapes from Chile in my North American supermarket, I know nothing about the grower and I am never likely to. And neither I nor anyone I know is likely to inspect that farmer's operation. Moreover, if I had concerns about the economic and environmental conditions of the Chilean farmer, I can do little about it. I cannot pay the farmer more or less and I can not encourage less pesticide use. I can only assume that, in a Southern country, the grower needs the employment and, hence, the grape sales and that the grower is getting a fair portion of the rents. Because I know little about Chileans' way of life, little about their economic or legal status, and little about their farm practices, I am at a cultural distance. As a result, I have no way of knowing if my consumption is supporting or undermining that farmer, economically or ecologically. With no feedback or with uncertain feedback I can only assume and, as with most of us, prefer to assume, that my purchases are supportive. I can only assume that all parties have entered the bargain voluntarily and that coercion or extreme dependency have not been the conditions of my enjoyment of Chilean grapes.[21]

In practice, as consumers most of us never give producers' welfare or environmental impact a thought. Japanese consumers of ivory, for example, never considered that their aggregate demand was driving many African elephant populations to extinction. In an ecologically rational world,[22] this lack of consideration would not be a problem. When elephant populations diminish or soil erosion accelerates, ecological and market signals would combine to relieve the pressure on the resource. But in the real world of incomplete information, strategic interaction, and shaded costs, lack of consumer consideration for environmental impact is a problem because cultural distance provides opportunities for some producers to ignore or avoid or misrepresent certain information, including that which reveals the long-term costs of a consumption decision.

Historically, cultural distance was often accompanied by economic mobility and coercion. English colonial tea planters bought forest lands for clearing from the government of British India and from private owners but also simply commandeered village commons.[23] Possibly equally destructive was the introduction of Indian workers from outside the region and from lower caste than the residents. These workers had been subsistence farmers but became totally dependent on tea plantations. Thus, outsiders, culturally distanced Europeans and Indians, assumed the land conversion role.[24]

In general, mobile outsiders, whether migrants (voluntary or coerced) or capitalists (foreign or domestic), view others' resources as frontiers, as opportunities to exploit and move on. These interveners treat others' property as open-access while ignoring or overriding with force or the lure of riches the closed access property regimes of residents.[25] Mobile outsiders relate to a resource as a mining opportunity whereas long-term residents tend to relate to it as a source of economic security.[26]

The primary effect of cultural distance is to block ecological feedback by inhibiting information flow from extraction to consumption decisions. To illustrate, distant-water, highly mobile fishermen fish until the financial return on effort diminishes sufficiently and then move to the next site. With efficient technologies, moving on may occur when stocks are severely depleted. Distanced production and consumption via networks of mobile fishing, processing, and distributing firms remains constant whether or not a given stock is depleted. By contrast, inshore fishers have long encountered periodic declines in fish stocks. They may not know the precise causes—overfishing or weather or predator-prey dynamics, say—but they have enough negative feedback to reduce fishing effort, increase other activities such as farming, and then resume fishing when all agree the stocks are replenished.[27]

In sum, as geographic and cultural distance increase, disparities in information, especially ecological information, increase. These first two dimensions of distance suggest that, as argued in chapter one, from an ecological and political economy perspective, the entire chain of decisions for a given resource and its use, i.e., from primary resource extraction to ultimate consumption, must be assessed. So-called "green production" is meaningless if only a few steps in the production chain are cleaned up. What is more, these dimensions suggest that even if, say, the primary extractors know they are depleting the resource, there are conditions under which they will continue to do so. Among these conditions, I argue next, are distance created by bargaining asymmetries and multiple agency.

Bargaining Power Dimension

To assess the impact of bargaining asymmetries, I begin with a single stylized buyer-seller relationship. Later, under agency, I consider the impacts of a series of buyers and sellers in a chain of production and consumption decisions.

Differences in monopoly power can be assessed in terms of alternatives the seller and consumer face.[28] In the local market, the seller has many customers and many competing sellers. It is truly a "competitive" market. Because producers come from roughly the same community (their geographic and cultural distances are low), they are unlikely to get away with lowering prices by externalizing costs—that is, to each other by drawing too much irrigation water or to their children via depleted soils. And, without market power, farmers are unlikely to shade costs via predatory pricing. With comparable bargaining power regarding critical resource decisions, they are unlikely to externalize costs to each other. In many farmers' markets, each farmer has an equal vote in the market's regulation. Producer cooperatives have similar structures, as do traditional caste-based resource systems in India.[29]

By contrast, the Chilean grape grower likely has only one buyer—the government or the shipper. This is because transnational operations, to be cost effective, must have high capital inputs (docks, ships, etc.) which only governments and a few large firms can afford.[30] Under these conditions, disparity in bargaining power can thus be considerable. With one buyer and many growers, it will be in the buyer's interest to promote competition to drive the price ever lower. This kind of "competition," unlike that in the self-regulated farmers markets, puts continuous pressure on producers to lower costs. Efficiency seeking may indeed be the first strategy employed. But as producers fully realize such efficiencies, cost externalization and shading become increasingly prominent, taking the form of pushing the soil beyond its regenerative capacities or applying herbicides and pesticides with unknown effects on the soil, water, workers, and the larger environment.

Because there are many buyers and sellers in the local farmers market, these farmers will not have this same pressure. Competition here works in two ways. It drives economic profits to zero (normal) and it promotes a degree of collusion, that is, collusion to keep prices above a level of desperation and collusion to prevent monopolistic control. When farmers are not desperate, when buyers will pay what is visibly a competitive price, the farmers' cost-cutting decisions are less likely to tip from a strategy of productive efficiency to one of cost externalization and shading. The assumption here is that cost externalization will increase when people are desperate (i.e., have few alternatives for basic survival) and when technologies and capital allow one group (say, traders) to internalize the benefits of a transaction while externalizing many of the costs to other groups.

In sum, bargaining asymmetry, especially the monopsony power of a buyer vis-à-vis the primary resource extractor, tends to tip a producer's cost reduction strategy from technical efficiencies to externalization. Externalization is carried out by the primary producer, not the buyer. The buyer gets a low price and the producer either bears the cost later or passes it on to others.

Multiple Agency Dimension

To this point, I have assumed one set of buyers and one set of producers, members of which interact directly with each other. This is literally the case in the local farmers market. The international grape market, by contrast, has many intermediaries—or agents—between the primary producer, the grape grower, and the ultimate consumer, the supermarket shopper.

As the principal-agent literature demonstrates, economic transactions are problematic when an agent's actions cannot be observed and when asymmetries in information and differences in incentives between the principal and the agent are prevalent. When questions of resource depletion are included in the analysis (which, to my knowledge, has not been done in the agency literature), one can predict comparable problems between, on the one hand, principals, the primary extractors and ultimate consumers and, on the other, the agents, the intermediate buyers, processors, packagers, distributors, and retailers. That is, only the primary producer, the farmer or fisher or forester, has the requisite intimate knowledge to ascertain threats to the resource and hence the point beyond which a resource cannot be pushed. Those in direct contact with the resource know when the land must lie fallow, when the fishery must be avoided, when the forest must be burned or thinned or left alone. Some of this knowledge may be scientific, but much is a product of experience and learning. All other actors along the production chain have relatively less of this information. And they have relatively less capability of responding to declines and less ability to monitor use of the resource. Moreover, incentives diverge. All want the best price and quality and, for buyers, reliability of supply. But the primary producer is dependent on the resource in a way no other decision maker in the production chain is.[31] The primary producer seeks long-term extraction capability whereas all others simply seek alternative supplies. Put differently, the primary producer is risk averse with respect to the resource whereas all others are risk seeking since it is not their resource and not their livelihoods that are at risk and, for them, alternatives either exist or are conceivable. Moreover, land and water, despite their commodification in capitalist economies, has ecological qualities entirely distinct from that of all other purchased supplies and products in a commodity chain. Central among them is the lack of alternative to regenerative soils or estuaries or groundwater pools. This ecological difference thus creates a discontinuity in multiple agency commerce that rarely gets represented in decisions up the chain.

Ecological information, monitoring capability, and incentives thus diverge among decision makers in a chain of production decisions. If intermediaries and final consumers find ways to lower prices and increase resource extraction, they have imperfect and delayed means of knowing whether doing so risks the regenerative capacity of the resource itself. Signals are rare and confused as one moves up the production chain. As a result, each intermediary serving as the agent for the next up to the final consumer can only make decisions based on price and observable quality.[32]

To this point, I have assumed that all producers—primary and intermediary—have a propensity to maintain the resource, albeit a declining propensity as one moves up the production chain. When strategic interaction is added to the behavioral assumptions, especially regarding those who are best positioned to maneuver strategically, namely, producers (primary and, especially, intermediary), the likelihood of severed ecological feedback and overuse becomes more apparent. To illustrate, the farmer in Chile never sees the entire chain of individuals who participate in getting the grapes to my supermarket. The multiple middlepersons compound any geographic, cultural, or bargaining distance by opening up opportunities to withhold information, emphasize some information, and outright misrepresent other information. Put differently, with multiple agents, growers, buyers, and distributors can easily shade costs and, as a result, pass them on to others. If the Chilean farmer wishes or is compelled to cut costs, that farmer, whether intentionally or not, can easily mislead me, the North American final customer. That farmer produces only what the middleperson needs which is, one, a low-priced grape and, two, a grape the next middleperson will buy. If the farmer wishes to cut costs by, for example, washing the grapes less thoroughly, the middlepersons are unlikely to know. And they may not need to know because their primary interest is to sell it to the next buyer who, until I the final customer pick it out in the supermarket, is similarly motivated. Most importantly, if the farmer or, for that matter, any middleperson along the way, can discover ways to reduce costs that are difficult to detect for the next buyer in line or that are irrelevant to that buyer, right up to me the customer, then that seller comes out ahead.

In a social sense, such stratagems break what, in the local farmers' market, is a direct link of *accountability*. Farmers may indeed shade costs in their transactions at the farmer's market. But, with respect to the product itself, they literally face the recipient of those shaded costs. They cannot be absolved from responsibility by, in effect, hiding behind the backs of numerous agents.

In an ecological sense, such stratagems break the feedback necessary to restrain resource use. In a chain of buying and selling, the first buyer may pick up signals from the grower about stressed soil and polluted water or changing predator-prey populations in a fishery. And that buyer may pass that information on to the next buyer. But, once again, the primary motive of each is to purchase a product that can then be sold and all at a remunerative rate. What is more, even if buyers' concern for the ecological conditions of the soil or forest or fishery exceeds their commercial concerns, they can do little about it. They and the government agencies that support them are organized to promote commerce, not to restrict it. Restraint in use of the resource (or its products) makes competitive sense only when the playing field is level, when no buyer gets a competitive advantage from others' attempt to restrain their use. But the politics of restraint and leveling are far more contentious technically and politically than the politics of market expansion. Only in extreme cases such as ozone depletion and the imminent loss of a popular species such as the elephant, do the politics of restraint win out and, then, often with only doubtful long-term success.

Severed ecological feedback and diminished accountability via the distance of multiple agency also separates *rights* for resource use from the *responsibilities* of that use. Literature on durably managed resources, especially the common property literature, suggests that a necessary condition for sustainable use is a tight marriage by key decision makers of the rights and the responsibilities of use. The distance of agency or bargaining power or both promotes just the opposite: a separation and an asymmetry of rights and responsibilities. An agent can readily acquire rights through purchase or coercion, but responsibility for long-term management is not easily had. In fact, investors are quick to point out that such responsibilities are properly held by others—the state and the nominal property owner. But states, as discussed, are notorious for abandoning resource responsibility when revenues are needed. And property owners are often owners only on paper and, in practice, only with respect to one value (often market value) of the full range of resource values.

Not only do agents have little ability to assume resource responsibility, they have strong incentives not to. Intermediaries in a commodity chain aim to maximize the difference between their selling price and their purchase price. For such an agent to assume true responsibility for the resource, that agent would, in effect, raise the purchase price. Responsibility also complicates transactions by requiring investment in institutional constraints on resource use, both of which are outside the objectives or capabilities of agents.

Responsibility is further inhibited by the psychology and ethical effects of multiple agency. Humans are cognitively limited to deal effectively with only some five major items, plus or minus two, at a time.[33] Given the multitude of consequences of everyday production and consumption decisions in modern life, individuals simply cannot give attention to, process, and, hence, be responsible for, any but a small fraction of these decisions and their consequences. Ethically, no established principle of distribution would hold individuals responsible for the consequences of their actions when they are so removed that the individuals can have no effect.[34] With the rare exception of a successful international consumer boycott, say, distanced consumption choices via multiple agents can neither hinder nor promote a primary resource use decision. The aggregate effects of such choices are, of course, real and consequential but each individual is necessarily and unavoidably absolved of responsibility.

In short, the greater the distance of agency, the less the responsibility for resource use decisions any actor in the production chain will want to have or is cognitively or ethically capable of having for the resource. A national or international political economy that promotes distance on any dimension is likely to be an economy—and a society—that claims rights but assumes few responsibilities for the resources it uses. Social norms for restrained resource use are likely to be replaced by norms for rapid, unregulated use, the result being overexploitation.

Trends in food production illustrate the distancing effects of multiple agency and asymmetric bargaining. Food production in North America and Europe is becoming increasingly subsumed under organized networks of wholesalers and retailers, many of which operate under tightly linked corporate structures controlling key production decisions from seedling to supermarket. These agents of production contract with farmers to supply them with specified products at specified times and prices. The farmers are effectively transformed from self-employed producers to contract workers. Because they contract with only one buyer, usually a large, integrated retailer, the distance of bargaining asymmetry is great. But decisionmaking itself largely shifts to the networks. Farmers, as workers, are told what crops to plant, which and how much fertilizer and pesticide to use, and what price they will receive. They are continuously compelled to increase output. What is more, they, not the wholesalers and retailers, accept the risks of crop failures and market downturns.[35]

As a result, the separation of primary production decisions from consumption decisions via large, often highly integrated corporate food networks results in continuous pressure to convert land, to specialize, to plant "fence to fence". Farmers are compelled to intensify production with monocultures and external inputs. The long-term costs include depleted soils, poisoned ground water, and, often, economic ruin for the farmer. At the same time, some owners of land and of processing and distributing systems do very well economically. Their risks are spread among multiple chains in these food production webs. From their perspective, a given farmer, now a contract worker, may go bankrupt but others can be hired.

This example suggests that the distance of multiple agency is, in part, one of institutional setting: low distance tends to have more built-in checks and balances, more reasons to restrain production and consumption and less reason to rely on external authorities. High distance, by contrast, requires institutional mechanisms that span great numbers of individuals, organizations, and cultural and legal settings. Producers in high distance webs of enterprise have few incentives to organize themselves up and down the chain of production. Regulation and restraint are both unlikely when primary producers have little political power and ultimate consumers are unaware of their impacts.

In sum, the distance of multiple agency is one of strategic interaction and of information and feedback. With the Chilean grape, each individual only has transactions with two others, a seller and a buyer, up to the ultimate consumer who interacts with only one seller. If the grower's production methods harmed me, the ultimate consumer, or if my consumption harms the Chilean grower (by, for example, paying a price that makes overexploitation irresistible), none of the intermediaries would ever know. Moreover, if my consumption, say, led to the degradation of one grower's soil, I would not know it until the grapes stop coming. Even then, the buyer, the intermediary would simply acquire alternative suppliers. Feedback between producers and consumers in a market with substantial distance on this or any set of dimensions, is minimal. All the grower and the consumer can respond to is price. And limited response implies, quite literally, limited responsibility. As has been repeatedly shown, price—especially under relatively non-competitive conditions and conditions where non-market values such as ecosystem health are prominent—is a poor indicator of resource maintenance and degradation. Under these conditions, cost externalization via shading and distancing will become increasingly attractive components of competitive strategies.

To summarize the concept of distance: The propensity to externalize costs through production processes that separate production and consumption decisions can be examined by contrasting two extremes, a farmers market and an international agricultural market. The dimensions of geography, culture, bargaining power, and agency divorce the primary production decisions (e.g., planting, fallowing, irrigating, fertilizing) from the selling and consumption decisions. One result is that, as key decisions are removed from primary producers and costs are externalized, feedback from the resource (e.g., land, fishery, forest) in both production and consumption is severed. Sustainable production requires effective feedback from all decisions in a production chain. When distance approaches zero in, say, a household or a self-sufficient community, that feedback is likely to exist.[36] But as distance increases, feedback diminishes and the need for accountability and governance increases, possibly exponentially.

A second result is that producers can self-regulate if the distance dimensions are low enough. Self-regulation is the social correlate of self-organization in complex ecosystems.[37] Fisherpersons can manage their own inshore fisheries and, much as traders in Chicago organized their own, non-governmental board of trade,[38] farmers can manage the conditions of sale in a jointly owned farmer's market. International traders largely cannot self-organize. Without self-regulation and without institutions that recreate ecological feedback, efficiencies may be achieved (in the short term, anyway) but costs will also tend to be externalized over space and time. These conditions, the absence of self-regulation and of effective institutions, arguably prevail in international markets and anywhere the dimensions of distance are significant.

A third result is that as distance increases along any dimension and, as a result, feedback is severed, restraint in resource consumption diminishes. Individuals restrain their consumption for a wide variety of reasons, including the promise of secure supply and ethical concern for farmland, natural habitat, and endangered species. But when cultural, bargaining and agency distances increase, the trade-offs between, say, restrained consumption and low prices, shift from the non-material (e.g., stewardship) to the material (immediate consumption). Responsibility for everyday production and consumption decisions diminishes as distance increases. Even the most committed environmental altruist and the broadest thinking global citizen cannot know of, or have influence on, production and selling decisions at a distance.

Conclusion

The framework developed here will be useful to the extent it aids analysts and activists in reinterpreting claims coming from the dominant belief system regarding growth, development, globalization, technological advance, and progress. By applying many of the same analytic tools business uses to promote the dominant political economy, I have made the hard case, namely, that, even on their own terms, firms and states undermine the very material and social basis on which their enterprise rests.[39] Market expansion and factor mobility increase distance on many dimensions rendering ecologically informed and ethically responsible decisions impossible. By insisting on economic relations that shade costs and separate resource use decisions from resource depletion impacts, firms and their representatives in government can simulate frontiers. But on a finite planet, market expansion without corresponding self-regulation or institutional control erases all true frontiers, no matter how much the economic activity appears to be, say, value-added or information based.

My approach has been much like that of business analysts, i.e., as a rational calculation of interest maximization in a strategic environment. The difference is that whereas the normative goal of business analysis is promoting a firm's (or a nation's firms') competitive position, mine is minimizing the tendency to externalize costs, including where that externalization is shaded or inadvertent and thus not responsive to conventional cost internalization measures. What is more, the questions I have posed center on ecological impact, especially irreversible damage, not on profitability, return on assets, or even long-term business survival.

Four specific propositions and two general implications emerge from this analysis of cost externalization and frontier simulation via shading and distancing.

As distance increases:

i. *Negative feedback loops break down.* This severs the information flow that connects the dynamics of ecosystem support to resource availability. In the process, users misperceive scarcity and irreversibility and tend to act as if resources are infinite or infinitely substitutable.

ii. *Stakeholders expand yet decisionmaking remains constant or contracts.* Complicated production processes including geographic dispersion and multiple agency allow a few actors to make critical decisions. At the same time, the persistence, dispersion, irreversibility, and intractability of many environmental problems increase the number of actors affected by the true costs. The combination allows for the skewing of the distribution of costs and benefits among such stakeholders.

iii. *Environmental problems are displaced.* Combining propositions i. and ii., displacement results in the appearance of solutions when, in fact, the problem is only shifted to other media, other ecosystems, or other, usually unrepresented, peoples.

iv. *The likelihood of shading and cost externalization increases.* As a firm attempts to lower costs it can seek technical efficiencies or pass on the costs. The trade-off between these two choices is in part a function of competitive pressures and the opportunities for technical efficiencies. It is also, however, a function of distance and the firm's ability to concentrate decisionmaking (ii.) and displace environmental problems (iii.). As distance increases, the choice will tip towards shading and externalization.

One general implication of shading and distancing is that uncounted costs and unaccountable actors exist not because economic agents are nefarious, ignorant, or short-sighted nor because a given economic system—capitalist, socialist, or feudal—is inherently exploitive. Nor do they exist because resource management techniques have been too primitive or too sophisticated or because an economy is too dependent. Rather, if shading is pervasive and distancing promotes that shading and other forms of cost displacement, then the problem is one of decision authority and location. When critical resource decisions are made by those who will not or can not incur the costs of their decisions, accountability will be low and what gets counted is likely to be financial capital, not social and natural capital. Decision authority must shift from those who, directly or indirectly, knowingly or unknowingly, overexploit the resource to those who receive negative ecological feedback and who have the capacity and incentives to act on that feedback.

In part, this shift can be made by lowering distance on any of the four dimensions. For example, international trade may at first appear to be a prime candidate for high-distance, overexploitative commerce. But some of the alternative trade arrangements between Northern food cooperatives and Southern communities of growers reveal that, although geographic distance can be high, cultural, bargaining, and agency dimensions of distance can be made quite low. The result is that decisionmaking rests largely with those who depend on the land and those who express a desire to consume sustainably produced products.

Where lowering distance is difficult or where shading is prevalent, institutions must be devised or re-oriented. By institution I mean both formal, governmental structures and informal social norms occurring at all levels from the local and self-organized (e.g., common pool resource systems) to large-scale, intergovernmental organizations (e.g., the World Trade Organization). To increase resource accountability, the purpose of institutions must be to direct primary decisionmaking to those who are most likely to receive and act on negative ecological feedback. To the extent such feedback occurs among those who interact with the resource (e.g., fishers and farmers), the local institutions must jealously retain decision authority and the larger, overarching institutions must strive to ensure such authority. From a sustainability perspective, all other institutional objectives—efficiency, growth, cooperation, equity—become secondary.

A second general implication is that the burden of proof for any economic intervention—a new trade relationship, investment, technology or retailing method—must be on the interveners. In the contemporary policy environment, such interventions are assumed to have net benefits and those who would promote sustainability goals must prove otherwise. The concepts of shading and distancing help delineate the conditions under which net benefits cannot be assumed, thus shifting the burden to the interveners. If, for example, the existing economy is based on subsistence agriculture and the proposed intervention involves conversion to cash export crops, the interveners must show that the dimensions of distance are low, that costs will not be shaded, and that institutions exist to retain decision authority among growers.

Reversing the burden of proof thus becomes a key component of a political economy of sustainability. It may contribute to a weak form of sustainability, but shifting the burden may be more tractable than establishing the sustainability of either the current system or the proposed intervention. It may also be empowering if those who would have their existing resource use pattern disrupted need only show, for instance, that institutional safeguards are not in place or that interveners are unable to assume full responsibility for their resource use. This is not to say that high-distance interventions should always be rejected. It is to argue that, although such interventions may be risky, they may be worth doing as small experiments with on-going evaluation. This would indeed slow the pace of economic expansion and change. But, as argued, it is precisely the rapidity of change and the mobility of capital and technology that makes institutional response so difficult. Internalization efforts thus become after-the-fact attempts to account for costs when the very structure of the economic intervention—high-distance and pervasive shading—makes costs impossible to count and interveners unaccountable.

6

Consumption and Environment in a Global Economy

Ken Conca

Not long ago I received an unsolicited mailing from the Sierra Club, containing a request in the form of a sales pitch. The letter urged me to become a member, using the problem of logging in America's national forests to illustrate why I should become involved. The letter presented a powerful explanation for the decline of America's old-growth forests: the destruction is driven by rapacious corporations; access roads on federal lands are a valuable subsidy for the loggers; I and "every honest, tax-paying American" have been subsidizing deforestation in this manner; the logging constitutes plunder of the public trust, at public expense, for private gain.

I was also offered a special, discounted membership rate of $15 and a free gift.

In other words, what started out as an urgent appeal for help had shifted into an opportunity to buy into the Sierra Club at a low price. The free gift, a backpack emblazoned with the Club logo, was pitched as "perfect for a hike in the woods or a picnic in the park" and something that would "help you show your support for all the Club's efforts on behalf of our environment." The promotional flyer—*Join Now* and you'll receive this *FREE GIFT*!—carefully described the backpack's special features, differentiating it from all the other backpacks presumably already owned by the recipients of this mailing.

Timothy Luke has argued that such practices expose much of today's mainstream environmentalism as a form of consumerism.[1] Luke argues that despite the genuine accomplishments of groups such as the Sierra Club, "Environmentalism, like corporate capitalism, increasingly is forced to pitch its messages in consumerist terms to win any widespread popular support."[2] For Luke, common practices—such as marketing calendars with glossy, surreal, and dehumanized images of nature, or privatizing landscapes in order to protect them, or promoting ecotourism to generate revenue for environmental protection—reveal mainstream environmental NGOs to be engaged in the commodification of nature.

This harsh criticism may be warranted—particularly for groups busy developing schemes to buy and sell slices of the Earth's atmosphere, or engaged in promotional giveaways of wild-animal trading cards with gasoline retailers. No doubt they would reply that these are necessary evils in the quest for real-world environmental protection. In any event, what struck me most about this episode was the absence of any recognition—in both the forest narrative and the backpack giveaway—that consumption patterns constituted part of the problem.

When environmental problems crowded their way onto the international agenda in the early 1970s, overconsumption was a central theme in the discourse of many environmentalists. The UN Conference on the Human Environment, held in Stockholm in 1972, framed the problem of environmental deterioration primarily as a by-product of affluence. The notion that human society was beginning to encounter "limits to growth" strongly informed ideas about the link between economic growth and environmental degradation.

This frame created substantial discord between the industrialized countries and the less developed world, which argued that the North's international environmental agenda could stifle development in the South. The resulting impasse forced a new look at the links among environmental degradation, poverty, and economic development, which in turn yielded some powerful insights: not all forms of economic activity do equal environmental harm; there is a pollution of poverty as well as a pollution of affluence; careful attention must be paid to the quality or "sustainability" of growth. These themes were stressed in *Our Common Future*, the influential report of the Brundtland Commission, which provided the framework for the 1992 Earth Summit.

To be sure, focusing on sustainability marked an advance over some of the more simplistic notions of ironclad limits to growth that marked the Stockholm-era debate. But one side effect of reframing the problem this way was to shift attention overwhelmingly toward the need for more efficient, cleaner production systems—and away from the consuming interests and needs fed by those systems. Thus, the Brundtland Commission saw no contradiction in offering a vision of global sustainability that presumed a fivefold increase in world economic output.[3]

The notion of sustainable growth has been a salve to the many environmental organizations and movement groups that have been reluctant to confront the values and practices underpinning the consumer society—preferring instead to channel their energies into regulatory and technological strategies of environmental protection. The ironic result is that, at a time when social theorists see consumption, consumerism, and the globalization of consumption patterns as increasingly important phenomena, the centrality of consumption to ecological worldviews and environmental activism seems to have faded.[4] This subordination of consumption is alarming because traditional supply-side strategies of technological and regulatory controls seem increasingly ineffective in the face of deepening trade integration, accelerating capital mobility, technological innovation, and changes in the organization of production—that is, in the face of globalization.

In this chapter I will argue that changes in the world economy are giving new importance to the question of consumption and its linkages to global environmental degradation, for two reasons. First, economic globalization is altering the nature of the "consumption problem." We tend to see the world in terms of an overconsuming North (with its pollution of affluence) and an underconsuming South (with its pollution of poverty). Yet perhaps the biggest environmental dilemma of globalization is the impact on consumption patterns in the planet's "sustaining middle"—the large but fragile segment of the planet's population that lives, works, and consumes in ways that most closely approximate genuine sustainability. Viewing economic globalization from the consumption angle calls attention to this eroding middle.

A second way that global economic change is giving new salience to consumption is political. Changes in the organization of production and the scope and complexity of international transactions are making traditional regulatory approaches to global environmental protection increasingly ineffective. Power in global production systems has shifted both upstream and downstream from the factory floor, where traditional environmental approaches have focused. If efforts to protect the planet and its peoples from environmental harm are to be effective, they will have to follow that shift in power. In particular, activism and advocacy will have to follow power downstream to the ideologies, symbols, relationships, and practices that drive consumption.

Understanding the ecological impacts of a changing world economy requires careful consideration of what we mean by *economic globalization*. Thus, the chapter begins with an overview of some important economic trends at the heart of what is often referred to by this term. In contrast to the more popular (but ultimately incomplete) view of globalization as more frequent cross-border transactions, I stress changes in the underlying organization and logic of production in the world economy. Two particularly important changes in the global organization of production are discussed: the proliferation and lengthening of genuinely global commodity chains and the continuing shift toward so-called post-Fordist modes of production, emphasizing competitiveness, adaptability, flexible specialization, and niche marketing.

The chapter then examines what I take to be the most important consequences of these changes for efforts to promote environmental protection. These consequences include the problem of greater spatial and social distancing between production and consumption, the threatened obsolescence of traditional approaches by environmentalists in the context of a new configuration of power in transnational economic activity, and the squeezing of the planet's "sustainable middle" via global economic restructuring. The chapter concludes with speculations on some of the ways that environmentalism as a global social movement will be forced to adapt in light of these changes and their consequences.

Changes in the Global Organization of Production

Some have suggested that talk of globalization constitutes little more than globalization hype or "globaloney."[5] One common criticism is that much of the analysis on which globalization claims are based lacks a sense of history. Critics are quick to point out that we have had a highly integrated capitalist world system since the age of empire and colonialism. In this view, transnational financial power, a global logic of capitalism, and the tight economic coupling of the international system do not represent anything particularly new. A lack of historical perspective can lead one to attach undue significance to recent trends. For example, introductory textbooks in international relations are fond of pointing out that several of the largest multinational corporations have annual earnings rivaling the gross national product of midsized nations—implying a shift in the balance of power between state institutions and private economic agents. This observation, while true enough as an analysis of trends over the past few decades, overlooks the history of the international system. During the age of empire most of the countries being compared today to corporations were not countries at all, but rather resource colonies and captive markets exploited by alliances of state and corporate power in the imperialist nations. We would do well to remember this history before concluding that economic transnationalism or the scope of the corporate reach are new experiences for most of the world's societies.

But qualifications of this sort should not be allowed to obscure the fact that very real changes are occurring in the world economy, creating new problems and challenges for social movements seeking ecological sanity. Economic globalization has become a catchall phrase to describe these changes. The term is used to refer to any or all of several macrotrends: an increase in economic transactions across national boundaries, the growing mobility of inputs across those boundaries in increasingly transnationalized production processes, or the growing insignificance of the boundaries themselves in the face of regional and global integration.

The most common way of describing globalization is also the simplest—as simply a deepening of the process of growing economic interdependence that has characterized the global economy since the end of World War II.[6] One hallmark of this transaction-based view of globalization is the accelerated growth of international trade. This growth, which has greatly outstripped the rate of world economic growth in recent decades, has been driven by continuing tariff reductions, aggressive attacks on such "nontariff" barriers to trade as health and environmental regulations, and the incorporation of previously excluded sectors such as agriculture, services, and intellectual property into the global free-trade regime. The other key process in this transaction-based view of globalization is growing transnational capital mobility. Discussions of capital mobility often commingle several different trends, including growth in the overall volume of transnational investment, the deepening integration of capital markets, and the growing speed with which transnational capital can relocate in response to short-term fluctuations in economic conditions.

These globalizing trends in trade and finance are often described as the product of technological innovation.[7] Information and communications technology have made it possible to link previously separate national capital markets, greatly enhancing the short-term flow of money across borders. Innovations in transportation have greatly reduced the costs associated with the movement of goods. Despite this emphasis on technology, it is also generally recognized that policy changes have played a key role in ushering in the growth of global transactions. Chief among these are aggressive trade liberalization and the widespread relaxation of restrictions on short-term capital movements.

But what the transaction-centered understanding of globalization overlooks is that trade has exploded because production has dispersed across borders, and that much of the capital that moves across borders is being attracted by, or merged into, new types of industrial organization. In other words, globalization means not only quicker and more frequent transactions among separate national economic units, but also changes in their relations of power and authority that derive from reconstitution of the units themselves. These changes have profound impacts on work, community, power, and place that cannot be captured by viewing the world economy as an "interstate system" marked by more frequent but still arm's-length transactions in trade and finance.

Two underlying changes in the global production system stand out. First is the rise of increasingly complex transnational commodity chains in products as diverse as automobiles, personal computers, athletic footwear, and garden vegetables. Second is the shift in many of these chains toward so-called post-Fordist forms of industrial organization, and with it the dispersal of the power to shape the chain's activities away from the most traditionally powerful nodes in the chain—a process I will refer to below as the upstreaming and downstreaming of power. These changes have enormous ramifications, both for the nature of the consumption-ecology problem and for the methods and tactics most likely to bring about effective environmental protection. Before turning to the consequences of these changes, however, it is worth considering their essence in more careful detail.

Transnational Commodity Chains

An increasing share of global production is organized into global commodity chains. A commodity chain consists of "sets of interorganizational networks clustered around one commodity or product, linking households, enterprises, and states to one another within the world-economy."[8] According to Gary Gereffi,

> Global commodity chains have three main dimensions: (1) an input-output structure (i.e., a set of products and services linked together in a sequence of value-adding economic activities); (2) a territoriality (i.e., a spatial dispersion or concentration of production and distribution networks, comprised of enterprises of different sizes and types); and (3) a governance structure (i.e., authority and power relationships that determine how financial, material, and human resources are allocated and flow within a chain).[9]

In other words, the commodity-chain lens brings into focus the technological, commercial, and organizational networks that link the various stages of production and exchange for a particular commodity—be it an automobile, an orange, or a pair of tennis shoes. The nodes or "links" in a commodity chain include those involved in extraction, component manufacture, assembly, packaging, marketing, advertising, retailing, and myriad associated services. Across many sectors of the world economy these nodes have become increasingly globally dispersed, causing the production process to cross multiple borders on the road to final consumption.

Viewing global production through this conceptual lens produces a very different image than the interdependence paradigm underpinning conventional understandings of globalization. In the interdependence model, the input-output structure is national production, the territory is the space within a nation's borders, and the governance structure is the state. From a global-commodity-chain perspective, in contrast, the input-output structure is inherently transnational, the territory is the spatial distribution of activities along the chain, and the governance structure is made up of the relations of power and authority among participants in the array of production and exchange activities up and down the chain. Thus commodity-chain analysis "highlights the need to look not only at the geographical spread of transnational production arrangements, but also at their organizational scope (i.e., the linkages between various economic agents—raw material suppliers, factories, traders, and retailers) in order to understand their sources of stability and change."[10]

The growth of international trade merely hints at a process of transnationalization that is at work across all stages of the commodity chain in a wide variety of sectors. Consider Latin America as an example. A transaction-centered view of globalization would begin with the growing volume of intraregional and extraregional trade, centered on NAFTA and MERCOSUR. It would also stress capital transactions, pointing to the instability of regional stock markets, the enduring problems associated with debt, and the chronic problem of capital flight.[11] But important as they are, these transactions must be understood in the context of a larger process of the transnationalization of production. Brazil and Mexico saw a net inflow of foreign direct investment of $14 billion and $12 billion, respectively, in 1997, with much of this money gravitating toward globally oriented firms and industries.[12] Foreign direct investment in the region as a whole increased from $33 billion in 1995 to $65 billion in 1997.[13] A key stimulus in this surge has been the privatization of publicly owned enterprises, which draws foreign capital to industries as diverse as telecommunications, energy, transportation, and manufacturing. Market liberalization has also stimulated the inflow of foreign capital, as local producers find themselves unable to compete.

Observers disagree as to whether the explosion of investment inflows represents a sustained trend or a one-time pulse driven by privatization. A mix of corporate strategies seems to be at work, including the search for transnational production efficiencies, market access, regional production platforms, raw materials, and market share in the opening of the previously untapped service sector.[14] In any event, the result has been to create more complex, transnationalized ownership structures and strategic alliances across most major industries in the region. Regional institutional arrangements, including NAFTA, MERCOSUR, and the Caribbean Basin Initiative, facilitate these processes of foreign investment and intrachain trade. Although typically framed as trade-liberalization initiatives, they are actually more significant as guarantees against the political risk associated with foreign investment. In assessing the impact of NAFTA on the auto industry, for example, the United Nations Economic Commission for Latin America and the Caribbean concludes that "the greatest impact of the agreement was to convince investors—both international and domestic—that the Government of the United States was willing to accept the Mexican economy as a manufacturing platform for entry into the North American market."[15] In this context, it makes little sense to focus narrowly on "international trade" as the engine of globalization. Far from being an arm's-length exchange involving a final product, "trade" occurs at virtually every stage along the commodity chain in these denationalized production systems. So do the key financial and technological aspects of production.

The reorganization of production affects both new and traditional economic sectors across the region. Obvious examples are the free-trade zones that have proliferated in Central America and the Caribbean, emphasizing assembly and light manufacturing of consumer nondurables and the low-skill end of hi-tech. But the transnationalization process can also be seen in more traditional capital-intensive sectors such as the region's auto industries, which have been the linchpin of industrialization in several of the larger Latin American countries since the 1950s. Foreign investment has been the dominant force in the industry's regional development since its inception, and remains so today. But whereas the emphasis was once on production by multinational subsidiaries for protected local markets, the new trend is toward integration into global production systems in the context of trade liberalization and global competitiveness.[16] Mexico's auto industry is illustrative. In recent years the industry has shifted from car production for the local market and the freestanding export of auto parts to a much more integrated role in the global commodity chain for motor vehicles. As a result, the domestic content of Mexican automobiles manufactured for export has fallen from 60 percent in the late 1970s and early 1980s to 30 percent today.[17]

These changes have been driven by a larger logic of global reorganization in the auto industry, stemming from the Japanese challenge to American and European producers in the 1970s and early 1980s. Japan's production efficiency and more competitive forms of industrial organization forced American and European producers to undertake dramatic restructuring, a key element of which was greatly enhanced foreign investment in selected developing countries. To be sure, policy choices by governments and corporations have altered where, how, and how quickly the process of transnationalization has proceeded in the auto commodity chain. Corporate responses have not been identical; among U.S. producers, for example, Ford and General Motors moved much more aggressively than Chrysler to build overseas plants in the early 1990s.

National policy responses have also varied; MERCOSUR has produced less dramatic transnationalization of the auto industries of Argentina and Brazil than NAFTA has for Mexico, in part because those two countries have disagreed on key policy issues.[18] Nevertheless, the overall regional trend is clear.

Post-Fordist and Flexible Specialization

Accompanying the transnationalization of production has been the rise, in many sectors, of so-called post-Fordist forms of organization at the firm and industry levels. *Fordism* is the term used to describe the system of mass production and consumption that became dominant in industrialized countries after World War II, with the Ford Motor Company being an early prototype of the organization of production underpinning the system. At its core, the term *post-Fordism* refers to changes that have replaced fixed capital, vertical integration, and mass markets with more fluid, decentralized alternatives captured under the rubric of "flexible specialization." Reynolds summarizes the post-Fordist argument:

> According to a great many authors, the Fordist model of production has broken down since the 1970s and is increasingly being replaced by a more flexible, post-Fordist pattern of production. Piore and Sabel argue that the new production model is based on flexible specialization—batch production in small firms that are linked through dense networks and produce for niche markets. They suggest that post-Fordist production can outcompete the Fordist model because of flexibilities in work organization, product specification, and marketing strategies. Many studies have found that large manufacturing firms are undergoing a process of vertical disintegration whereby production is increasingly undertaken by small specialized firms linked through production contracts.[19]

Consider, in this context, the difference between Ford and Nike as prototypical global businesses. Ford symbolized the classic American multinational of the 1960s and 1970s—a vertically integrated oligopoly whose power within the auto commodity chain resonated from control of the manufacturing process and the factory floor. Nike, in contrast, owns not a single production facility, relying instead on shifting short-term contracts with a diverse array of suppliers. The firm's key assets are not manufacturing facilities or production technologies but rather a powerful marketing symbol (the Nike "swoosh"); a set of marketing, retailing, and advertising relationships; and the public lives of Michael Jordan and Tiger Woods. Production is farmed out to multiple sources via shifting, short-term contracts.

Again, the shift toward flexible specialization has been at work across a wide variety of sectors—not just the labor-intensive, export-oriented manufacturing activities that sustain Nike. As Reynolds points out in her analysis of the agricultural sector in Latin America, "All firms, whether they be subsidiaries of very large transnational corporations or modest firms established by entrepreneurial capitalists of domestic or foreign origin, have had to cut costs and institutionalize flexible production systems in order to remain competitive under changing world economic conditions."[20] Even Ford itself (and more generally, the prototypically Fordist auto industry) has moved in a post-Fordist direction.

Claims about the causes of this ongoing transformation remain controversial. Some observers conceptualize post-Fordism as the latest in a series of "long waves" of world capitalist development, with the shift driven by technological innovation as well as a crisis of accumulation in the dominant spheres of the previous wave. Others see post-Fordism not as a historic inevitability but rather as the product of strategic choices made by firms and industries in the context of changing competitive dynamics and regulatory regimes.[21] Regardless of origin, the rise of post-Fordist flexible specialization, combined with the globalization of commodity chains, is creating new configurations of power and authority within global production systems—and thus changing what is required to rein in the forces behind pollution and ecosystem degradation.

Consequences of Globalization: Distancing, Downstreaming, and Destruction of the Sustainable Middle

The changes at the heart of globalization have important implications for a global political ecology of consumption. First, the polarizing effects of globalization force us to rethink the essence of the "consumption problem" in that they draw attention to the erosion of the planet's "sustainable middle"—the diverse array of relatively sustainable communities and lifestyles that are often hidden from view between the excesses of a billion overconsumers and the desperate struggles of a billion marginalized poor. Second, to the extent that global economic restructuring increases the spatial and social distance between production and consumption, it does further harm to the already-damaged negative-feedback mechanisms that are an important dimension of sustainable economies. Third, by rearranging the balance of power among the economic agents in a commodity chain and between those agents and the regulatory state, globalization renders regulatory and technological approaches to environmental protection much less effective. As a result, these traditional tactics are increasingly likely to target the weakest nodes in the chain rather than the strongest.

The Problem of Spatial and Social Distancing

As commodity chains grow longer, more complex, and more deeply transnationalized, one result is to deepen the spatial and social distance between production and consumption. Princen defines *distance* as

> the separation between primary resource extraction decisions and ultimate consumption decisions occurring along four dimensions—geography, culture, bargaining power, and agency. At one extreme, zero-distance is production and consumption by one household or individual; at the other extreme, it is global, cross-cultural, and among agents of disparate abilities and alternatives.[22]

As Princen points out, these diverse forms of distance can block feedback effects by inhibiting knowledge, information, and contextual understanding of the production process feeding one's consumption decisions. As a result, consumers lack the information and incentives to behave in a more sustainable fashion even if they are otherwise disposed to do so. As Princen suggests,

> The propensity to externalize costs through production processes that separate production and consumption decisions can be examined by contrasting two extremes, a farmers market and an international agricultural market. The dimensions of geography, culture, bargaining power, and agency divorce the primary production decisions (e.g., planting, fallowing, irrigating, fertilizing) from the selling and consumption decisions. One result is that, as key decisions are removed from primary producers and costs are externalized, feedback from the resource (e.g., land, fishery, forest) in both production and consumption is severed.[23]

Of course, large and complex modern economies already exhibit dramatic distancing effects. This may lead some to conclude that globalization does little to worsen the distancing problem, whereas it may promise offsetting efficiency gains in resource use. By this logic, Detroit is already sufficiently distanced from, say, Nebraska as to effectively sever any meaningful ecological feedbacks. Outsourcing auto manufacturing from Detroit to Mexico does nothing to increase the effective "distance" for

Nebraska consumers, but may yield global efficiency gains through a better mix of technologies and resource inputs. This presumes, of course, that globalization enhances the efficiency of resource use—as opposed to merely worsening the externalities embedded in market distortions and national policies.[24]

But even allowing for the existing distances between production and consumption, important aspects of the distancing problem do appear to be worsened by globalization. Domestic aspects of geographic distancing may or may not already be large; globalization supplants both minimally and extensively "distanced" suppliers. It is difficult to envision distances being shortened by globalization, except in the extreme case of border regions—and border regions are often the most subsumed by global as opposed to local production systems, making the shortening of geographic distance irrelevant.

More important, distance is multidimensional; it exists along axes of culture and power as well as in the literal, geographic sense. Cultural distance—that is, cross-cultural barriers that inhibit the flow of information, understanding, or sympathetic identification—is clearly increased by globalization; only the most naive variants of "global village" idealism would argue otherwise. As commodity chains grow longer and more complex, and production systems more dynamic, it becomes harder to contextualize production in terms of its social and ecological ramifications. The feedback-distorting effects of multiple agency are also deepened, given the extra layer of intermediaries that typically accompanies transnational transactions. Finally, the concept of distancing can also be applied to growing asymmetries in bargaining—again, with greater distance having deleterious effects for sustainability. Princen suggests that "bargaining asymmetry, especially the monopsony power of a buyer vis-à-vis the primary resource extractor, tends to tip the producer's cost reduction strategy from technical efficiencies to externalization."[25] In principle, it is an open question whether globalization creates such monopsony power or undercuts it by creating access to more potential buyers in more countries. The answer to this question will vary across commodity chains. But to the extent that power is shifting downstream to retailing, marketing, and advertising nodes of the chain in a post-Fordist world (see below), the effect seems likely to be an increase in power-based distancing.

Finally, and crucially, distancing affects not only consumers and producers but also the advocates for and crafters of environmental regulations. Distancing undermines several of the favored international techniques of international environmental diplomacy, including technology transfer and the codification of centralized rules. It becomes ever more difficult to craft rules and technologies with universal applicability and effectiveness, or even to know whether those rules and technologies are having the desired effects across diverse locales. For example, compare the international regime to protect the ozone layer to ongoing talks to build a regime on climate change. Seen through the conceptual frame of the distancing problem, the climate talks are infinitely more complex in that distance is far greater along all the relevant dimensions of the concept.

The Changing Balances of Economic Power

A second important effect of globalization is that it shifts the balance of economic power. The most obvious and widely discussed dimension of this power shift is the eroding capacity of the regulatory nation-state in a globalized world economy.[26] Simply put, transnational commodity chains resist the reach of national regulation. To be sure, state power has not played a purely defensive, catch-up role in the process of globalization; the application of that power has done much to accelerate globalization processes through trade liberalization, regional economic integration, the weakening of labor, and the gutting of environmental, health, and safety regulations. Moreover, the application of state power is not limited to framing the institutional architecture of economic globalization. Export-credit agencies in the OECD nations provide market-distorting guarantees by subsidizing foreign investment in the global South at levels that exceed the World Bank's annual budget for development assistance.[27] If for no other reason than to rein in these pushes from the state, the struggle for state power remains a critical arena.

But state power is not what it used to be. Greater capital mobility shifts bargaining leverage away from the regulatory state, flexible specialization facilitates the rapid shifting of activities among a transnationally dispersed myriad of potential suppliers, and the resulting competitive pressures undercut regulatory initiatives (environmental and otherwise). Reorienting state power to play a more constructive role in global environmental protection is problematic, because the bounded geographic spaces known as states too often cannot or will not control that space in isolation from broader developments in the global production chains that pass through those spaces.

Globalization also alters balances of power among key economic agents. The traditional vertically integrated manufacturing multinational of the 1960s and 1970s enjoyed dominance through economies of scale, barriers to entry, and technological dominance concentrated at the manufacturing node of the commodity chain. But globalization exerts both an upstream and a downstream tug on that concentrated power. Upstream lies the explosion of global finance capitalism, shifting power from fixed industrial capital to institutional investors and well-heeled global speculators. Their choices increasingly dictate both national policy and long-term investment decisions. The instability in global financial markets that generated the so-called Asian flu in late 1997 was symptomatic of the upstreaming of power; an entire array of East Asian industries previously lauded as models of international competitive efficiency found themselves plunged into crisis by choices made upstream.

Less obvious than the upstream power of institutional investors, but just as important, is the downstreaming of power in increasingly "buyer-driven" global commodity chains. Buyer-driven commodity chains include "industries in which large retailers, brand-name merchandisers, and trading companies play the pivotal role in setting up decentralized production networks in a variety of exporting countries, typically located in the Third World." They are distinct from producer-driven chains, which involve "industries in which transnational corporations or other large integrated industrial enterprises play the central role in controlling the production system (including its backward and forward linkages)."[28]

As Gereffi points out, there is an obvious affinity between flexible specialization and the rise of buyer-driven chains. The decentralization and dispersal of the manufacturing node of the chain—a central element of the flexible specialization model—weakens the power of manufacturing vis-à-vis other nodes, for the simple reason that it becomes easier to substitute alternative suppliers. Profits tend to be highest, and power most concentrated, in segments of the chain most readily monopolized through barriers to entry.[29] In buyer-driven chains, power resides downstream from manufacturing, in the hands of the retailers, brand-name firms, marketers, and advertisers who "add value" to what otherwise would be just a toy, a shoe, a grape, or a spark plug. Flexible specialization and the decentralization of suppliers has created new sources of power for these downstream nodes, which control "the means of consumption" through advertising, marketing, retailing, and brand-name product identification.

How far this trend extends and how quickly it may be at work remain unclear.[30] The shift is particularly obvious in labor-intensive consumer-goods industries. But to the extent that post-Fordism is a more generalizable trend, so too may be the downstreaming of power. The shift in global demand from materials-intensive to knowledge- and symbol-intensive industries as varied as telecommunications, media, fashion, and entertainment clearly strengthens the hand of downstream nodes.

These power shifts—privatizing, upstreaming, and downstreaming—raise serious questions about traditional environmental tactics grounded in national regulations and technological approaches. Simply put, such approaches risk targeting the weakest nodes in the chain rather than the strongest. They focus on the application of public power in a world where power is more extensively privatized, and on the production point source of pollution or ecosystem degradation rather than the diffuse, societally based attitudes, ideologies, and consuming practices stimulated by increasingly influential upstream and downstream nodes.

Consider the example of the semiconductor industry in the United States, as documented in a carefully detailed study by Jan Mazurek.[31] The American approach to toxic pollution in this industry and more generally has focused on the classic regulatory strategy of monitoring and enforcement at the firm and production-facility level. EPA's Toxics Release Inventory showed a slight emissions decrease from semiconductor manufacturers between 1988 and 1995. Touting this as evidence of an improving, self-policing industry that deserved regulatory relief to boost its competitiveness, the Clinton administration struck an agreement on "flexible" regulation with America's largest chip manufacturer, Intel, in 1996. Under this deal, Intel was no longer required to seek prior EPA approval for "routine" changes in production processes, so long as a production facility's overall emissions record was acceptable. This would allow for quicker adjustment to the constant stream of new chemicals used in semiconductor fabrication, often cited as a critical condition for international competitiveness.

Through the traditional lens of pollution control via technology-forcing regulation applied at the level of the firm and the factory, the case looks like a success story. Yet as Mazurek shows, even these modest gains seem illusory, in that they owe much to bookkeeping artifacts, inadequate monitoring, and the reorganization of production. One decidedly new-economy feature of microchip manufacturing is that much of it has been outsourced. Intel and other Silicon Valley manufacturers have shifted much of their production to foreign facilities that lie beyond EPA's reach, or to smaller factories in the American Southwest that fly under the radar screen of EPA monitoring. Not surprisingly, the tracking devices at the chronically underfunded EPA have little hope of keeping pace with the hundreds of new chemicals now in widespread use or the small-scale, specialized fabricators in the United States and abroad to which Intel now farms out production.

This episode is not to suggest that Intel or the semiconductor industry cannot be regulated to promote environmental protection, or that tackling the consumption side of the equation is the only remaining option. On the contrary, one appropriate response would seem to be to move upstream from the factory to the boardroom, perhaps through the approach of "cradle-to-grave" corporate responsibility. But the story does illustrate important aspects of the disconnect between traditional environmental approaches and new configurations of power in a globalizing economy. In particular, it shows the growing inability of state regulatory capacity to keep pace with a dynamic, increasingly globally dispersed industry, and the problems of production-focused solutions in a world of decentralized, flexibly specialized, "just-in-time" manufacturers.

The power shifts inherent in globalization should not be read as creating a world where power is so diffuse as to reside nowhere. On the contrary, power is in many cases being reconcentrated—but at new points along the commodity chain. The challenge is to identify and map the shifting locus and currency of power, rather than to fall back on simplistic assertions of its dispersion.

Rethinking "The Consumption Problem": The Squeezing of Sustainable Communities

Perhaps the most important ramification of globalization is its effect on communities exhibiting relatively sustainable consumption practices.

Table 6.1
Durning's consumption strata

Category of consumption:	Consumers (1.1 billion)	Middle (3.3 billion)	Poor (1.1 billion)
Diet	meat, packaged food, soft drinks	grain, clean water	insufficient grain, unsafe water
Transport	private cars	bicycles, buses	walking
Materials	throwaways	durables	local biomass

Source: Alan Durning, *How Much Is Enough? The Consumer Society and the Fate of the Earth* (New York: Norton, 1992), p. 27. Original source: Worldwatch Institute.

These communities have been largely absent from global discussions of the consumption problem, or of environmental degradation more generally. Prior to Stockholm, the dominant frame for understanding global environmental problems recognized only a single dimension to the problem of sustainability—overconsumption and the pollution of affluence in industrial society. Post-Stockholm, a second dimension was added to our understanding, emphasizing the aspects of global environmental degradation associated with poverty and underdevelopment. But the resulting "North-South" construction provides a misleading picture of current global consumption patterns and obscures the impact of globalization on those patterns. Perhaps the greatest challenge of global sustainability today is to engage a third threat that is largely hidden from view by conventional frameworks—preserving the planet's fragile middle in the face of the wrenching changes of globalization.

Despite the popularity of dualistic representations such as North-South and rich-poor, most of the world's inhabitants consume in ways that fall somewhere between the excesses of a billion overconsumers and the desperation of a billion marginalized. Alan Durning, the author of *How Much Is Enough?*, suggests that we think not of two global consuming strata, but rather of three, which he labels consumers, middle income, and poor.[32] Indicators such as an individual's mode of transportation, sources of dietary protein, and use of materials provide the basis for this categorization (table 6.1). Durning identified roughly 1.1 billion overconsumers who eat meat, packaged food, and soft drinks, transport themselves in private cars, and use throwaway materials. In contrast, some 1.1 billion marginalized poor survive as best they can on protein-poor diets and unsafe water, transport themselves by walking, and derive their materials from local biomass.

What lies between these extremes? According to Durning's schema, it is the bulk of humanity. Between the poles of overconsumption and underconsumption, he identified some 3.3 billion people occupying the planet's ecological middle stratum. (Durning wrote in the early 1990s, when the global population was approximately 5.5 billion.) According to Durning, these midrange consumers tend to eat a reasonable diet without the ecological and energetic absurdity of grain-fed beef, they enjoy substantial personal mobility without the planetary burden of the single-passenger private automobile, and they hold a modest but sufficient stock of durable goods.

To be sure, much of the world's ecological peril derives from overconsumption by the upper-income stratum and—to a lesser but still troubling extent—from the economic and ecological marginalization of the world's poor. But it is the large yet fragile middle stratum that provides the models for sustainable lifestyles—lives rich in rewarding social relations, meaningful work, and enjoyable leisure activities, but low in material throughput, energy use, and environmental degradation. This middle stratum is far too diverse to be romanticized as a planetary middle class. Indeed, that diversity is its greatest significance: the rich array of practices, norms, and livelihoods it contains provides the bulk of the world's social germ plasm for alternatives to the pollution of affluence and the pollution of poverty.

If this middle stratum is indeed full of models for sustainable livelihoods rich in human relations, leisurely reflection, meaningful work, and intergenerational connectedness, then preventing its erosion from both ends is arguably the planet's biggest ecological challenge. Tellingly, this challenge is not even seen through the traditional lens of North and South, with its emphasis on national levels of output. But a country such as Mexico—where Coke is now the most popular drink, even as a third or more of rural people lack safe drinking water—spans all three of Durning's consuming strata. So, increasingly, does America, where gas-guzzling sport utility vehicles have become a middle-class fashion statement even as countless small towns victimized by corporate downsizing and relocation have lost access to interurban bus or rail service.

What are the impacts of globalization on the sustaining middle, wherever it is found? Not surprisingly, little in the way of systematic research exists on a question that is rarely even recognized. But there are ominous signs that the effect of global economic restructuring is to erode the sustaining middle from both ends. On one end, the globalization of production, trade, communications, and capital linkages appears to be sucking the upper reaches of the middle stratum into the ranks of the overconsumers. Many of the world's sustainers live in the so-called big emerging markets of the global South, which are being dragged into the world economy through capital mobility and trade liberalization.

At the same time, a large expanse of the sustaining middle faces the constant threat of marginalization in a hypercompetitive, turbulent world economy of constantly adjusting flexible specialization. All too often, "flexibility" in this context means downsizing, corporate relocation, and outmigration. These forces may raise aggregate national incomes—but they also rend the fabric of community and social capital on which sustainable livelihoods so often depend.

The Challenge to Environmentalism: Rethinking and Retargeting Consumption

Not all of the consequences of globalization are ecologically harmful. Some of the industries, technologies, and practices obliterated in global economic restructuring are no doubt polluting and resource inefficient—although one wonders whether this effect can outweigh the damage to sustainable practices, lifestyles, and communities. Too, many observers hold out hope that the shift from a goods-based to a service-based economy and the rise of "e-commerce" will have ecologically beneficial effects—although a skeptic would point out that Silicon Valley, home of the information revolution, has more toxic-waste sites on the U.S. Environmental Protection Agency's "Superfund" list than any equivalently sized area of the United States. Perhaps the strongest strand of hope is in the new possibilities for transnational networking and activism in response to these problems.

Whatever its precise ecological effects, globalization demands refocused attention on the consumption side of the global ecological problematic. Technological and regulatory strategies to improve the efficiency of throughput and reduce the harmful by-products of production will always have their place. But they must be accompanied by new strategies that start from the consumption side of the new equation of globalization. We must respond to the problem of greater spatial and socioeconomic distancing, through policies and activism that relink distanced consumers to the effects of their consumption. We must follow power downstream, from the traditional vantage point of the producing sources of pollution to the newly powerful actors, ideologies, and interests that stimulate overconsumption. Finally, we must learn to see the world not simply as a static array of rich Northern countries and poor Southern ones, but rather as a horseshoe-shaped distribution of sustainable and unsustainable consuming practices, livelihoods, and communities. Seeing the world this way, we are challenged to develop strategies to defend the large but fragile sustaining middle from the twin perils of overconsumption and marginalization.

7

The Distancing of Waste: Overconsumption in a Global Economy

Jennifer Clapp

Most residents of North America are well aware of the growing mountain of waste plaguing the planet.[1] Newspapers regularly report on the "landfill crisis," and the local TV news seems fond of covering the politics of recycling and waste disposal. But most North Americans do not experience waste firsthand beyond their own garbage, which is conveniently taken away once a week. When they do think beyond the boundaries of their own trash can, they tend to associate the problem with industrial inefficiency, lack of recycling services, or population growth. This is not surprising, because these issues dominate both popular thinking and the academic literature on waste. Yet looking at these issues alone misses key elements that challenge the long-term sustainability of processes, products, and by-products. Looking at waste through a consumption lens helps identify those missing elements. Utilizing this lens helps us get at questions that confront consuming itself: What is the consumer link to the waste problem? Where is the "away" place to which waste is taken? Why does the waste go there and not elsewhere? Who is affected in these "away" places? Can such practices continue on a finite planet?

An important dimension to the waste problem that becomes apparent when asking these questions is that of waste distancing. In today's world, there is a growing distance, geographically as well as mentally, between consumers and their waste. A focus on waste distancing highlights two important components of the consumption angle on waste. First is the fact that when decision makers have little knowledge of the ecological and social impacts of the wastes associated with goods they produce or purchase, they have little incentive or ability—as producers or consumers—to change their habits based on waste considerations.

Second is the fact that such distancing consumes waste-sink capacity—both the social capacity to deal with waste in a manner that minimizes harm and that is fair, and the ecological capacity to assimilate waste that avoids toxic contamination and intergenerational effects.

In this chapter, I map out several factors that contribute to waste distancing. I examine distancing with respect to the gulf between consumers and waste—both postconsumptive garbage and the waste associated with the production of consumer items. I argue that the growing scale of industrial life, economic globalization, and economic inequality are the central drivers of the distancing of waste. This distance leads to excessive waste output and associated waste-sink demand, which are played out on a global scale, taking advantage of economic inequalities and resulting in environmental damage as waste-sink capacity is exhausted. Addressing the waste problem in a way that is ecological and fair, then, involves more than environmental education or improved consumption choices. It calls for political action to address the broader forces that make waste distancing a normal and accepted pattern of everyday industrial life.

Beyond Waste Management and Population Control: The Consumption Angle

The global waste crisis tends to be defined primarily as a problem of production. Production processes are inefficient and wasteful—too many materials used, not enough recycling of by-products. Strategies for dealing with the global waste crisis from this angle focus on improving industrial waste-management policies such as prevention and minimization, source reduction, better waste treatment, and enhancement of recycling opportunities.[2] Through a production lens, the issue is not one of "too much"—that is, too many items consumed per person, or just too much consuming in the production process. Instead the focus is on how to *manage* waste. Neo-Malthusians also chime in on the waste discussion. Through a population lens the amount of waste in production is "normal," but there are simply too many people, all demanding products. The issue from this angle, then, is not even one of improving waste management but, rather, of reducing demand by reducing the number of people.

If the analyst turns the waste issue around, however, and interprets it as a consumption problem, different issues must be explored. As Princen argued in chapter 2, the consumption angle changes the focus from adding value to subtracting value, from producing goods and producing correctives to consuming ecological and social capital. The consumption angle calls attention to the contribution of "consuming" to the generation and distribution of waste. It probes the broader waste implications of the consuming of materials that go into, and the by-products that come out of, the manufacture and consumption of products. It examines where and why waste ends up where it does, and who is affected by it. Further, it explores the implications of the current overconsumption of waste sinks, the biophysical capacity of soil, bodies of water, and the atmosphere to assimilate wastes.

A useful analytic means of identifying the key elements of waste generation and distribution from a consumption angle is the concept of distancing. In chapter 5, Princen defines distancing as the separation of primary resource-extraction decisions from final consumption decisions. The greater the distancing on any of several dimensions, the greater the likelihood ecological feedback will be severed and a resource overused. But whereas Princen locates the environmental impact of distancing primarily at the resource-extraction end of the commodity chain, I locate it at the waste end. That is, by following resource use all the way downstream one finds two sets of end points: waste from production at each node of the chain and waste from consumption at the end-use node. Distancing and its predicted environmental impacts occurs when key decision makers—producers and consumers—are separated from the decisions that directly lead to the generation and disposal of waste.

In highly distanced waste chains, some information is inevitably lost and accountability diminished. Part of the separation occurs along a geographic dimension. With economic globalization, especially, the physical distances between producers and consumers as well as between consumers and wastes, expand. But part of distancing occurs along other dimensions as well. These include a cultural dimension (the ways consumers lack information about the specific environmental impacts of wastes), a bargaining-power dimension (asymmetries among decision makers and other stakeholders over the siting of waste disposal), and an agency dimension (the role of middle agents between consumption and waste-disposal decisions). All of these dimensions of "distance" create not just a geographic distance, but also a mental distance, what I will call an "understanding gap"—a gulf of information, awareness, and responsibility between consumers and wastes.

Significantly, as distance between consumers and waste increases, more waste tends to be generated. When people have little understanding of where their postconsumptive wastes actually go, and have even less understanding of where the waste associated with the production of their purchases ends up, they tend to make decisions that perpetuate the generation of waste. Once waste is out of sight, people tend to forget about it, assuming that it is someone else's responsibility. This "someone else" may be firms, the federal government, or local municipalities. The understanding gap and geographic distance between consumers and their waste sever ecological feedback loops, thus perpetuating otherwise undesirable consumption choices. This distancing is present between the consumer's act of "throwing away" and the ultimate fate of that garbage (what I will call *postconsumer waste*), and between consumption and the waste generated in the production of consumer items (what I will call *preconsumer waste*).

Distancing is also crucial for understanding another important and closely related aspect of the waste crisis. With more waste being generated, more must be disposed of; it must go *somewhere*. As more and more waste is disposed of, the waste-sink capacity—that is, the capacity of the earth, both socially and environmentally, to absorb this growing mountain of waste—is being consumed. There is little if any frontier for waste disposal left on the planet. Most of the waste-sink capacity in rich industrialized countries has already been used up. With the ecological impacts of waste dumps better understood today, fewer communities are willing to accept them. There are fewer landfills available in the United States today, for example, than 20 years ago. For this reason wastes are traveling longer distances, including to poorer, less industrialized countries. The transport to and disposal of wastes in developing countries is perpetuating the consumption of waste-sink capacity, only now on a global scale. It creates an illusion that there are enough places for wastes to go. But such transport creates grossly unfair situations where the poor must bear the burden of the wastes from the rich. And it cannot go on forever. Eventually the waste-sink capacity of the Earth will be used up. Indeed, some might argue that we are already there, because there is just no place for certain wastes to go (see below).

A better understanding of the root causes of waste distancing should help to promote policies to reduce it. Three factors promoting waste distancing stand out. One is the extremely large scale of modern industrial life. Seemingly ever-expanding systems for production, consumption, and waste collection have developed, beginning with the industrial boom experienced at the end of the Second World War.[3] To achieve economies of scale and enhance growth prospects, industrialists, planners, and government agencies promoted ever-larger production and consumption systems. Increasingly, a growing number of items used in daily life are manufactured on a large scale in factories and brought to consumers via globally connected distribution systems (chapter 6). Such mass-production and mass-consumption systems contribute to distance between consumers and waste because of the sheer volume of goods, which makes it nearly impossible to know much about the wastes associated with their production. At the same time, a mass waste-collection system, for both postconsumer and industrial wastes, has developed, making waste disposal easy and cheap, something producers and consumers need not think about once the waste is taken away.

A second factor is economic globalization. The drive toward economic liberalization since the 1980s, in North and South alike, has contributed to this increasingly global organization of production, trade, and consumption. In the North this process has been largely driven by ideological beliefs about the benefits of a global free market. The North then presses policies on the South that promote a global free market via "structural-adjustment" reforms prescribed by global economic institutions such as the International Monetary Fund and the World Bank.[4] Liberalization of trade, investment, and financial policies under these reform programs bring these countries into the global-scale systems for production, consumption, and trade. Concomitantly, globalization leads directly to waste distancing. The waste-collection business itself, particularly the private industrial waste-disposal industry, has become ever more global in scale over the past few decades, facilitated by trade.[5]

A third factor promoting waste distancing is economic inequality. On both the local and the global scale, inequality can lead to situations where some communities have little choice but to accept others' wastes, despite the environmental and social problems that such wastes create. These communities accept wastes because they are desperately poor; they cannot turn down the jobs and financial remuneration offered. Economic inequalities both within and between countries have in fact become more pronounced over the past 30 to 40 years. Between 1960 and 1989, for example, the economies of the countries with the richest 20 percent of the world's population grew almost three times faster than the economies of the countries with the poorest 20 percent. The result is that over that same time period, the poorest 20 percent's share of global income dropped from 2.3 percent to 1.4 percent, while the richest 20 percent's share of global income rose from 70.2 percent to 82.7 percent.[6] Indeed, some argue that economic globalization is a contributing factor to global inequality.[7] Along with this growing inequality, waste brokers have increasingly sought out disadvantaged communities, people who are the least able to refuse the jobs and revenues. And the brokers use the channels of the global economy to get the wastes there. The Not in My Backyard (NIMBY) syndrome with respect to the siting of waste dumps has meant that some communities keep dump sites out of their neighborhood while others are paid to take them. A number of studies have shown that such dumps tend to end up disproportionately in poor communities.[8] This displacement of waste disposal from rich to poor occurs within local communities, within countries, and globally.

The Volume of Postconsumer Waste

Not only is waste distancing increasing, but the sheer amount of postconsumer waste generated is growing. Americans, for example, dispose of some 720 kg of garbage per person every year, more than people in any other country in the world.[9] In 1998 this amounted to some 4.5 pounds per person per day, up from 2.7 pounds per person per day in 1960.[10] The OECD country average for 1998 was 500 kg per person per year, a level some 30 percent higher than several decades ago.[11] This contrasts with the 100 to 330 kg per person in developing countries (and, for instance, the estimated 20 kg per person per year waste disposal in Nigeria) in the early 1990s.[12]

Garbage—that is, the kind of postconsumer waste that cannot be reused or composted—is almost exclusively a twentieth-century phenomenon.[13] Many of our grandparents in North America do not remember what happened to their garbage as children because there was so little of it. A century ago, nearly all products used in daily life were fabricated from natural materials, and, if left to the elements, would biodegrade. The use of synthetic materials, those that do not break down so quickly, has risen remarkably over the last 50 years, making the waste-disposal problem more pressing.[14] Postconsumer waste today—that is, the waste that consumers toss into garbage bins—accounts for up to 20 percent of the waste stream in North American landfills.[15] Although this is not the dominant source of waste landfilled, the role of postconsumer waste is indeed significant. A number of factors are commonly cited to explain the emergence of the "throwaway society" and the resultant growth of postconsumer waste. These factors include the rise of a consumer culture, increased difficulty repairing items, growing obsolescence of technologies, and more extensive packaging. These developments are related to and interact with the broader forces outlined above: large-scale production and consumption systems, economic globalization, and growing economic inequalities.

The rise of consumer society, particularly in Northern industrialized countries, is a well-known phenomenon.[16] An expanding array of products have become accessible to consumers through the proliferation of shopping plazas, retail outlets, television ads, newspapers and magazines, telemarketing, and Internet sales services. In this environment, consumers are purchasing more and more "stuff," and throwing more of it away when its relatively short useful life is over. The quantity and availability of "stuff" that makes this culture of consumption possible, however, are also intimately linked to the mass-production and consumption systems facilitated by economic globalization. And increasingly skewed global economic inequality has exacerbated the situation; the "haves" increasingly "have more," and often "buy more," resulting in "overconsumption."[17] Not surprisingly, as incomes rise, material consumption and its corresponding wastes also tend to increase.[18]

Another factor behind the increased amount of postconsumer waste is a growing inability of consumers to repair items themselves, as well as the high cost of repair services. Increasingly less durable products, produced in large scale and sold in the name of affordability and consumer convenience, are at the root of this trend. Affordability and convenience have overtaken durability as major selling features in the marketing of products.[19] Systems for mass production have directly contributed to this problem, because it is far easier to make goods in large quantities on automated assembly lines than it is to repair them. Making repairs requires an intimate knowledge of how a machine works and an ability to creatively craft a new part.[20] Large-scale production systems do not encourage workers to acquire this type of knowledge or skill, and thus the availability of repair services has dropped and their price has risen. In this context, it is often less expensive to purchase a new item than to repair it.

Increasingly rapid technological obsolescence is another factor associated with growing postconsumer waste. Obsolescence, more and more often planned by producers, is a product of the growing scale of production and consumption systems and the mentality of the perpetual-growth economy.[21] Consider the ubiquitous personal computer (PC), for example. PCs become obsolete so fast that it makes more sense, and is cheaper, to buy a new one every couple of years than to upgrade them. In 1998 alone, some 20 million PCs were rendered "obsolete."[22] Less than 10 percent of those computers were recycled.[23] Some obsolete PCs are donated to schools, churches, or charities. But giving them away is increasingly difficult, because no one seems to want an out-of-date computer. And the recycling option is increasingly costly, even for those trying to get rid of them. The majority of obsolete computers end up in storage or in landfills, or are incinerated. When left in a landfill or incinerated, these computers pose a serious hazard because of their high levels of heavy metals such as lead and cadmium. Indeed, they are considered toxic waste in the United States. It is estimated that there will be 315 million obsolete computers in the United States by the year 2004.[24] Rather than protesting this situation, consumers in rich industrialized countries seem quite happy to acquire new computers on a regular basis. Some 130 million PCs were sold in the United States alone in 2000.[25]

The increased volume of postconsumer waste is also intimately linked to the growth of packaging waste, very much linked in turn to large-scale systems for production and consumption.[26] About one-third of U.S. municipal solid waste is discarded packaging.[27] To service consumers on a global scale, goods are packaged to enhance their ability to travel long distances, to give them uniformity of size for efficient distribution, to keep them sanitary, to increase their convenience of use, and to make them more appealing.[28] Producers claim that they are only trying to meet consumers' needs by, for example, increasing the use of plastic packaging, because it is lighter than glass and thus easier to carry. Much of the packaging we see today is ostensibly designed for "consumer convenience," especially with respect to "disposable" items and the packaging of foods, fast food in particular.[29]

There are thus many factors working to increase the amount of consumer waste, and these are interconnected with the factors that contribute to the distancing of waste from consumption. The growing mountain of postconsumer waste is distanced from consumers through a variety of channels. What is not compostable is typically dumped in landfills, enters recycling programs, or is incinerated. In the United States over the past few years, for example, of the 220 to 380 million tons of municipal solid waste generated (statistics vary), 50–60 percent of it was disposed of in landfills, 25–33 percent was recycled, and 7–17 percent was incinerated.[30] Below I discuss each of these outlets of the consumer-waste chain and its distancing implications.

Most people in North America have a sense of where their local landfill is, though they probably have never visited it. Wastes are conveniently taken away each week by municipally organized waste-collection systems. Most cities in the United States began to set up municipal systems for waste collection around 1900, in an attempt to clean up the streets where waste was regularly tossed. This waste was sent to local dumps, and since the 1930s, has increasingly been disposed of in so-called "sanitary" landfills. The large-scale organization of municipal waste collection and disposal, however, has only been available in the industrialized countries since around the 1950s.[31] Today municipal solid-waste-collection services are available to over 95 percent of the population in OECD countries.[32]

But not all municipal solid waste finds its final resting place at the landfill. Postconsumer waste often travels a significant geographic distance with a large understanding gap. Some cities in the United States and Canada, for example, regularly send their municipal garbage to other communities that have more capacity or, at least, greater willingness, to dispose of it. The geographic distance traveled could be within the city limits, within the state or province, or even further afoot. The

New York "garbage barge" that sailed down the East Coast of the United States in 1987, loaded with over 3,000 tons of garbage from Long Island looking for a place to unload, is a reminder of the global potential of city trash. After being rejected by the state governments on the U.S. coast, the barge continued on, seeking permission from Belize, Cuba, and Mexico. Finally, after two months at sea and many rejections, the barge finally returned to Long Island and the waste was reluctantly dumped there.[33] In this case other communities, even those in need of the economic resources that would come with accepting the waste, were unwilling to take it, indicating an exhaustion of social capacity to absorb the waste. But just as often wastes sent afar are successfully dumped somewhere. Within cities and local regions, waste dumps are located in neighborhoods that are economically depressed. A large environmental justice literature has analyzed this phenomenon. It is widely recognized that poor neighborhoods and communities, including those made up largely of people of color or Native peoples, tend to be targeted for the siting of landfills and other toxic activities.[34]

It is not just municipal waste destined for landfills that is increasingly distanced from consumers. Most communities in North America have a local recycling program in which recyclable postconsumer waste is collected weekly together with garbage. When people "recycle" their aluminum cans and foil, plastic containers, and paper products, they may feel as if they are "doing their bit" for the environment and for society. But most people know little about the conditions under which these wastes are recycled, or whether the environment is better off for it. Reducing the amount of recyclable material that ends up in landfills may only increase the amount of landfill space available for other wastes.[35] What is more, once the recyclable material is sorted, there may not be enough demand to ensure that it is actually recycled into other products.[36]

Much of what is collected for recycling in the United States is recycled domestically, though a growing amount is also exported. Recyclables are usually sorted locally, but they travel to recycling facilities located some distance away. There exists a global trade in, for example, used plastics, obsolete computers, and used lead-acid batteries, which serve as raw materials for recycling operations around the world. In these cases, it is often cheaper to recycle such items in poorer countries where environmental regulations for recycling are more lax. Latin America and Asia are key locations for the recycling of lead-acid batteries, while Asia seems to dominate the recycling of both computers and plastics.[37]

The plastics that go into local recycling programs in North America and Europe are a stark illustration of this distance that many people are unaware of when they diligently rinse out their used plastics and place them in recycling bins. For example, some 100,000 to 150,000 tons of plastic waste collected in local recycling programs made its way from the United States and Germany to Indonesia every year in the early 1990s. Only about 60 percent of this waste was actually recyclable, however, and the rest, some of which was toxic, was disposed of in landfills. The plastic recycling operations in Indonesia operate under extremely unsafe conditions. Plastic wastes often contain residues from their original contents, such as toxic cleaners, pesticides, and fertilizers. Protective clothing is seldom given to employees, mainly women and children. Moreover, the importation of plastic wastes has been implicated in diminishing the livelihood of some 30,000 to 40,000 scavengers of local plastic waste (supporting some 200,000 dependents) who supply local, small-scale plastic recycling operations.[38] In part because of these problems, the Indonesian government banned the import of plastic wastes in 1993. But other countries in Asia remain large importers of postconsumer plastic waste, including China, the Philippines, Korea, India, and Taiwan, to name just a few.[39] Indeed, the Asian market for postconsumer plastic waste has been a big part of the economic viability of plastic recycling in the United States, with the United States exporting to Asia some 200,000 metric tons of plastic waste for recycling in 1995.[40]

Another major outlet for waste is incineration. Although in North America incineration is not as prevalent a waste outlet as are landfills and recycling, it is still significant and represents yet another set of distancing issues between consumers and waste. Support for the incineration of waste emerged in the 1970s and 1980s as a way to solve two pressing problems: lack of landfill space and the energy crisis. The combustion of waste reduces the volume taken up by the waste, thus saving landfill space. And in the process of combustion, energy can be recovered.[41] Environmental campaigns in the 1980s sought to stop the incineration trend, however, on the grounds that the ash left over from the combustion stage is highly toxic. Moreover, the smoke from incinerators, particularly when they are not burning at the correct high temperature (which is extremely difficult to monitor), releases toxic substances into the air.[42] These toxic air pollutants, containing persistent organic pollutants (POPs) such as dioxin and furans, can travel thousands of miles. They typically make their way to ground level in cold climates. Incinerators in communities as far south as the Southern United States and Mexico have been implicated in the POP contamination problem of many Arctic native communities.[43]

The solid waste remaining from the incineration process—the fly ash, which is usually highly toxic—has also made its way to vulnerable communities. The fly ash is normally disposed of in landfills, which, as mentioned, are themselves often sited according to economic and ethnic criteria. Economic globalization has also had an impact on the distancing of this ash. In some cases, particularly in the 1980s before an international outcry, there were exports to developing countries of toxic fly ash from municipal incinerators. The now legendary voyage of the *Khian Sea* is a stark illustration. In 1986, loaded with 14,000 tons of toxic incinerator ash from Philadelphia's municipal waste incinerator, the ship sailed to the Caribbean Sea to unload. After several months the cargo was unloaded in Haiti under the label of fertilizer. The government ordered it removed when the true content was revealed, but an estimated 4,000 tons were left on the beach where it had been dumped. After failed attempts to offload the remainder of the cargo in Africa, Europe, the Middle East, and East Asia, a voyage of some 27 months, the ash mysteriously disappeared from the ship in Southeast Asia.[44] The waste left in Haiti sat on the beach leaching toxins for ten years before a deal was brokered with the United States for its removal. It was finally removed from the beach in Haiti in early 2000 and made its way by barge to Florida, where the Florida state Environmental Protection Agency agreed to dispose of it. The state revoked the offer several days later and, as of October 2001, the waste was still on the barge off the coast of Florida, awaiting its fate.[45] This case shows that the social as well as environmental sink capacity for such wastes, even in economically disadvantaged countries, has been exhausted. There is no "away."

The Fate of Preconsumer Waste

A second important dimension of the consumer-waste distance is the understanding gap and geographic space between, on the one hand, consumption of items and, on the other, the fate of wastes associated with the production of those items. In other words, consumers do not know where the waste by-products from the manufacture of consumer items they buy end up, or even what that waste is composed of. As in the case of postconsumer waste, the disposal of this *preconsumer waste* is taking place on a global scale, made possible by the globalization of the industrial waste-disposal industry. Taking advantage of economic inequalities, waste traders often seek out communities willing to accept waste for disposal or recycling. Moreover, with the increase in scale of industry, spurred by the globalization of trade and investment, a much larger percentage of consumer items available these days in North America are produced in another country, making the distancing of industrial waste often much larger than the distancing between consumers and their postconsumer waste.

The global generation of toxic industrial waste is huge and growing, up from 300–400 million metric tons per year in the 1980s, to some 400–500 million metric tons per year over the past decade.[46] Most of the world's toxic waste is generated in the United States and Western Europe, though a growing proportion of it is being generated in the developing world.[47] To assess the total waste picture, though, it is important to also consider the extraction of raw materials, the necessary inputs to industrial production, in addition to toxic industrial waste. Mining, oil extraction, and logging all produce immense amounts of waste, some toxic and some not.[48] In Canada alone, for example, nonfuel mining generates over 650 million tons of waste material per year, more than 20 times the amount of municipal solid waste and industrial production waste combined.[49] As in the case of the growing amounts of postconsumer waste, this preconsumer waste is expanding due to a number of forces, including those that contribute to waste distancing.

The push for economic growth over the past 50 years has played a large role in the mounting industrial and raw-material waste problem. Production is counted as economic growth: if more growth is good for the economy, then more production is good for the economy, too. And because the waste-disposal industry generates economic activity, adding to growth, it too is seen as good for the economy. In today's global economy, this has led to government subsidization of industrial growth and pressure to relax environmental regulations and/or enforcement as a means of attracting foreign investment. This trend, dubbed a "race to the bottom," is a highly contested notion. But it is increasingly recognized that, regardless of whether firms actually do seek out pollution havens or take flight from highly regulated economies, the threat that they might do so does indeed have an impact on how governments impose and enforce environmental regulations.[50] In this climate, there has been much less emphasis among governments on promoting "clean production" methods that reduce the generation of industrial waste than there has been on creating jobs and attracting investment. Economic globalization only contributes to this trend, as countries strive to compete in global markets for consumer goods by creating economic efficiencies and producing items cheaply on a mass scale. Waste-disposal companies then can clean up the mess left behind, themselves contributing to economic growth.[51]

The growing use of synthetic materials, including chemicals and plastics, in the manufacture of consumer items is a major contributor to the increasing volume of industrial waste generated.[52] These components of the industrial process generate particularly toxic by-products. They are used in the manufacture of items with especially high consumer demand, such as electronics goods (including televisions, cell phones, and computers), textiles (dyeing), furniture (finishing), and automobiles, to name a few. The mass-production and mass-consumption systems that have developed over the past century exacerbate this trend. To produce electronics items in mass scale, for example, the use of synthetic materials such as plastics keeps them portable and lightweight, reducing the cost of materials and transportation. This enables them to serve a global market more easily. Similarly, the use of chemicals and synthetic products in other industries has enabled more economies of scale and reduced weight, making them easier to transport around the globe via mass consumer-distribution channels.

By buying a product "made in the USA," or anywhere else for that matter, one cannot be sure that the industrial waste created by the production of that item is disposed of on the site of the factory where it was made, or even in that country. Just as mass-production and consumption systems have developed in ways that have increased the generation of toxic waste, mass systems for its removal, disposal, and recycling have also developed, and in this case the toxic-waste-disposal industry has become increasingly global. Waste-disposal companies are setting up operations in more than one country, wastes are traded around the world, and transnational corporations tend to dispose of their wastes in the country in which they operate.[53]

Like municipal garbage dumps, toxic-waste-disposal facilities are often located in poor communities, taking advantage of economic inequalities. At the domestic level, particularly in the United States, industrial wastes from the production of consumer items also tend to gravitate to poor and disadvantaged communities.[54] Though a number of factors determine the location of such industries, wealthier and more educated communities appear to be more able to block toxic-waste-generating industries, or at least to be able to require them to clean up their operations.[55]

The export of hazardous wastes to developing countries also became a thriving business in the 1980s and early 1990s. The first cases that came to the world's attention were in Africa, such as the infamous case of toxic-waste drums exported from Italy to Koko, Nigeria, and stored on a farmer's land in exchange for a small fee.[56] But Africa was not the only recipient of such wastes. In the following years, countries in Latin America, the Caribbean, South Asia, East and Southeast Asia, as well as Eastern Europe were targets of waste traders.[57]

The 1989 Basel Convention, which came into force in 1992, sought to control this trade. The impetus for this environmental treaty was the public outcry in both rich and poor countries against hazardous-waste export to developing countries. But soon the export of industrial waste for recycling emerged, in large part as an attempt to circumvent the Basel Convention and other rules to control the global hazardous-waste-disposal industry. Developing countries were recipients of toxic wastes destined for recycling operations throughout the 1990s. Eventually the parties to the Basel Convention adopted an amendment in 1995 that bans the trade in toxic waste for both disposal and recycling purposes between rich and poor countries. This amendment, however, is not yet in force as of October 2001, and thus is not yet legally binding. Despite the public outcry against the trade in wastes from rich to poor countries, there have been hundreds of cases of hazardous-waste exports to poor countries over the past two decades. Though the number of such incidents has declined since the mid-1990s, when the Basel Convention came into force and the Basel Ban Amendment was adopted, they still occur often enough to raise concern.[58]

A recent example of this phenomenon is the late-1999 shipment from Japan to the Philippines of 2,700 metric tons of wastes destined for disposal in Manila. This shipment was in direct contravention of the Basel Convention. Labeled as paper for recycling, the wastes were in fact a mix of hazardous medical and industrial wastes.[59] When the Philippine government discovered the wastes after they had been abandoned at the Manila Harbor, it protested vigorously to Japan. Embarrassed by the incident, the Japanese government agreed to remove the waste within 30 days, as required by the Basel Convention. In one account of this incident, a reporter speculated that the exports may have been the result of a tightening of Japanese laws for dioxin emissions, which had shut down the incinerators of the firm responsible for the waste shipment.[60]

Most of the export of toxic industrial waste from rich to poor countries these days is destined for "recycling" operations. Though recycling implies environmental stewardship of hazardous wastes, in most cases, especially in developing countries, it has been as harmful to the environment as dumping. The recovery of products from imported hazardous wastes in developing countries is, in almost all cases, carried out under unhealthy and environmentally dangerous conditions. A large proportion of these wastes cannot actually be recycled. And hazardous byproducts are often left behind that must be disposed of.[61] Once again, poor communities, attracted by the economic benefits of handling others' wastes, are further disadvantaged environmentally while the wastes are increasingly distanced from consumers.

The recent attempted export of spent mercury wastes from a decommissioned U.S. chemical plant to India illustrates the ongoing problems associated with the global toxic-waste recycling industry. The HoltraChem Manufacturing company, which used mercury to produce chlorine and other chemicals for use in the paper industry, closed in September 2000, leaving behind 260,000 pounds of mercury waste. The governor of Maine asked the U.S. Department of Defense to put the waste in the national stockpile of mercury, but was refused. HoltraChem said it would sell the waste to a broker who planned to ship it to India, already the largest recipient of mercury exports from the United States.[62] News of this planned shipment sparked a huge controversy in both the United States and India. Mercury-recycling operations in other developing countries in the 1990s, such as the Thor Chemicals recycling operation in South Africa that recycled mercury from the United States and Europe, raised concerns about the safety of recycling in India. In the South Africa case, mercury wastes from the recycling operation caused severe poisonings among workers and polluted the local environment before the plant was ordered closed.[63]

The U.S. government has claimed that because mercury, despite its highly toxic properties, is considered a traded metal, rather than toxic waste, it is not covered by the Basel Convention. Environmentalists and the Indian government disagreed. With a shipment of 18 tons of the waste on its way to India in January 2001, it was rejected by the Indian government and returned to the United States.[64] The waste was then slated to make its way to a mercury-recycling facility in Pennsylvania. But the recycling facility pulled out of the deal because of the bad press surrounding the mercury waste. The waste returned to the closed plant in Maine. The company ceased operations in April, 2001, raising questions about who would be responsible for the waste.[65] If the waste is eventually recycled, it will be resold, and could well make its way back to India for use in the manufacture of thermometers or in chemical plants. Even so, there is, once again, no "away" for such materials.

There is also a growing amount of toxic waste generated *in* poor countries. These wastes stem from industries producing consumer items that serve global markets, such as electronics and chemicals. When wastes are both generated and disposed of in developing countries, the distance from consumers in rich countries is huge, largely because the information is even less available to them than is the case when those wastes are produced in rich countries. The developing world's share of toxic-waste production from manufacturing is in fact growing, a trend that has been widely recognized.[66] Whether or not firms are moving to developing countries to take advantage of more lax environmental regulations, they do tend to take advantage of the low-cost conditions once there.

The *maquiladora* firms in Mexico are a stark example. These U.S.-owned industrial factories located just inside the U.S.-Mexico border in Mexico were set up in the 1960s to produce goods for export to the United States. In the early years these plants, concentrated in the garment-assembly sector, were not generators of large amounts of industrial wastes. By the 1980s, however, plants were being set up in the electronics, chemicals, and furniture sectors, all of which are large generators of hazardous wastes.[67] By the early 1990s, nearly four-fifths of *maquiladoras* were generators of toxic waste, while the number of such firms also climbed significantly.[68] Technically this increased generation of toxic waste in Mexico among the *maquiladora* factories should not have affected that country's environment. The La Paz agreement, made between the United States and Mexico in 1983, as well as Mexican law, required the toxic waste generated by the *maquiladoras* to be returned to the United States, where the materials originated, for disposal. This agreement explicitly aimed to reduce waste distancing. In the early 1990s, however, only 2 to 3 percent of firms were returning their waste.[69] At that time, both the United States and Mexico admitted to not knowing the amount of toxic waste generated on the border.[70] Improved monitoring and tracking of wastes were put into place following the adoption of the North American Free Trade Agreement (NAFTA) in 1994. Although recent figures show that the return of hazardous wastes to the United States has risen to 25–30 percent with increased monitoring since NAFTA came into effect,[71] it is widely recognized that accurate figures on toxic-waste generation on the border are elusive, and that illegal dumping of toxic waste continues.[72] Though the geographic distance to the U.S.-Mexican border may not be all that large for many in the United States, the understanding gap among Americans regarding the fate of toxic wastes associated with consumer items on the market there is significant.

The growing export of industrial waste *among* developing countries adds yet another dimension to waste distancing. The case of hazardous industrial plastics wastes exported to Cambodia from Taiwan in December 1998 demonstrates this clearly. When Formosa Plastics Group (FPG), a Taiwanese firm, had difficulty obtaining permission to dump its mercury-contaminated plastics waste locally, it hired a waste-brokerage firm to get rid of it for them. The waste broker exported the waste to the Cambodian town of Sihanoukville. Neither country is a party to the Basel Convention. Moreover, the Basel Ban Amendment does not prohibit such transfers among poor countries, only between rich and poor countries. Increasing the understanding gap, the waste was wrongly labeled as "polyester chip," and the shipping papers called it "cement cake." At least six deaths have been attributed to the waste, two from contact with it and the bags in which it was shipped, and four from a stampede out of the town when it was announced that the waste dumped there was hazardous.[73] FPG agreed to remove the waste, but it did not intend to return it to Taiwan.[74] Instead, it attempted to ship it to an industrialized country for treatment. The United States, France, and Germany were approached, but all three refused. By mid-2000, FPG agreed to reimport the waste and dispose of it in Taiwan.[75] It is virtually impossible for consumers purchasing plastic products marked "made in Taiwan" to know whether such products are linked to incidents such as this, illustrating the enormous distancing, along both geographic and cultural dimensions, of industrial waste from the end consumer.

Conclusion

Looking at waste disposal through a consumption lens highlights the mounting problem of distancing. Wastes are being separated from end consumers in increasingly new and incomprehensible ways. Unbeknownst to consumers, both postconsumer and preconsumer wastes are making their way around the world, with detrimental environmental and social impacts. The mass-production and mass-consumption systems developed over the past half-century give little information about wastes. At the same time, global collection and disposal systems help keep consumers in the dark about the fate of their garbage. Wastes often follow the channels of the global economy as vehicles to disadvantaged communities, taking advantage of economic inequality to "solve" the waste problem. In these processes, global waste-sink capacity is consumed, and at such a rapid pace that there is now a growing number of waste shipments—of both postconsumer and industrial waste—that no one will take, in either rich or poor countries.

Attempts to reduce the distancing of waste via education may make consumers more aware of the ecological implications of consumption choices. These attempts may improve ecological feedback loops so that more environmentally responsible consumption decisions can be made. But as Maniates argues in chapter 3 of this book, focusing only on the domain of private consumption choices serves to "individualize" responsibility for the environmental crisis through purchasing practices, rather than challenging more powerful structures and forces. Efforts to tame the forces causing waste distancing in the first place—large-scale industrialization that has brought mass production, mass consumption, and, now, mass waste collection, as well as economic globalization and economic inequality—must be undertaken. Such efforts are obviously much more difficult than merely educating consumers about the impact of their consumption choices. They are, indeed, political, and deeply resisted at many levels.

Some initiatives have taken on this task, though, despite the enormity of the challenge. At the international level, in the context of the Basel Convention, environmental NGOs and developing-country governments pushed Northern countries to ban the trade in hazardous waste between rich and poor countries. This ban, based on the precautionary principle, aims to put a damper on the exploitation of economic inequalities in the globalized businesses of hazardous-waste disposal and recycling. Once the Basel Ban Amendment comes into force it should further reduce transfers of toxic waste from rich to poor countries. Further efforts will still be needed, however, because transfers of toxic waste between rich and poor communities *within* countries continue and toxic waste continues to be generated within poor countries and traded among poor countries. The Basel Convention does encourage waste minimization and the proximity principle for disposal, but these are not enforced. Efforts to strengthen these components would go a long way toward improving the contribution of the Basel Convention to reducing waste distancing.

Another important step on the part of the international community is the adoption of the Stockholm Convention on the Elimination of Persistent Organic Pollutants (POPs). This treaty employs the principle of zero discharge as well as the precautionary principle to phase out the generation of the most serious, bioaccumulative persistent pollutants, including dioxins and furans that are the toxic by-products of waste incineration. This treaty will probably not come into force for several years, but it promises to significantly reduce waste-related pollution inequalities. These efforts at the international level mainly ameliorate the negative waste consequences associated with economic globalization and inequality, but they do not tackle the problems directly.

At a more local level, there are efforts to reduce the scale of industrial life through the localization of production and consumption. One example discussed in this book (chapter 11) is local exchange and trading systems (LETS). LETS attempt to reduce the large scale and global nature of industrial production and distribution systems by encouraging small-scale local production, service provision, and exchange. Though LETS do not focus directly on waste generation and distribution, they do have implications for waste issues. Extensive use of LETS in some communities can help reduce the scale of production and consumption for consumers who participate, thus increasing awareness of production methods and pressuring local producers to reduce waste. But because these systems are small and, to some extent, still operate within a global economy, there is little assurance that wastes generated either by producers or consumers within LETS will stay within the community. Recognizing this problem, some communities have debated whether to institute a proximity principle for their waste. The city of Toronto, for example, seriously debated whether it should be sending its municipal waste great distances for disposal. A proposal was floated to send its wastes to an abandoned open pit mine in a community 600 km to the north. In the end the NIMBY syndrome won out over the proximity principle, and a deal was struck to send that city's waste to Michigan.[76] But now Michigan is having second thoughts, and the debate over the proximity principle for waste may be reopened.[77]

Extended product responsibility (EPR) initiatives are another promising effort aimed to reduce waste distancing along the entire commodity chain. EPRs assign responsibility to producers and consumers all along the chain in an attempt to reduce the environmental impact of a product at all stages of its life cycle—that is, from the extraction of raw materials that go into a product, through to its production, distribution, and disposal. EPR initiatives originated in Western Europe, where the concept is now embodied in packaging laws in a number of countries, including Germany, the Netherlands, and Sweden.[78] These laws require producers to take back products at the end of their useful life, a provision that encourages producers to design products for easy reuse, repair, and recycle. They also encourage consumers to make more environmentally sound purchase choices. In the United States, most EPR efforts are only voluntary on the part of firms, and consequently in place mainly where there are clear economic benefits. To reduce waste distancing, more government oversight in the United States, including the enactment of laws to require EPR measures such as product take-back, could spur reluctant firms to get on board.[79]

To date the efforts to reduce waste distancing have been on a relatively small scale or have only partially addressed broader structures. Increasing efficiencies and expanding opportunities within the global economy are still the main organizing principles for national and local economies. Large-scale production, consumption, and waste disposal systems follow, along with inequality. Measures to reduce distancing still largely exist within this broader economic framework. Whether such measures will be successful within the global economic context depends in large part on whether they explicitly connect their aims to a revision of the broader structures and whether they are emulated more widely. Institutionally, the precautionary principle, zero discharge, waste minimization, and the proximity principle must override risky free-trade principles such as open markets, capital mobility, and centralized production.

8

Environmentally Damaging Consumption: The Impact of American Markets on Tropical Ecosystems in the Twentieth Century

Richard Tucker

Once one begins to wonder where the tropical products go, who uses them, for what, and how much they are prepared to pay for them ... one is asking questions about the market. But then one is also asking questions about the metropolitan homeland, the center of power, not about the dependent colony, the object and target of power. And once one attempts to put consumption together with production, to fit colony to metropolis, there is a tendency for one or the other ... to slip out of focus.
—Sidney W. Mintz[1]

In the half millennium since Europeans began to fling their nets across the planet, Western Europe and the United States have built their prosperity and power to an important extent on the appropriation of tropical and colonial natural resources. Many of the agricultural and forest products that have contributed to northern affluence have been grown in tropical climates, where they have damaged or displaced natural ecosystems. Remarkably, the causal links between the economic expansion of Europe and North America and the environmental degradation of the tropics have not been studied with any precision, for they cross the lines of both geographic and disciplinary specialization. As Sidney W. Mintz observes, it is not a simple task to keep in mind both arenas—consumption and production, consumer markets and agroecosystems—and beyond that, to comprehend the complex causal links between them. Yet awareness of those connections is vital for our understanding of the modern world and the relation of human culture to its biosphere home. In today's era of global free trade, with its insatiable demand for expanding material production and consumption, the issue of the environmental consequences of expanding consumption demands scholarly attention as never before. As social critic Robert D. Sack has written, "Our economies are geared to increase people's desires for goods. Mass consumption powers the economy, but it does more. Since mass consumption transforms nature and since the economy of mass consumption dominates the world, these transformations have a global reach."[2]

Historians have barely begun to explore the specifics of how the rise of mass consumerism has resulted in the depletion of the natural resources of distant and "less developed" lands. Perhaps even this assertion is inadequate: some social scientists assert that the study of consumerism itself is grossly neglected, especially as regards the varied values that different cultures ascribe to material consumption. As social historian Grant McCracken emphatically puts it, "The history of consumption has no history, no community of scholars, no tradition of scholarship."[3] In recent years leading social theorists have similarly begun to assert that the blindness to the links between the cultural and economic world of material goods and the living world of ecology has been a similar fundamental failing in their disciplines.[4] Among historians the gap is equally wide.[5] Historians of the American industrial and consumer economy, for example, have not adequately probed the environmental transformations that accompanied the rise of urban, affluent North America. Nor have diplomatic historians concerned themselves with the environmental aspects of the global reach of the American imperium.[6]

Thomas Princen observes that in order to describe land-use change and social dislocation in less developed countries, it may be sufficient to document those changes along with the activities of transnational corporations there. Or if the object is to explain the evolution of production patterns and the international division of labor, it is probably sufficient to trace foreign direct investment, trade flows, and relevant government policies. "But if one wants to explain the ever-increasing use and abuse of resources and waste sinks, the strain on ecosystem services from the local to the global, then one must do more," Princen urges. "In particular," he continues, "one must ask not just how the commodities are produced, but how they are consumed: how distribution, marketing, advertising, pricing, retailing, and government policies and programs affect consumer demand."[7]

This chapter argues that the issue of ultimate importance is ecological decline, especially the reduction of the biologically diverse ecosystems of the tropical world. It also insists that to understand what has caused that decline, one must focus on both production systems in the tropics and the ways by which the consumer economies of the North have increased their ability to extract resources and simplify ecosystems. One must be concerned for the past five centuries as a whole, but focus on the massive acceleration of scale in the past century or so. The analysis centers on one type of commodity consumption that established direct links between rising affluence in industrial societies and environmental damage as much as half a world away in the tropics. Concentrating on the rise of American consumerism since the late 1800s, when the American economy became the world's largest importer, it surveys the weakly understood ecological impacts of American consumption of bananas, coffee, and tropical timber.

The market production of these crops entailed replacement of forest or prior patchworks of subsistence farming and natural vegetation. In many instances what emerged were large monocrop plantations. In other instances relatively stable perennial crops became interspersed with the unstable frontier agriculture that had previously displaced primary forest. For each commodity specific links can be traced from markets back to tropical and subtropical production locations, processes, and ecological impacts.[8]

Consumerism as a dominant set of values is rooted in the emergence of the colonial world economy in the era of mercantilism, which created links between colonial production locations and metropolitan centers of investment and consumption. Its first locus was the emergence of mass markets in eighteenth-century England. There is an extensive literature on the history of the links between mercantilism and colonialism in the era before free trade became hegemonic in the nineteenth century. There is also a growing literature on conspicuous consumption in the era of European commercial capitalism, including the rise of European markets for mass food and fiber brought from great distances, including sugar, cotton, and other crops.[9]

Mass consumption, as the core of economic development, emerged in early modern Europe, and then was surpassed by the United States in the years around 1900. It evolved within the framework of the longer history of Northern imperialism. From the sixteenth century onward, European wealth and power increasingly reflected Europe's ability to command the natural wealth of the tropical world. The environmental history of the tropics cannot be understood except in its subordinate role in the modern "world system."

The growth of cities, London above all, was crucial to this societywide transformation. In the 1600s London became a booming market for mass-produced cotton cloth from India, shipped westward by the East India Company.[10] A specific link between European consumption and its subtropical source area was forged.[11] What impact this early colonial trade had on forest and field systems in cotton-production areas of India has not yet been clearly measured for the period before the mid-1800s.[12] From the late 1700s onward, consumption accelerated in England at an unprecedented rate in association with the early industrial revolution, a rapid increase in British population, migration to urban areas, and the rise of mass marketing. Some historians go so far as to insist that the consumer revolution preceded and fueled the industrial revolution. As an authoritative analysis puts it,

> There was a consumer boom in England in the eighteenth century. In the third quarter of the century that boom reached revolutionary proportions. Men, and in particular women, bought as never before. Even their children enjoyed access to a greater number of goods than ever before. In fact, the later eighteenth century saw such a convulsion of getting and spending, such an eruption of new prosperity, and such an explosion of new production and marketing techniques, that a greater proportion of the population than in any previous society in human history was able to enjoy the pleasures of buying consumer goods. They bought not only necessities, but decencies, and even luxuries.[13]

McKendrick, Brewer, and Plumb's analysis provides a prelude to the even-greater consumer market in the United States, which surpassed European economies around 1900 as the most powerful force for transforming economies and ecosystems throughout the tropical world.

The Hegemony of Consumerism in the United States

There is a large literature on the rise and significance of mass consumerism in American social and economic history. In this field, though, as in others, little effort has been made to relate domestic American history to its impact—both human and ecological—in parts of the world that provided important elements of the material wealth Americans consumed. A brief survey of themes in this field will provide a context for this chapter's cases, cases that trace some consequences of the modern American consumption engine.

The urbanizing and industrial transformations of American society brought into being a culture of consumerism in the late 1800s. As social historian Richard S. Tedlow describes it, "The United States during the course of the past century has been a nation of consumers. A higher percentage of the population has been able to purchase a greater variety of goods and services than even the most visionary dreamer in the mid–nineteenth century would have imagined possible."[14] Tedlow defines three eras of mass marketing. Before about 1880 the domestic market was fragmented among many small producers for hundreds of localities. This was determined largely by weak transport systems in the era before the national network of railroads was constructed. Then in the 1880s the unification era emerged, enabled by the transport and communications revolution, especially railroads and the telegraph. In cities and towns across the country, inexpensive consumer goods shipped in from distant sources appeared in the new department stores. This revolution in general merchandise retailing was led by John Wanamaker, followed by the great chain stores, Montgomery Ward and Sears Roebuck.[15] Simultaneously the spread of mass-circulating newspapers and magazines signaled the rise of the modern advertising industry, which was ultimately responsible for shaping the national culture of material craving, and with it an insatiable appetite for goods from around the world.[16]

The 1920s saw the third era in the corporate consolidation of commerce, as individual products came to be dominated by one or a few companies, and national brands came to dominate the advertising industry. As James D. Norris argues, "Separated by both time and distance from the producer, consumers became more and more dependent on brand names to ensure real or perceived quality."[17] In a society where social mobility was becoming the norm, status depended more and more on acquisition of its symbols. Norris concludes, "In connecting consumption to the achievement of social acceptance and status, . . . advertising simply put the old wine of materialism and conformity into new bottles seemingly available to everyone."[18]

Supermarket chains appeared as the regional and finally national distribution networks of corporate food giants. The new A&P chain led the centralization of grocery retailing in the 1920s, closely followed by Kroger's. This trend featured technological breakthroughs and corporate consolidation in food processing as well,[19] which became important for the marketing of tropical products like coffee and rubber.

Then, after the Great Depression and World War II, a burst of pent-up consumer desire produced the full florescence of American consumerism. Based in mass auto-mobility, it featured suburban sprawl, shopping centers, and the fast food complex, with such innovations as instant coffee.[20] Consumer sociologists Richard W. Fox and T. J. Jackson Lears describe "the potent ideal that underlay the consumer society of the 1950s: a world of goods produced in profusion, packaged and marketed by advertisers with subtle skills, avidly acquired on the installment plan by middle- and working-class Americans."[21]

Consumerism had already become closely associated with the global hegemony of the American economy after World War I, through the rise of mass marketing of imported tropical commodities. This market sector, and its immense ecological consequences, had several features associated with the American economy's global reach. By the 1890s U.S. industrial output greatly exceeded domestic consumption. This led to a push for overseas markets, which was linked to new capital investment abroad, especially for industrial raw materials and tropical consumer products. Domestic and foreign operations were closely linked, since marketing them involved the same transport, communications, advertising, and retailing networks that peddled domestic products. But in international trade, even more than in the domestic economy, monopoly or monopsony became the characteristic pattern. Because local governments were not always cooperative, social conditions at production sites were unstable, and logistical arrangements were expensive, there was little room for head-on competition among North American and European corporations.[22] (Timber imports, in contrast, had no parallel dominant firm or two. Hence the example of mahogany consumption necessitates a contrasting analysis, as we will see.)

The efforts of American corporations were backed by an official ideology of global progress through the production of mass (i.e., democratic) consumer goods. Elihu Root, Secretary of State from 1905 to 1909, declared:

> The people of the United States have for the first time accumulated a surplus of capital beyond the requirements of internal development.... Our surplus energy is beginning to look beyond our own borders, throughout the world, to find opportunity for the profitable use of our surplus capital, foreign markets for our manufactures, foreign mines to be developed, foreign bridges and railroads and public works to be built, foreign rivers to be turned into electric power and light.[23]

He might have added to his list tropical agricultural and forest products for American consumers.

Other federal agencies were as involved as the State Department in Washington's support of the new extractive enterprise. In the 1920s the Department of Commerce greatly expanded its international operations under Herbert Hoover, Secretary of Commerce from 1921 to 1929, who dominated foreign economic policy. Hoover's Bureau of Foreign and Domestic Commerce staffed 50 foreign offices and organized commodity divisions specializing in single items of foreign trade to advise American companies abroad, and it assisted with the lobby on Capitol Hill that successfully pressured Congress to reduce tariff barriers, making the flow of imports easier.[24]

In sum, the first third of the twentieth century was the formative era for American imports and consumption of the agricultural products that replaced tropical forests. A small number of emerging American multinational corporations, supported by the power of the imperial state, set in motion forces that enriched inhabitants of the northern temperate zone of the planet at the expense of the tropics. However, and this must be emphasized, the engine of consumerism did not in itself determine the precise forms of environmental impact. That was the province of the managers on the production end. The three cases that follow illustrate the wide range of impacts. They reflect the biological nature of particular varieties of bananas, coffee, and timber exploited in each instance, the state of agronomic and logging knowledge in each era, as well as the division of power between North American and local land managers.

Bananas: The Classic Corporate Duopoly

The most famous—or to many, notorious—multinational corporate giant of all was the United Fruit Company, formed in 1899 by a merger of two early American operations in the Caribbean basin. For many years its sole American competitor in the banana business was the Standard Fruit Company. Until World War II the two rivals centered their operations in the banana republics of Central America. Until the 1960s brought large-scale use of agrochemicals to the industry, banana plantations required virgin soil from primary rain forest if they were to yield profitably. The expansion of banana production for American markets required the clearing of tropical forest, reducing the biotic richness of the rain forest to a commercial monocrop.

Until the two Yankee giants succeeded in dominating both production and shipping of the fruit, local small-scale producers had combined bananas with a variety of other crops for both local consumption and distant markets. But bananas were delicate, easily damaged during transport or turning overripe and useless if they were delayed in reaching retail markets. By their very nature they demanded efficient, large-scale production and vertical integration of the entire process from the place of production to the checkout counters of distant stores.

The agribusiness giants cleared forest by the thousands of acres and built the infrastructure of an entire economy in the former forest zone, importing a concentrated labor force to areas where human population had previously been sparse.[25] As early as 1913, United's first corporate biographer extolled the company for vanquishing tropical Nature, bringing chaotic, unproductive rain forests under the disciplined management of corporate agribusiness.[26] Growing a delicate product far from its market, United and Standard had to create an unprecedented system of vertical integration. They cleared and sold hardwood timber; they established agronomic research stations and built entire towns and port facilities. United organized cross-Caribbean shipment by the refrigerated ships of its own Great White Fleet.

The first banana had reached the wharves of New York in 1804 from Cuba. By midcentury various shipping companies that traded with the Caribbean were loading bananas as an exotic curiosity, most often from the British colony of Jamaica. Consumption accelerated in New England in the 1870s when the Boston Fruit Company began regular imports of Jamaican bananas. In 1884 the U.S. government provided important support to the trade by canceling import duties on bananas. Tariff manipulation became (however inadvertently at the time) one of the most powerful tools for directing the development of tropical export agriculture, and thereby its ecological impacts. In 1892 the banana importers carried more than 12 million stems of bananas to American consumers.

Significantly, these buyers were increasingly successful in instructing growers in the islands and along the Caribbean coast of Central America to grow only one variety, the Gros Michel, to concentrate yields in the spring months, and to grow the largest possible stems, the nine-hand stems. The sole criterion was to maximize profitability. Efficient, rapid, reliable marketing became assured in the late 1800s, when the national railroad system of the United States and the new era of oceangoing steamships made possible prompt delivery of bananas to their ultimate consumers. Responding to the new opportunity for corporate consolidation, United Fruit came into being in 1899 as an amalgamation of Boston Fruit, the largest distributor, and the production network controlled by the American entrepreneur Minor Keith in Costa Rica and Panama. A year later United organized the first national distribution chain when it established a subsidiary, the Fruit Dispatch Company.[27] From then on, no other firm, American or European, could match United's capital resources or its control of the entire process from tropical forest clearing to American dining table.

Consumers could now be encouraged to shift from seeing bananas as an occasional exotic treat to adopting them as a regular item on their tables. Environmental historian John Soluri shows that "the expanding market for bananas coincided with the rising availability of other fresh fruits, including citrus, that were increasingly viewed by doctors, public health officials, and nutritionists as important components of daily diets."[28] Cookbooks and home economics manuals featured bananas for their sweet taste and nutritional value. But most important to the cultivation of a mass market, bananas became the cheapest fruit available on the carts of street peddlers in poor neighborhoods, and on the shelves of general stores and finally in A&P and Kroger supermarkets for the middle class. As the industrial labor force expanded, and its housewives scrambled to produce meals more quickly, bananas became one of the nation's first "fast foods." Soluri concludes, "the banana, both as food and symbol, crossed boundaries of social class and racial identity."[29] Its mass availability at low prices ensured that the consumer class could expand indefinitely.[30]

The advertising industry, now in full swing as the primary shaper of consumer tastes, took up the work of expanding consumer demand for bananas. As radios became pervasive in American life, the new song, "Yes, We Have No Bananas," became a national hit in 1923, giving a new level of popularity to tropical fruit. When the Depression descended on the economy six years later, the banana companies searched for additional ways of generating demand. Importing bananas became a tool to tie Latin American regimes and their economic foundations closely to American strategic interests throughout Latin America. In 1939 the theater impresario Lee Schubert discovered Carmen Miranda, the Brazilian nightclub entertainer, and took her to New York to star in a Latin revue. A year later Darryl Zanuck and Twentieth Century Fox launched her in the first of a series of popular Hollywood films. Her famous hat piled with tropical fruit and flowers, she became Chiquita Banana; everyone north of the border soon knew her theme song.

United Fruit captured Carmen/Chiquita in 1944. United had set up a Middle American Information Bureau a year before, publicizing the delights of tropical imports such as bananas, coffee, and hardwood timber. Launching its new brand name, Chiquita Banana, United achieved one of the greatest corporate successes in creating American consumer loyalty to one brand of a standard mass product. A corporation as powerful as United could take over a national icon, using it to shape demand for its tropical product. Until she died in 1955 (by then a Brazilian national heroine), Miranda served corporate interests well.[31]

The successes of marketing and consumption had direct environmental consequences at production sites in the tropics. The banana giants' gamble with tropical Nature was a dangerous one, because monocrop plantations virtually guaranteed massive attack by pathogens. Shortly after 1910 Panama disease began destroying entire plantations. A decade later Sigatoka disease also began ravaging the plantations. The two microorganisms forced the companies to move onward every ten years or so into virgin soils, devouring additional rain forest as they left old plantations either to their former workers and scrub cattle, or to revert to secondary woodlands. During the 1930s the companies moved to the Pacific coast of the Central American countries, opening a second front in the destruction of the Mesoamerican rain forest. By the 1940s they spread even farther, making Ecuador the world's largest banana ex-porter. In 1955 the American market imported 434,000 tons of bananas from Ecuador, 290,000 tons from Costa Rica, and 180,000 tons from Honduras, out of a total of 1,389,000 tons from Latin America as a whole.[32]

By the late 1950s the two companies, United and Standard, were able to return to their old haunts on the Caribbean littoral, using a new variety of bananas, the Cavendish, which was not susceptible to Panama disease. Intensive use of arsenic sprays succeeded in keeping Sigatoka disease under control. This produced a new generation of tropical corporate agriculture: the pseudosustainability of the agrochemical era, in which long-term production has been maintained only by massive applications of chemical fertilizers and pesticides, with severe damage to soil, water, and human health. In recent years, although banana plantations have no longer had to move every decade, both workers' health and downstream ecological health have been seriously damaged by intensive pesticide use.[33]

Coffee: Latin American Growers, Marketing Oligopolies, and American Consumers

Americans' century-long taste for tropical fruit has been paralleled by the two-century history of industrial America's craving for caffeine, primarily in the form of coffee. Coffee is the largest in money value of all tropical agricultural commodities produced for global markets. It has replaced both natural forests and subsistence multicrop farming in some 40 countries of the tropical and subtropical world. These transformations have occurred in a different ecozone than where bananas grow: hill regions wherever the climate is moderately moist and beyond the reach of frost. In sharp contrast to banana production in tropical moist lowlands, the extremely hardy coffee bean can be grown in any number of production systems, both as a large monocrop plantation and as the cash crop on small multicrop farms. Since it is grown under complex and varied arrangements, its production has had correspondingly variable social and environmental consequences and degrees of sustainability.[34] Here too, as with bananas, vast market demand has been the engine of changing tropical agroecology, but in itself it has not determined the specific ecological consequences.

In contrast with the environmental impacts of banana consumption, the saga of American coffee consumption has depended on importers and retailers rather than corporate managers of tropical lands. American coffee managers did not reach onto the tropical land itself to own or manage the plantations. Instead, Yankee buyers purchased coffee in tropical ports, led by São Paulo's port of Santos in southern Brazil[35] and Colombia's north coast ports. In both of these countries local elites controlled the land, the production and primary processing of the product, and the internal transport of the beans to the coastal ports, where American and European buyers maintained offices.[36] The plantation owners' methods of land management determined the specific environmental impacts, but the coffee economy would never have thrived without the pull of the American market. That is the end of the trail where the analysis must begin.

The marketing of coffee in the United States began in the late 1700s; imports were largely from South America and mostly flowed through the port of New York. For decades the trade was characterized by small-scale roasters, who retailed the beans in bulk to shops in towns and cities throughout urban America. The market grew steadily, reflecting both population growth and rising per capita consumption, until by the mid-1800s it permanently surpassed Europe's coffee imports. Gradually, economies of scale began to be decisive in the American industry. The New York Coffee Board was established in the 1870s to centralize and standardize the import, roasting, and wholesaling of coffee throughout the eastern states.[37] Just as with many consumer products in the late 1800s, coffee marketing came to be consolidated in fewer, more highly capitalized firms. Arbuckle Brothers of Pittsburgh became the largest processor, packer, and distributor of coffee that poured in through New York Harbor. After 1920 San Francisco emerged as a major competing center for the Western states, as Hills Brothers became a West Coast regional powerhouse. Coffee was now a standardized product, advertised aggressively in daily newspapers and nationally circulating monthly magazines. The major brands were instantly recognized by a wide spectrum of consumers.[38]

Historian Michael Jimenez has shown that the expansion of the American coffee market closely reflected the process of industrialization of labor and the increasing domestic responsibilities of women. The advertising industry was largely responsible for creating the national demand for coffee, teaching women to define their households by the coffee they served, and encouraging industrial workers to raise productivity by taking coffee breaks to sustain their working energy. By the 1930s an oligopoly of coffee distributors controlled mass distribution of Brazilian and Colombian coffee through supermarket chains, as well as the advertising of universally recognized national brands: Chase & Sanborn, A&P's Eight O'Clock, Hills Brothers, and others. Colombia serves to illustrate how dominant the U.S. market could be for the export economy of a tropical country. In the mid-1860s the United States imported only 26 percent of Colombia's exports; the rest went to Europe. But by the late 1920s the United States took 92 percent of Colombia's coffee.[39] (Figures for Brazil are roughly similar.)

In the years after 1945, instant coffee and the fast food era led to another explosive round of expansion in American purchases of South American coffee, particularly the robust Brazilian product. In 1960 the United States imported 563,000 tons of Brazilian coffee and 261,000 tons of Colombian coffee, by far the largest figures for any importing country, from any region of the world.[40] The advertising industry took on a leading role in this promotion, but American firms were by no means the only major players. The Colombian Coffee Board invented Juan Valdez, a male parallel to Chiquita Banana, the personification of the happy, prosperous, independent small farmer. His clothes were always neat, his donkey always well fed, and his hills always green. Consumers were not to worry that their thirst had any deleterious effects a continent away.[41] The sanitizing of consumption has rarely been more charming—or more misleading, because the reality on the Colombian (and Brazilian) landscape was very different from the advertising image.

The environmental impacts of the American coffee market were massive but varied, depending on the cultivation system. Southern Brazil has been the world's dominant coffee producer throughout the past 200 years. In the mid–nineteenth century coffee became the leading edge of Brazil's economy; it was grown on large slave plantations in the rolling hinterlands of Rio de Janeiro and São Paulo. The slave owners became a landholding aristocracy that dominated Brazil's government, selling large tracts of government-owned forest land for virtually nothing. The voraciously expanding American market meant that over the last six decades of the twentieth century, high coffee prices guaranteed big profits to speculative owners. With labor and land costs low and profits high, landlords allowed rapid soil erosion to occur over wide areas of Rio de Janeiro and São Paulo states, then moved on to clear additional forests, rather than practice any kind of soil conservation. The result was the most massive erosion of hill slopes that ever resulted from world coffee production.[42] For a century Brazilian estate owners kept forcing the frontier of forest destruction onward, planting trees in rows up and down hillsides without interplanting shade trees whose root systems might inhibit soil loss and preserve some degree of biodiversity. Only after 1950, when cheap frontier forest land was no longer readily available, did the planters begin efforts to reduce soil erosion.[43]

In the middle hills of the Colombian Andes, perennially the second largest coffee-producing region, both frontier peasants and large landowners grew coffee from the 1840s onward. By providing a cash crop for squatter peasants, coffee was the engine that removed tens of thousands of patches of Andean forest. Even that process of deforestation on a moving frontier, fueled by Northern markets, could not prevent endemic social violence from flaring repeatedly in the coffee region. Colombian society came to be the most convulsively violent in all of South America.[44]

No study has yet determined with precision the dynamics linking the international coffee market, the social structure of coffee production, and the environmental consequences in the Colombian Andes. Local soil conditions and production systems varied widely in scale, soil management, and use of shade trees. Hence land degradation in the coffee regions has been somewhat less notorious than in Brazil's first century of coffee expansion. But any steep slope, once stripped of its vegetative cover, rapidly loses its soil; the Colombian coffee zone was no exception. T. Lynn Smith, an American observer who traveled at length in Colombia in the 1940s, reported seeing devastating erosion—land degradation accompanying social degradation—throughout wide areas where coffee had been the cutting-edge crop of frontier settlement.[45]

The hill region of Central America became a third major site for the coffee boom from the 1840s onward. The volcanic hillsides of the Central American republics produced a mellow, aromatic bean that was exported largely to England, France, and Germany until World War I. During the war, when European markets were cut off, American buyers took up the slack, assisting in the strategic work of displacing European capital from Latin America. After the war ended, the New York market remained for a while the largest buyer in El Salvador and Guatemala, and briefly in Costa Rica as well. Production systems varied from the aristocratic estates of El Salvador to the small farmers' multicrop system in Costa Rica.[46] But all of them used more careful mulching and shade trees than in Brazil, so individual coffee groves sustained coffee production more effectively. This has changed recently, as growers have adopted new varieties that bear more heavily but do not use shade trees, thus requiring intensive chemical fertilizing.[47] Where coffee grows, forest is gone, but Central American coffee has generally been a more stable substitute for natural vegetation than its South American counterparts.

Mahogany and the Economic Value of the Tropical Forest

Tropical hardwood timber contrasts with crops grown as a replacement of forest, not only in production processes but in marketing and consumption patterns as well. Yet the hardwood timber trade—and consequently its ecological impact—can be traced with similar accuracy from retail purchaser back to the source region. Mahogany is the classic case of northern imports of tropical hardwood. There are three closely related species in the *Swietenia* genus, the dominant hardwood throughout the moist lowlands of the Caribbean, southern Mexico, and Central America, as well as in parts of the Amazon basin.[48] Mahogany grows as a dominant tree, but it is only one species in the intricate complexity of the rain forest. Thus mahogany forests have always been cut selectively, never clear-cut. But the work of reaching each prime tree has resulted in severe damage to others around it and along the extraction routes.

Beginning in the sixteenth century, mahogany and Spanish cedar were exported from the Caribbean islands and mainland lowlands to European colonial centers, especially Spain, France, the Netherlands, and England.[49] In Britain,

> The golden age of mahogany, between 1725 and 1825, saw the development of furniture styles by master craftsmen who were directly influenced by certain characteristics of the wood. The large size of the mahogany planks made possible a greater flexibility in the designing of furniture than had been possible when using other woods. The richness and warmth of the reddish-brown wood, the variety of figure, and the fine texture appealed to the British people.[50]

Chippendale, Adam, Hepplewhite, and Sheraton styles of cabinetry boomed. By its peak in the years around 1850, the trade imported 50,000 tons of mahogany annually through London and Liverpool.[51]

Following the lead of British styles, the urban centers of the United States became important markets for luxury furniture by the early 1800s.[52] Mahogany filled a separate niche from what domestic hardwoods could satisfy, so imports steadily expanded as the consumer class prospered. In the course of the 1800s it spread from the wealthy few to a much broader spectrum of middle-class buyers. In the years around 1900, imports varied between 10 and 20 million board feet annually. By 1920 American importers dominated the international market, reflecting American displacement of Europe's trade with Latin America generally.[53] Moreover, hardwood furniture is a high value-added industry, based on skilled craftsmanship. Mahogany furniture has been produced at many factories in several regions of the United States, in a generally decentralized industry.[54]

The prestige hardwood flowed into New Orleans, New York, Boston, and several other American ports from West Indian ports such as Santiago, Cuba, and Gulf Coast ports from southern Mexico to Nicaragua. During the 1920s average American imports were over 32 million board feet; then they gradually fell off by one-third by the late 1950s, because accessible sources were shrinking.[55] The degradation of the great rain forest resource was proceeding. The great trees were harvested throughout the Caribbean islands as well as mainland shores and riverbanks. In contrast with agricultural monocrops, which replaced forest cover entirely, mahogany loggers since early colonial times had extracted only the finest trees, thereby degrading but not destroying the forest.

After 1900 some logging was associated with forest clearance for export monocropping; the banana companies introduced some of the most technically advanced logging. As easily accessible mahogany supplies ran low and transport costs rose, furniture manufacturers began to worry about their sources.[56] After World War I new technologies—trucks, bulldozers, and more powerful milling machinery—enabled loggers to penetrate ever deeper into rain forests.[57] As Caribbean, Mexican, and Central American sources of mahogany became depleted, loggers searched for additional sources of the elegant wood in South America and Southeast Asia. The greatest New World rain forest, the Amazonian basin, presented such formidable challenges to hardwood extraction that corporate capital and lumbering technology succeeded in penetrating Amazonia on any significant scale only after about 1960.[58]

In the Western United States the market for tropical hardwoods flourished after World War I, just as coffee, banana, and rubber markets expanded. Importers from Los Angeles to Seattle found an equivalent resource in lauan, the dipterocarp timber species from the Philippines, calling it *Philippine mahogany* for their markets. Exports from the Philippines, almost entirely to the United States, rose from 252,000 board feet in 1907 to a high of 196 million board feet in 1936.[59] But Philippine supplies were severely battered by the Japanese wartime occupation from 1942 to 1945, and then by a new round of commercial logging in the years after Philippine independence in 1946. By the 1960s the country's exports, still primarily to the United States but increasingly to Japan, began an irreversible decline, which reflected the decimation of its hardwood forests.[60]

Domestic U.S. prices for mahogany products rose in that era of mass affluence, motivating the search for Amazonian mahogany. The industry also turned increasingly to substitutes such as *Danish modern*, the styles that used teak from either natural or plantation forests in Southeast Asia. And additional species of tropical hardwood began to appear on American markets. Consumers, their tastes shaped and broadened by the industry, were learning how to digest an ever-increasing range of tropical timbers. Whether this market could impart positive money value to the tropical forest, in competition with the plantation crops that destroyed the forest, was unclear. Whether selective hardwood logging in the rain forest can be sustainable in any circumstances is still unclear.

Conclusion

The three products surveyed in this chapter illustrate specific links between the American market and its environmental impacts in tropical source locations, in the era when that market became the world's largest for tropical commodities. A full list of those products would include cacao, cotton, tobacco, citrus fruit, more recently coca, and other agricultural and forest products—both annual and perennial crops, grown in both monocrop and multicrop systems. Their impacts have been complex and varied, and there have been a wide variety of ecological processes and transformations in source areas. In each case the two keys to the pattern of environmental impacts have been the changing size and character of the consumer market, and the specific agroecological production system. One general production pattern has been large-scale clearing of forest, by both foreign corporations and local landlords, under both slave and free-labor systems. A second general type of landscape change has been patchwork, under multicrop systems, usually produced by smallholder farmers, but these systems too have fallen increasingly under corporate control of transport and marketing. (A third dimension of the environmental impact of all these systems, their ripple effects into surrounding areas of forest and settlement, lies beyond the local production area. This dimension is still more diffuse and complex, but in the total picture the regional ramifications of all the expanding export systems must not be overlooked.) As Mintz reminds us, a full understanding of each commodity trail takes consumption and production equally into account. Economy and ecology must both be considered.

As a major element of the global environmental economy, export agriculture has been driven most fundamentally by oligopolistic corporations: the production, processing, and marketing giants such as United Fruit and Standard Fruit for bananas; and Arbuckle, Chase & Sanborn, Hills Brothers, and now Starbucks in coffee marketing. Their power to develop, shape, and manipulate markets has been decisive in determining the consumption end of tropical agricultural commodities. In contrast, the import of mahogany and other tropical hardwoods has remained smaller in scale, more varied, and more specialized. These timber products have been transported and manufactured into furniture and other products by a wide range of relatively small firms, and their products have been purchased by a smaller, more affluent circle of purchasers than the mass markets for tropical fruits and coffee.

These briefly summarized cases illustrate that the backward links from ultimate consumption, through the corporate economy of processing, advertising, and marketing, and ultimately to environmental change at the sources of tropical commodities, can be traced directly in many cases, though they rarely have been, by either consumers or analysts. If these specific cases were replicated for dozens of others, the emerging jigsaw puzzle would show many and variable impacts of Northern commercial investments and consumer affluence on tropical ecosystems. Each would remind us that the cost of that prosperity, however far removed from the ultimate consumer, has been damage and reduction of tropical Nature's bounty.

III

On the Ground

9

In Search of Consumptive Resistance: The Voluntary Simplicity Movement

Michael Maniates

It's quiet, countercultural, potentially subversive, but also mainstream. It flies low, usually hidden amidst reports of increasing productivity, rising consumer confidence, expanding personal debt, and the dizzying array of new products promising to make life easier, faster, more productive, and more rewarding. Unpromisingly rooted in an apolitical and consumerist response to social ills, it also sows the seeds of collective challenge to fundamental dysfunctions of industrial society. Focused as it is on the quality of work and quest for personal control of one's time and one's life, it resonates with the American deification of individual freedom. But inevitable connections to questions of environmental quality, workplace control, and civic responsibility lend it more complicated hues.

Some call it *simple living*, evoking images of earlier, more prudent times. Others prefer *downsizing*, *downshifting*, or *simplifying*. The popular press and many scholars know it as *voluntary simplicity*. Call it what you will, but don't lose sight of the irony: at a time when policymakers, pundits, and corporations around the world embrace ever more deeply the assumption that consumption and happiness are joined at the hip, frugality appears to be back in fashion. Through their words and by their deeds, a seemingly large and growing number of people are claiming that "we can work less, want less, and spend less, and be happier in the process."[1]

According to Rich Hayes of UC Berkeley's Energy and Resources Group, these people generally "value moderation over excess, spiritual development over material consumption, cooperation over competition, and nature over technology."[2] Historian David Shi is more pointed—to him, practitioners of the simple life "self-consciously subordinate the material to the ideal."[3] Alan Durning, a former analyst with the Worldwatch Institute, an environmental think tank, quotes a voluntary simplifier who says that "simple living has come to mean spending more time attending to our lives and less time attending to our work; devoting less time earning more money and more time to the daily doings of life."[4] Noted sociologist Amitai Etzioni brings a finer-grained analysis to bear by offering three categories, each more intense, of "voluntary simplifiers": (1) "downshifters," who reduce their consumption and income without deeply altering their way of living, (2) "strong simplifiers," who significantly restructure their lives, and (3) "holistic simplifiers," whose consistent rejection of consumerism flows from a coherent philosophy.[5] And Cecile Andrews, whose book *The Circle of Simplicity* is often credited with catalyzing the "voluntary simplicity movement" (VSM) in the United States in the 1990s,[6] advances an understanding of voluntary simplicity that zeros in on the pace of life:

> A lot of people [are] rushed and frenzied and stressed. They have no time for their friends; they snap at their family; they're not laughing very much. But a growing number of people aren't content to live this way. They are looking for ways to simplify their lives—to rush less, work less, and spend less. They are beginning to slow down and enjoy life again.
>
> There's a movement associated with this—it's called the voluntary simplicity movement. Around the country, thousands of people are simplifying their lives. They are questioning the standard definitions that equate success with money and prestige and the accumulation of things. They are returning to the good life.[7]

Andrews's estimate of "thousands of people" is likely conservative, though no observer of the U.S. simplicity movement really knows for sure. Writing in 1981, Daniel Elgin, author of the simplicity bible *Voluntary Simplicity: An Ecological Lifestyle That Promotes Personal and Social Renewal*,[8] claimed 10 million dedicated converts to the cause. But this figure, an estimate for the United States alone, struck some as "optimistic."[9] And yet a later study commissioned by the Merck Family Fund[10] concluded that from 1990 to 1995, 28 percent of Americans (or over 60 million) voluntarily reduced their income and their consumption in conscious pursuit of new personal or household priorities. In 1998, Harvard economist Juliet Schor found that some 20 percent of Americans (roughly 50 million people) have, in the past several years, permanently chosen to live on significantly less and are happy with the change. Most of these "permanent downshifters" are not rich, says Schor; half reside in households with annual incomes of $35,000 or less prior to any downshifting.[11] More recently, Gerald Celente at the Trends Research Institute claimed that 15 percent of the nation's 77 million baby boomers were significantly engaged in the simplicity movement.[12]

Perhaps because of their growing numbers, simplifiers have become the subject of increasing media attention. In 1993, for example, readers of major U.S. newspapers[13] would have learned little if anything about the VSM, since relevant stories rarely made it into print. This had changed by 1996; that year, an average of just over two articles or features per paper appeared.

By 1998 the number of stories or features had jumped threefold to fivefold, depending on the newspaper, and as of this writing (mid-2001) there is no indication that this pace of coverage is slowing. Strikingly, relevant articles are finding their way into marquee venues. The *Washington Post Magazine* ran a cover story on "voluntary simplicity" just before Christmas 1998. The *New York Times* ran four pieces, three in its prestigious Op-Ed section, that connect directly to the subject in three months between November 1998 and late January 1999. Major West Coast newspapers ran feature pieces on simplicity and frugality throughout 1999 and into 2000. (Ubiquitous "resolutions for the new millennium" pieces frequently highlighted the need to simplify and "slow down.") The U.S. Public Broadcast System aired two programs—*Affluenza* and *Escape from Affluenza*—on consumerism and its cure.[14] The producer of the programs characterizes the outpouring of public interest in them as "astonishing" and "completely unexpected." The distributor of the videos describes them as the biggest production his company has distributed in years.[15] *Affluenza* has been translated into four languages for distribution to 17 countries of the former Soviet Union, and similar plans are in the works for *Escape from Affluenza*.

Curiously, the media's fascination with simplicity has not been matched by attention in more scholarly circles. One reason, I explain shortly, is that voluntary simplicity butts up against mainstream environmentalism's understanding of "sustainability" and "the consumption problem" and is marginalized as a result. By its very existence, the VSM insists that real reductions in consumption—at least for some, framed in particular ways—bring real net benefits to be enjoyed rather than sacrifices to be endured. But mainstream environmentalists—not to mention policymakers, planners, and many academics—find it difficult to entertain and give voice to the distinct possibility that doing with less could mean doing better and being happier. Locked in a calculus of sacrifice, activists and academics and policymakers alike tend first to romanticize the VSM, and then to marginalize it to the domain of "fringe" activity that is oddly interesting but fundamentally irrelevant to the practical politics of sustainability.

This is unfortunate because, as I later argue, the relevance of the VSM to larger struggles for sustainability remains ambiguous. Simplicity is teetering between a self-absorbed subculture looking to effect social change by, say, using soap more sparingly, and a nascent social movement capable of fostering lasting change in how work, play, and consumption are organized in industrial society. Writing off voluntary simplicity as the newest incarnation of yuppie "self-help" claptrap[16] runs the risk of slighting a potentially defining element of an alternative politics of environmentalism. On the other hand, framing voluntary simplicity as the next major trend to sweep industrial society[17] dangerously glosses over simple living's worrisome deficiencies of politics and analysis.

Simplifiers are apparently a diverse group, drawn to simplicity for myriad reasons and networking in novel ways. The VSM is a kaleidoscope, and thus answers to reasonable questions become both straightforward and confusing. Is voluntary simplicity a growing social movement for reducing consumption and fostering environmental sustainability, as many of its adherents maintain? One could make a convincing argument. Is it instead a passing fad, the contemporary manifestation of a long-observed pattern of yearning for simplicity that is then followed by a burst of hyperconsumption? This case can also be made. Does the VSM stand as evidence of broadly shared disquiet with the direction and pace of industrial society; is it, in other words, a subversive attempt to strike at the heart of consumer capitalism? Or is it an irrelevant subculture, one that has enjoyed more than its share of media attention, which brings together two social groups: burnt-out urban professionals earning $200,000 year whose choice to "downshift" to a "simple" lifestyle has them trading in the Mercedes for a Honda, and everyday citizens hiding from the rough-and-tumble world of environmental politics by quietly focusing on simple acts of frugality like clipping coupons at home?

One way of slicing through this confusion is to return to the organizing themes of this book: production as consumption; the chain of material provisioning and resource use; the myth of consumer sovereignty; the politics of consumption and the struggle to build new institutional mechanisms that more fully communicate the full range of the costs of consumerism, commodification, and overconsumption; the consumption juggernaut. When applied to simple living, what do these concepts tell us about the strengths and weakness of the VSM and its potential as a social movement that confronts consumption? The concluding paragraphs of this chapter take on this question.

Some of what follows is necessarily speculative—anecdotes about the VSM abound where one would hope for hard data. This chapter, though, seeks less to be a definitive treatment of the VSM than, ultimately, a reconnaissance of what is said about the movement and why, and an argument for taking seriously the simple-living phenomenon that will not, for reasons both encouraging and troubling, go away. Simple living in an age of instant gratification and globalized mass consumption deserves our attention.

Tyranny and Fever, AIDS, and Inoculations

As the introduction to this book reminds us, "sustainable development" is the defining lens through which contemporary environmentalism sees the world. Though this lens brings into sharp focus many issues of importance, it remains blind to deep questions of "consumption." Because the prevailing definition of a sustainable society is one that "satisfies its needs without jeopardizing the prospects of future generations,"[18] sustainable development becomes economic development that meets the *expanding* needs of the present global order while maintaining, even enhancing, regional and global ecological processes crucial to the well-being of future generations. Present needs, and future ones too, are assumed to be large, growing, and exogenous. The task becomes meeting them through the enlightened use of high-efficiency technologies paid for by a rapidly expanding globalized economy. Whose needs, why these needs grow or diminish, how the desire for more is fanned and fueled—these questions are never raised, either in conventional policy circles or among environmentalists. This docile acceptance of ever-expanding wants and needs leaves unchallenged what economist Robert Frank calls "luxury fever," and what Nepalese environmental scholar and activist Dipak Gyawali labels "the tyranny of expectations."

Robert Frank lives in Ithaca and teaches at Cornell University, while Dipak Gyawali calls Kathmandu his home. From opposite sides of the planet, living in very different worlds, both point to the same dynamic: satisfaction with one's material life is significantly influenced by how much one spends and consumes *relative to others*. Beyond some threshold of minimally acceptable consumption, deprivation and abundance become relative conditions, easily altered by the consumption levels and decisions of benchmark social groups.[19]

Examples abound in the industrial North. A cartoon from the *New Yorker* appears in Frank's book, for example, in which a commuter, calling home from a roadside stop, says "I was sad because I had no onboard fax until I saw a man who had no mobile phone." A silly cartoon, yes, but on the mark too regarding relative satisfaction. Products constantly are becoming more "luxurious"—faster computers, larger cars, more spacious houses, smaller cell phones, better backpacks, higher-performance athletic shoes. Their introduction into the marketplace makes previously acceptable consumption choices pedestrian, and sometimes unacceptable, by comparison. "Recent changes in the spending environment," writes Frank,

> affect the kinds of gifts you must give at weddings and birthdays, and the amounts you must spend for anniversary dinners; the price you must pay for a house in a neighborhood with a good school; the size your vehicle must be if you want your family to be relatively safe from injury; the kinds of sneakers your kids will demand; the universities they'll need to attend if you want them to face good prospects after graduation; the kinds of wine you'll want to serve to mark special occasions; and the kind of suit you'll choose to wear to a job interview.[20]

Part of this dynamic is inescapably structural: if everyone else is wearing an expensive suit to that job interview, you have got to as well if you hope to send the right message, regardless of your feelings about consumerism or your leanings toward frugality. Likewise, the presence of more behemoth vehicles on the road ratchets up the arms race on the highway; standing a fighting chance against one in a collision requires buying a larger vehicle yourself, no matter how much joy or dismay driving it brings. But a big chunk of the problem is rooted in less obvious changes in the "benchmark" group against which much of the middle class compares itself.

"Keeping up with the Joneses" used to mean measuring up to the consumption choices of neighbors of similar means and aspirations. More and more of the middle class, however, are now gauging their material prosperity against the top 10 percent of consumers nationally—the yardstick has taken a quantum leap upward. Many interlocking forces are responsible, including easy credit, vigorous marketing, the increasing concentration of income, the marked rise in casual mixing among economic classes as corporations embrace "open offices" and "team-building" management techniques,[21] a spate of television shows and movies that portray consumption levels of the top 10 percent of Americans as those of an "average" middle-class family, and real-life consumption by the upper economic strata that is conspicuous and seemingly guilt free.[22] At work too, Juliet Schor reminds us, is the "Diderot effect,"[23] whereby a new acquisition—a new bookcase for the living room, for instance—suddenly makes the couch look rather shabby and that, when then replaced, highlights the sudden inadequacy of the curtains. Together, this trinity of forces—the structural, the rising benchmark, the Diderot-driven "upward creep of desire"—conspire to keep the consumer escalator moving upward. It is luxury fever: more people are consuming more, in terms of both quality and quantity; they are going more deeply into debt as a result; and if analysts like Robert Frank or Robert Lane (whose book *The Loss of Happiness in Market Democracies* marshals an impressive array of data[24]) can be believed, they are finding life less fulfilling and less secure as a result of this relentless ratcheting up of standards.

Take this "luxury fever" and extend it to the so-called Third World, and you get Gyawali's "tyranny of expectations."[25] Globalization brings sophisticated advertising to the Third World, spreads Western television programs that glamorize consumption ("Dallas," "Melrose Place," "Santa Barbara"), and facilitates the impossible-to-miss emergence of a conspicuously consuming Third World elite more at home in New York, Los Angeles, or London than in Delhi, Salvador, or Nairobi. Masses of poor in these countries feel poorer by comparison and come to expect more. Though their economies have a hand in fueling rising expectations, these economies cannot satisfy people's expectations quickly, if at all—huge numbers of people are involved, after all, whose understanding of what minimally constitutes the "good life" is rising exponentially. These expectations, heightened by media and elite behavior, go unmet; an enduring sense of "missing out" and being left behind takes root; and frustration, dissatisfaction, and social instability surface. Ironically, as prudent societies of the Third World are transformed into consumer societies, mass frustration and an overall *decline* in felt material satisfaction ensues, even as real incomes among the poor slowly inch *higher*. For scholars like Canadian historian and filmmaker Gwynne Dyer, this still-unfolding dynamic is nothing less than "a bomb under the world."[26]

This is not an argument to keep poor people poor because "we" somehow know that increased consumption will not make "them" much happier. Many around the world *must* be lifted from the depths of poverty, and nothing in Gyawali or Frank's analysis should be heard to suggest otherwise. But if the 4 billion or more global underconsumers are to raise their consumption levels to some minimally rewarding and secure level, the 1 billion or so global overconsumers will first have to limit and then reduce their overall level of consumption to make ecological room. At this point, mainstream environmentalism stalls out, for it has not yet focused on developing a politically practical language for analyzing and confronting the explosion of wants, much less for thinking about how to foster the "downshifting" of democratic industrial societies. Indeed, the sustainable-development lens, which removes from view any political or sociological analysis of consumption, has become the metaphorical AIDS of the environmental movement. The movement's ability to recognize and respond to the core threats to the environment has been compromised because it cannot engage an ethos of frugality.

What instead emerges is simplistic moralizing about consumption that little advances the intellectual analysis or collective action necessary for taking on the consumption question. This moralizing takes three forms: rhetorical lambasting of advertising, condemnation of the immorality of overconsumption, and a rosy-eyed, apolitical romanticization of the joys of simple living. Though each response makes some sense, in that each echoes credible analyses of the current human condition (corporations *do* spend egregious amounts on advertising, gluttony *is* fraught with ethical implications, frugality and self-denial *can* bring personal satisfaction), none alone hold much promise for undermining luxury fever or thwarting the tyranny of expectations. Here's why.

Advertising

In 1992, Alan Durning observed that

> communications in the consumer society is dominated by the sales pitch, by the unctuous voices of the marketplace. Advertising is everywhere, bombarding typical members of the consumer class with some 3,000 messages a day. Ads are broadcast by thousands of television and radio stations, towed behind airplanes, plastered on billboards and in sports stadiums, and bounced around the planet from satellites. They are posted on chair-lift poles on ski slopes, hung on banners at televised parades and festivals, piped into classrooms and doctors' offices, weaved into the plots of feature films, and stitched onto Boy Scout merit badges and professional athletes' jerseys.[27]

That was in 1992. Durning's litany of advertising excess, written to shock, might seem tame by today's standards. Advertisements are not just plastered in sports stadiums any longer: companies now buy whole stadiums and name them. Every sports story reporting on stadium events advances corporate name recognition. Product symbols are not just posted on billboards; they appear on computer screens, in rented videos, on the backs of Olympic athletes. Messages are not just piped into classrooms; major corporations now compete for "product placement" and "informational advertising" in the hallways, the playgrounds, the cable feeds, even the stationery of elementary and secondary schools.[28] It is a brave new world of advertising, one where tried-and-true strategies of name branding and lifestyle mass marketing mix with sophisticated technologies of communication to create heretofore unimaginable powers of persuasion.

Insofar as it cultivates needs, advertising is anathema to those who would have us rein in consumption. Calls, then, for "public control" of advertising or a redirection of advertising dollars to more socially productive ends are not surprising or uncommon. Realistically, however, this agenda stands little chance of success. Protected by guarantees of "free speech" common in some countries, and often more powerful arguments for "free markets" common to most, advertisers have the full weight of Western liberal tradition in their corner. To be antiadvertising is to be antidemocratic, if not antimarket. It is no surprise, then, that the invasion of public space by advertisers is growing rather than abating, as too is the psychological sophistication of the ad (see chapter 10). Railing against advertising may be morally satisfying, and is altogether consistent with impressive scholarship documenting the cultivation of needs and the escalation of "cola wars." But such objections fail any test of political efficacy. Advertising—ubiquitous, insidious, ever more global in its production and distribution (ads produced on computer screens in Bangalore or Kuala Lumpur are finding their way into the living rooms of United States, Canadian, and European consumers)—is here to stay. Those wishing to blunt its ability to stimulate acquisitive desire would be better off working to increase people's resistance to advertising than struggling for regulations that would severely limit the scope and freedom of advertising itself.

The Immorality of Consumption

The same rhetoric that takes dead aim at advertising also asserts moral claims about the ethics of consumption. Oddly, these claims often cast consumers as both the victims and the agents of their own moral destruction. By fueling feelings of guilt (about one's own behavior) and anger (at an industrial system that manipulates consumers), such "moral" arguments, like the one made by Ronald Sider, an advocate of rethinking Christian moral teachings in light of today's materialism, aspire to convince people to consume less:

> Many of us actually believe that we can barely get along on thirty-five, forty-five, or sixty thousand dollars that we make each year. We are in an incredible rat race. When our income goes up by another $2,000, we convince ourselves that we *need* that much to live—comfortably.
>
> How can we escape this delusion? Perhaps it will help to be reminded again that thousands of children starve every day. That over one billion people live in desperate poverty. And that another two billion are very poor. The problem, we know, is that the world's resources are not fairly distributed. North Americans, Western Europeans, and rich elites around the world are an affluent minority in a world where half the people are poor.
>
> Why are we so unconcerned, so slow to care? . . . We have become ensnared by unprecedented material luxury. Advertising constantly convinces us that we need one unnecessary luxury after another. Affluence is the god of twentieth-century North Americans, and the adman is his prophet. We need to make some dramatic, concrete moves to escape the materialism that seeps into our minds via diabolically clever and incessant advertising. . . . The more we make, the more we think we need in order to live decently and respectably. Somehow we have to break this cycle because it makes us sin against our needy brothers and sisters.[29]

Sider certainly has a point—consumption levels among the affluent can appear obscene when compared to the deprivation of the impoverished. The problem, though, is that moralists like Sider typically proceed from the assumption that "the affluent minority" does not care or remains unconcerned about global inequality, or "the adman," or the rat race within which it finds itself ensnared. In fact, though, many in the affluent world *do* care and *are* concerned; they know about global inequality, they intimately understand the invasiveness of modern advertising, they experience daily the enervating effects of the rat race. They just have little idea, given the daily options and constraints they face, about how to meaningfully oppose such forces and structures. Rooted as they so often are in assumptions about the immorality or callousness of the affluent, well-intentioned homilies like Sider's that urge "dramatic, concrete moves to escape materialism" too often overemphasize individual culpability for materialism at the cost of frank talk about the political and economic structures that manufacture desire and lock us into patterns of overconsumption. Ultimately, such preaching does little to help people of conscience bridge the gap between their morals and their practices.

Simple Living

Mainstream environmentalism's public articulation of "simplicity" only complicates matters. All too often, influential environmental publications bind "simple living" to the spiritual clarity that comes from time in the "wild." "Simplicity" becomes a set of simple, highly individual steps certain to ease life's complexities and save the planet in the process. It is an attractive formula, reminiscent of Thoreau's time at Walden or Muir's wanderings through the Sierra Nevada: time in the woods allows one to see the unnecessary complexity of life, and a list of simple consumerist steps to simplicity enables one to act on this newfound clarity on return to civilization. Alas, it is out of step with the experience of most people, who do not frequent the wilds for insight and clarity, and it wildly simplifies and stereotypes the life events and processes that bring people to "downshift."

Environmentalism's story of simplicity—what it is, how to find it, how to practice it—ends up telling us more about the transcendental and technocratic roots of environmentalism than the contemporary practice and politics of simple living. Consider Scott Russell Sanders's essay[30] (titled "Simplicity") in an issue of *Audubon*, the magazine of the National Audubon Society, which begins like this:

On our last night in Rocky Mountain National Park, after a week of backpacking near the continental divide, my son, Jesse, and I sat on a granite ledge overlooking a creek just below our campsite. The water crashed through a jumble of boulders, churning up an ice mist and turning the current whiter than the surrounding snow.... In the waning light, the trees along the banks merged into a velvety blackness, and the froth of the creek shone like the Milky Way. Waves rose from the current, fleeting shapes that would eventually dissolve—like my own body, like the mountains, like the earth and the stars. I blinked at my son, who rode the same current. Our time in the mountains had left me feeling cleansed and clarified.

After speaking a bit more to the details of his time in the wilderness, Sanders offers that:

Whenever I return from a sojourn in the woods or waters or mountains, I'm dismayed by the noise and jumble of the workaday world. One moment I can lay everything I need on the corner of a poncho, tally my responsibilities on the fingers of one hand. The next moment, it seems, I couldn't fit all my furniture and tasks into a warehouse. Time in the wild, like time in meditation, reminds me how much of what I ordinarily do is mere dithering, how much of what I own is mere encumbrance. Coming home, I can see there are too many appliances in my cupboard, too many clothes in my closet, too many files in my drawers, too many strings of duty jerking me in too many directions. The opposite of simplicity, as I understand it, is not complexity but clutter.

Sanders's lyrical essay is arresting, but what does it tell us about changing our lives and our world? Sanders speaks of the need for restraint, both personal and collective, lest unrestrained growth in production and consumption further undermine critical ecological processes. He argues for the evolution of cultural practices that foster within us a capacity for restraint, speaks eloquently to the urgent need to challenge the juggernaut of exponential growth, then concludes with his own story of pushing aside his mountains of work one night to accept a friend's invitation to star gaze, in the hope of recapturing the clarity and peace he found in the Rocky Mountains. He did; the essay ends with these lines:

The Milky Way arched overhead, reminding me of froth glimmering on the dark surface of a mountain creek. I know the names of a dozen constellations and half a dozen of the brightest stars, but I wasn't thinking in words right then. I was too busy feeling brimful of joy, without need of any props except the universe. The deep night drew my scattered pieces back to the center, stripped away all clutter and weight, and set me free.

We best go about confronting cultural and institutional impediments to personal restraint, in other words, by engaging in profoundly personal acts of reflection and communing with nature. Just in case the point is not clear, the editors of *Audubon* include a sidebar with Sanders's essay titled "12 Ways to Simplify Your Life and Save the World." They are:

> 1. avoid shopping 2. leave the car parked, 3. live in a nice neighborhood (that will allow you to walk to stores or easily access public transit), 4. get rid of your lawn, 5. cut down on your laundry, 6. block junk mail, 7. turn off the TV, 8. communicate by e-mail, 9. don't use a cellular phone, 10. drink water rather than store-bought beverages, 11. patronize your public library, and 12. limit the size of your family.

Despite this apolitical and individualistic response to problems that are fundamentally political and institutional (see chapter 3), mainstream environmentalism (the Audubon Society is far from alone in trumpeting this interpretation of simple living) is actually onto something important. Essays like Sanders's begin to voice the questions that come to mind whenever a news report about "simplifiers" and "downshifters" makes the newspapers or gets on television: What (to put it bluntly) is with these people? What, in other words, accounts for their ability to step back and ask tough questions about consumption and personal satisfaction? What has inoculated them against luxury fever and imbued them with a certain "consumptive resistance"? Are they just better people, or maybe just better off? Or has some combination of cultural, political, and social forces come together in their lives to extricate them from the tyranny of expectations? And could this combination conceivably be fostered, and even recreated, for a significant portion of industrial and overconsuming North America?

On the whole, mainstream environmentalism does not deal with these questions terribly well. It remains captured by a pastoral theory of simplicity that imagines consumptive resistance flowing from close contact with nature. It suggests that we best challenge the large social forces that gnaw away at our capacity for restraint by going stargazing. It implies that we save the world and make our life happier in the process by giving our lawn over to native species, giving up our cell phone, or giving away our TV. Without perhaps quite realizing it, mainstream environmentalism ends up locating voluntary simplicity to the affluent margins of society who consider themselves "environmental" but not actively political—if you are a white, economically comfortable, ecotouring suburbanite with a big lawn, e-mail, a cell phone, and the flexibility to search for "a nice neighborhood," and you want to save the world in 12 easy steps, simple living may be for you. Others need not apply. In this simplistic view of simplicity, the VSM becomes a curious little phenomenon lacking any real promise of molding the terrain of contemporary environmental politics. And members of mainstream environmental groups with real interest in "saving the world" are left with 12-point programs that steer well clear of confronting institutions and interests that erode the capacity for restraint.

Fortunately, a closer look at the VSM paints a more complicated picture, one rich with potential. Most North Americans, for example, do not seem to be pushed toward simple living by close contact with nature, nor are their reasons for downshifting centrally environmental. And while many members of the VSM are highly educated, they do not disproportionately represent the economic elite that is so often the target of environmentalism's urgings to live simply. Many of them, it seems, do not have cell phones or large lawns or big families. Job stress emerges as a principle source of consumptive resistance (complicating the conventional view that dissatisfied workers in search of solace make good consumers), though other factors come into play as well. Contrary to what historians might have predicted, moreover, the current wave of simple living has not waned, despite the consumption binges of the 1980s and 1990s. All this together points to a set of hopeful possibilities for simplicity. If it is not as elite and as dependent on close contact with nature as mainstream environmentalism tends to paint it, it could prove relevant to a new politics of sustainable consumption.

Focusing in: The VSM

A Surprise from the Field

The United States does not hold a monopoly on voluntary simplicity, but it might as well. Although offshoots of the movement thrive in some corners of Western Europe, especially in the United Kingdom, and its core tenets are preached and practiced in much of Asia, notably among those in India and Sri Lanka deeply influenced by Gandhian ideals, the United States remains the site of some of the most broadly felt and keenly expressed yearnings for simplicity in the world.[31] The city of

Seattle, in Washington State, is often characterized as "ground zero" of the movement, given the thousands of practitioners reputed to reside in the city's environs.[32] But simple living is not particular to the Pacific Northwest. It is scattered throughout the United States to a degree unfamiliar to the industrial democracies of Japan, Australia, Western Europe, and Canada.

What about American life might yield such a distinctly American response? Environmental activist turned environmental scholar Rich Hayes offers some hunches:

> Some observers link (the renewed interest in voluntary simplicity) to the resurgence of environmental activism that began in 1988. Others believe it was a pragmatic response to the recession and economic restructuring of the early 1990s. For still others it was a response to psychic stress associated with rapid technological change, social fragmentation, and the relentless expansion of consumerist values into all domains of human life. Today the prospects for the voluntary simplicity movement are uncertain. With strong and steady economic growth, a shortage of ecological disasters, and growing acceptance of high-tech social norms, the movement might fade once more. If any of these conditions should become problematic, interest in voluntary simplicity could grow.[33]

This passage is from Hayes's introduction to his analysis of the "first national gathering of the voluntary simplicity" movement, held at the University of Southern California (USC) in late September of 1998. To my knowledge, Hayes's assessment, which informs his doctoral dissertation in Energy and Resources from the University of California, stands as the most comprehensive attempt to make sense of the meeting, which remains the single largest gathering of the "movement." Like Hayes, I made my way to Los Angeles in the fall of 1998 to attend the day-long gathering, billed as "No Purchase Necessary: Building the Voluntary Simplicity Movement." Together with the interviews I had scheduled with many of the principals of the movement the day before, I expected the affair to shed light on the sociology and vitality of the VSM. What I learned and witnessed surprised me greatly. Hayes, who was a savvy, no-nonsense Sierra Club activist before returning to graduate school, confessed to being "astonished."

The gathering was scheduled for a Saturday, in a conference hall at USC's Davidson Conference Center. Representatives of Seeds of Simplicity, a VSM networking organization taking the lead on the event, were nervous. The gathering had been announced months earlier, but was advertised haphazardly, primarily through announcements posted on e-mail networks and postcards sent to a handful of mailing lists. Ultimately, Seeds of Simplicity and its many conference cosponsors[34] were counting on word-of-mouth advertising. Because there was no preregistration process (the conference was free, and staff resources were inadequate to process preregistrants), it was anybody's guess as to how many people might actually show up.

What if you call for a gathering of simplifiers and nobody comes? The question must have weighed heavily on the minds of the conference organizers. Looking onto the empty auditorium the day before, the 500-person-capacity conference hall looked cavernous and daunting; what seemed like a suitable venue months earlier assumed the appearance, in the hours before the meeting, of a monument to hubris—and the picture-perfect weather forecast for Saturday, inviting Los Angelinos (not known for their leanings toward simplicity and restraint) to head for the beach rather than hunker down inside a conference hall, only heightened the anxiety. "Maybe we'll get, at best, 200 people trickling in throughout the day," hoped one organizer. "Which wouldn't be too bad," she continued, "except for the size of this hall. Two hundred people will look puny."

Saturday dawned to a predicted 80°, partly cloudy day. ("Why couldn't it be raining," muttered one.) People steeled themselves for disappointment, not knowing what to expect. With the gathering scheduled to formally begin at 9 A.M., many were in the hall by 7:15 A.M., hurrying about with last-minute setup. A mixture of hope and anticipatory despair was in the air.

Surprisingly, by 7:50 A.M., many attendees had already trickled in, more than an hour early. The trickle soon turned into a steady flow, then a deluge. By 8:30 A.M. there was not an empty seat left in the hall. By 8:50 A.M. some 700 people were jammed into the hall; at least another 150 more were pressed in the entry hallway hoping to enter. Organizers tried to start promptly at 9 A.M., but introductory remarks from the stage were quickly disrupted by yells and chants from the now 200 or so pressing to enter, which brought the session to a standstill and sent people scurrying to locate an "overflow" room where many of those seeking entry could be accommodated.

Confusion reigned for more than 30 minutes as organizers sought to cope with problems they had never thought possible: too many people, too much interest, an unwillingness by "latecomers" (i.e., those who arrived 30 minutes *early*) to go away quietly. As some of the conference organizers searched for overflow rooms, others took to the stage to inform those 150 or so standing in the back of the room and in the aisles that they would have to exit or face eviction by the fire marshal, who evidently was threatening to shut down the entire affair. No one budged. Perhaps they had heard, as many of us came to learn later, that the USC campus police was at that moment turning away another 500–700 people from USC parking facilities. Having a space inside the hall, even an "illegal" standing-room-only space, was looking good by comparison.

The program finally got underway in earnest around 9:45 A.M. and continued to 5 P.M., with a lunch break and a few shorter breaks interspersed throughout the day. There was little audience attrition. All went as planned, with plenty of opportunity for audience feedback, though there were some surprises. The first was the general response to Kalle Lasn, founder of The Media Foundation and publisher of *Adbusters*, who spoke just after lunch. He described the history and work of his organization, then turned to the 800-pound gorilla in the room that nobody as of yet wanted to acknowledge: the seeming apolitical, self-absorbed, and escapist qualities of the VSM. Lasn spoke powerfully to the need for simplifiers to look beyond the household and band together for broader social change. Part stern patrician, part cheerleader, part union organizer, he argued in a tone both scolding and empathetic that "simplifying your life is important, but we need a pincer strategy if we are to succeed. Individual change *and* collective action geared to enforcing corporate responsibility is the *only* way that we will together achieve our goals."

The audience responded with a standing ovation, the only one of the day.

Another surprise was the lukewarm response to speakers from the major environmental groups, and from the Sierra Club in particular. Folks in the room were not unsympathetic to environmental concerns, but as the day wore on it became clear that simplicity, for them, was not about saving the planet, although limiting environmental degradation was certainly a side effect of frugality that should not go ignored. Nor, in many cases, did their curiosity about or practice of simplicity flow from any deep connection to nature. Some were at the gathering because, though they were successful and secure in their work, they felt that the cost of this success had become too high. Others—many, many others, in fact—told stories of ongoing workplace insecurity, periodic layoffs, and little wage growth. Still others spoke of doing too much overtime and of not being able to "catch their breath." Others came in search of some collective vehicle for addressing broader concerns about irresponsible advertising and the abuse of corporate power.

Representatives of the major environmental groups were like fish out of water. They came to the gathering assuming that "downshifters" had experienced some transcendental epiphany, and were rejecting consumption out of deeply held concerns about wilderness and pollution, only to find their envirospeak falling flat. Conversation at the conference was not about the environmental benefits of taking a three-minute shower rather than a five-minute one, or of letting your lawn go to weeds. It was "radical talk" about the state of work and the maldistribution of income and flat wage rates and corporate irresponsibility and ways that ordinary people caught in the midst of it all might reasonably react to preserve their sanity, their families, and their sense of self-esteem. The Sierra Club representatives never knew what hit them.

Trite as it might sound, the most consistently voiced reason for interest in simplicity was the felt need to "get my life back." There was give-and-take across different experiences and perspectives, to be sure, but *the* enduring theme of the meeting was that as currently organized, work and shopping—production and consumption, in other words—are not organized to meet the full range of human needs. Powerful actors will not intervene to correct this condition. Consequently, one must take control of one's own life, both by consciously moving one's lifestyle toward frugality and simplicity *and* devoting some of the independence and spare time that simplicity buys to collective struggle aimed at transforming institutions that drive consumerism and overconsumption.

In this respect the national meeting was true to form. Although some scholars understand the simplicity movement as a reaction to the psychological or spiritual emptiness of plenty,[35] the more compelling view appreciates the VSM as a kind of passive resistance to the legacy of "Reagonomics." The 1980s in the United States was a decade of impressive economic growth and ostentatious accumulation of wealth. But it was also a period when wage growth was stagnant and the proportion of households with two full-time wage earners was on the rise. Indeed, almost all growth in household income for the middle class in 1980s and early 1990s was from household members working more hours. And most of this new income was quickly dissipated on goods and services—child care, a second car, new work clothes, more expensive prepackaged meals that can be prepared and served on the fly—critical to the functioning of two-income households.[36]

Firmly established by the end of the 1980s, this middle-class treadmill of work and spend accelerated in the 1990s. First, the recession of the early 1990s produced waves of layoffs and "downsizing" that left those still on the job responsible for more, and more complex, tasks. The economy then kicked into high gear. As unemployment dipped to record lows and labor markets tightened, American employers came to demand more and more overtime from their employees. Workers, moreover, were not just spending more time *at* the job; the proliferation of personal computers, home fax machines, pocket pagers, and the overall rise in "home offices" meant that Americans were spending more time *on* the job as well, even when they were not physically in the office.[37] The result: overwork, stress, information overload, and growing doubts about the benefits of running the "rat race." "People are toiling overtime," writes Kristin Davis of *Kiplinger's Personal Finance Magazine*, "to pay bills or to pick up the slack in a downsized work force. Two-income families are cramming the job of running a household into a frenzied Saturday of errands and chores. And the array of choices in everything from finances to leisure time is exploding."[38] Is it any surprise that the overwhelming reason generally cited for downshifting, and expressed at the national conference too, is "wanting more time, less stress, and more balance in life?"[39]

Perhaps the lasting revelation is not what was said at the national simplicity meeting or who said it, but rather that in response to haphazard publicity and shoestring organizing, at least 1,200 people from 14 states decided to "show up" on an enticingly warm autumn day at a meeting that advertised itself as a place to engage the VSM. It is one thing to read study after study documenting the "time crunch" confronting Americans and analyzing the economic treadmill on which they run. It is quite another to witness, one after another, articulate and spontaneous expressions of frustration and resistance, and watch as people with little formal political experience or understanding of community organizing grope for an appropriate set of responses to their growing sense of life out of control.

Toward the end of day, *Washington Post Magazine* essayist Bob Thompson, whom I happened to be standing next to at the water fountain, leaned over and asked, somewhat incredulously, "could this be for real?"[40] For those looking for robust repositories of "consumptive resistance" to tap and expand, that remains the compelling question.

A Skeptic's Response

Despite its seemingly sudden burst on the scene, voluntary simplicity is nothing new. Some, like University of California at Santa Cruz scholar Marcy Ann Darnavsky, trace its origins to the counterculture and environmental movements of the late 1960s and early 1970s. Others, especially the noted historian of simple living David Shi, locate the movement in deeper waters of an American yearning for self-sufficiency and plain talk that reaches into the seventeenth century. Shi makes a compelling case that intense periods of interest in "simplicity" punctuate social life in the United States every 20 to 30 years; evidence of recent bursts can be found, for example, in the work of essayists of the 1930s, many of whom argued the merits of simplicity in ways little different from contemporary simplifiers. And lest we conclude that voluntary simplicity is solely an American invention, historian Peter Gould reminds us of the intellectual and political debt downshifters owe to nineteenth-century activists in Britain.[41]

Historically, the politics of simplicity has been anything but political. With few exceptions, simplifiers shunned heated debate and explicit struggle, choosing instead to let their frugal actions speak louder than their words. Their ire was reserved for profligate lifestyles, not institutional structures and public policies that fostered or accommodated overconsumption. They often joined with major industrialists and retailers of the time who found it advantageous to preach the virtues of thrift and frugality, of hard work and "making do."[42] In the end, simplifiers never proved terribly successful in recognizing, much less undermining or reforming, dominant institutions of consumption and consumerism.

To a skeptical observer of the 1998 "national gathering" at USC, little, at first glance, would seem different today. There are policy battles to be fought for frugality—a luxury tax on excessive consumption, an end to hidden subsidies of fossil-fuel-intensive production, a better measure of true social and personal costs of economic growth than gross national product (GNP), flextime and shared jobs and new ways of thinking about work—but the simple-living folk do not seem terribly interested. No simple-living lobby is at work in Washington, DC, advancing these policies, and no network of voluntary simplifiers is inundating key congresspeople with letters and e-mail. Whatever political backbone infuses the simplicity movement appears limited to the household, where family members struggle with one another and themselves over what to buy and how to live and what to do without.[43] And, like their predecessors, contemporary downshifters are finding allies in the corporate sector. With the support of CEOs and personnel managers, programs and seminars on "work balance," "reducing stress through simplicity," and "lifestyle management" are cropping up all over the modern workplace.[44] It is hard to miss the resemblance between the call to frugality some 70 years ago by major U.S. corporations eager to cap wages, and simplicity and stress workshops sponsored by today's major corporations who increasingly *require* workers to put in overtime.

It is thus easy to understand why those on the political Left, and labor too, would answer Bob Thompson's hopeful question of "could this be for real?" with skepticism tinged with contempt. Tom Vanderbilt, a contributor to the weekly magazine *The Nation*, captures better than most the problems many on the liberal end of the political spectrum have with simple living:

> The voluntary-simplicity authors like to underscore the remarkable courage and resolve it took for them to step off the roller coaster of getting and spending. But their advice, which includes such widely applicable gems as "Sell the damn boat," "Get rid of your car phone," and "Work less and enjoy it more," must come as a slap in the face to those who were pushed off the ride, or to those who never got on and probably never will. How many people actually shared the simplicity advocates' "frenetically paced lifestyle and rampant consumerism of the eighties?"... There's a vaguely insidious tone of resignation to the whole movement, implying that we accept an age of diminished expectations, steeling us for some future round of corporate downsizing. Perhaps AT&T could include a copy of Duane Elgin's *Voluntary Simplicity* in the same package that holds

the retraining and relocation stipend it is giving the 72,000-plus managers it is shedding.[45]

That was Vanderbilt as political economist. He ends his essay—which, again, captures well the leftist critique of "simple living"—as an analyst of the deterioration of participatory democracy:

> Luckily, the voluntary simplicists have a solution (to the continued concentration of wealth and consequent erosion of democracy in America): Stop reading the newspaper, stop answering the door when the doorbell rings, cancel your magazine subscriptions. But all these suggestions of "hard-core simplicity" speak of a conscious withdrawal from the convolutions of modern life. Albeit with different motives, they resemble the "off the griders" who inhabit the fin de siècle America, people who have torn up Social Security cards, stopped paying taxes and pulled themselves out of what they see as the oppressive data banks of central authority. It may well be enough to live "the American dream on a shoestring," as the voluntary simplicists have it, but when the string begins to fray, it may take more than tidying up the closet and giving up the BMW to mend it.[46]

Vanderbilt does not speak for the entire American Left, of course, but his essay captures well the essential leftist critique of the VSM, which has three elements. The first is that the capacity for restraint exhibited by simplifiers is principally a function of wealth and the power it confers. Consumptive resistance flows to those who can limit their purchases without pain (because they already have so much), and who command power over their economic lives sufficient to rearrange work schedules, change jobs, and say no to expanding workplace demands without fear of retribution. "Voluntary simplicists" hardly serve as a model for the rest of the world; they attained their current state of "simplicity" by walking the path of profligate overconsumption only accessible to a few. Such hypocrisy would not be so offensive if not for the second element: simplifiers are oblivious to the plight of those saddled with *involuntary* simplicity; they too readily renounce their responsibility as *citizens* to challenge unjust or dysfunctional institutions, perhaps because their privileged position rests on the economic deprivation of others. The third element is that the VSM is for real; we imagine its irrelevance at our own peril.

Voluntary simplicity: it is elitist, it lies somewhere between accommodationist and escapist, it looks to be spreading, and it promises to further undermine our faltering sense of collective civic responsibility. These are the charges leveled against the VSM. Just how seriously should they be taken?

Elitist?

The VSM is commonly assailed as a movement of "rich kids" tiring of life in the fast lane. The accusation makes for good copy (an "exposé" of simplicity titled "Cashing Out" takes the cake[47]) but falters under scrutiny.

The median income for U.S. households in 1997 was nearly $40,000, while the average household income of the middle 20 percent of American households for the same period was about $44,000. Rich Hayes, based on survey data he collected at the September 1998 voluntary simplicity conference, estimates the average household income of those attending the meeting (people interested in simplicity, as opposed to already practicing downshifters) to be around $40,000, with the median figure likely lower. Juliet Schor pegs the median household income of the downshifters she studied at around $35,000 before their changes. Linda Pierce's study of 211 simple-livers places 66 percent of her respondents in households with income levels of $50,000 or less. Sixty percent of these simplifiers live in dwellings of 1,300 square feet or less (typically two to three bedrooms), and 38 percent are renters.[48] Unlike the Hayes and Schor numbers, these data are for people who claim, via their responses to an online questionnaire, to have already downshifted.

Voluntary simplicity is not a poor person's movement. But neither is it the domain of the rich. All available data indicate that those expressing strong interest in simplicity, or who claim to have downshifted, hail predominantly from the middle 20 percent of American households. It makes sense that they would be. Those living in poorer American households are, after all, already practicing an involuntary form of simplicity and are not likely to express deep yearnings—to poll takers or anyone else—for a more frugal, less cluttered lifestyle. And, clearly, the onerous complications of owning a BMW, maintaining a quarter-million-dollar home, or keeping track of time-share property in the Caribbean have been grossly overstated by those who imagine the simplicity movement as a cynical "cashing out" of the affluent. Materially, downshifters and potential simplifiers are nothing special. They hail from the mainstream, hold or held typical white-collar jobs in corporate management, administration, customer relations, or computing, are often laboring under significant debt (especially high-interest credit-card debt), and are struggling with oppressive workplace stress or lack of personal fulfillment.[49]

Where they do stand out from the crowd is on education. Pierce reports that a whopping 54 percent of her 211 simplifiers had completed some postgraduate work, and 19 percent had an undergraduate degree—but, again, Pierce's study, though useful, does not purport to be a scientific sample of the larger population. Schor's study, a random survey of 800 adults nationwide conducted in November 1996, is more reliable—and it reveals that 30 percent of downshifters have "some college" under their belts, 25 percent possess a four-year college degree, and 14 percent command a postgraduate degree. For his part, Hayes notes that the attendees at the simplicity conference were "generally highly educated and middle-aged." These data stand in marked contrast to March 1997 figures for the United States as a whole: only 25 percent of U.S. adults had "some college" and 24 percent had a bachelor's degree or more.[50]

Those attacking voluntary simplifiers as agents of the economic elite look to be unduly influenced by environmentalism's one-dimensional characterization of simplicity. (Vanderbilt's annoyance with suggestions to "get rid of the car phone" or "work less" could not be any more pointed if he were directly confronting Scott Sanders's "Simplicity" essay in *Audubon.*) It is a bit ironic that those who consider themselves defenders of the economically beleaguered middle class are often the first in line to bash "elitist simplifiers," not realizing that—for better or for worse—it is the besieged middle class that is driving the current expansion of simplicity in the United States.

This does not mean that voluntary simplifiers are the salt of the earth. Educationally, they *are* among the select few, even though they have not translated their higher education into six-figure salaries. Indeed, the income-to-education ratio for downshifters and potential simplifiers must be among the least impressive of any social group in the country—which is both interesting and important, for at least two reasons. First, it again suggests that the capacity for restraint does not principally derive from profligate levels of consumption. The engine of the VSM is not the "decline of marginal benefits" of consumption to which economists refer, but rather workplace pressures from which workers estranged from union organizing or other forms of formal workplace politics are seeking escape. Second, the overriding "normalness" of simplifiers (they work regular jobs, they make average salaries, they live in typical houses, and—like almost 70 percent of Americans who say that they would be more satisfied with their lives if they were able to spend more time with family and friends[51]—they find themselves far too pressured) is compelling. Simplifiers are not fringe—what they are thinking and feeling and beginning to do is most likely representative of a large chunk of the American public.

So what makes simplifiers different? Perhaps they have a lower tolerance for job frustration. Maybe they are more natural risk takers. Certainly, by virtue of their higher educational levels, they can, as Schor notes, better "manage the world around them. They have social and personal confidence, know how to work the system, and have connections to powerful people and institutions. Unlike the traditional poor, they have *options*."[52] And while Schor may overstate the case—narratives of the lives of voluntary simplifiers do not reveal a pattern of connection to powerful people, for example—her observation points in a provocative direction. If simplifiers are like most other stressed out (typically but not always white-collar) workers, *except* for the social power they wield to renegotiate their work lives, might not the path to more sustainable levels of consumption best be paved by policy measures that would confer on others the same power over work that voluntary simplifiers are themselves now asserting?

Escapist?

The VSM has been slow to consider such policy measures, despite the many voices—some old, some new—urging it to do so. Robert Frank, the Cornell University professor who gave us "luxury fever," speaks appreciatively of the VSM but wonders how a focus on personal lifestyle choices can chip away at larger socioeconomic processes that are ratcheting up "acceptable" levels of consumption. Likewise, University of California sociologist Arlie Hochschild observes with both surprise and admiration the precipitous growth of the movement, but then notes that "a more daunting yet ultimately more promising approach to unknotting the time bind requires collective—rather than individual—action: workers must directly challenge the organization, and the organizers, of the American workplace." She argues that rather than voluntary simplifiers who withdraw from the workplace, the country needs a new kind of political activism, one that would create a *time movement.* For the truth of the matter is that many working parents lack time because the workplace has a prior claim on it. It solves very little to either adapt to that claim or retreat from the workplace. The moment has come to address that claim, to adjust the old workplace to the new workforce. As history has shown us, the only effective way to bring about such change is through collective action.[53]

Surprisingly, one of most influential framers of the contemporary VSM is firmly in Hochschild's camp. She is Cecile Andrews, a former community college administrator of adult education who became involved in voluntary simplicity in the late 1980s. Her widely read 1997 book *The Circle of Simplicity: Return to the Good Life* is a warm but challenging invitation to earnest reflection and focused study about the dynamics and possibilities of voluntary simplicity. The book, at its core, is about "simplicity circles," which are small groups of people (typically six to ten) who meet regularly to critically analyze study materials and explore how they connect to one's own struggle with simplicity. Reflecting her adult-education experience, *The Circle of Simplicity* is less a one-dimensional celebration of simplicity than a handbook, syllabus, history lesson, and political-action primer geared toward facilitating tough, meaningful inquiry within a small-group setting. Andrews jokes that her simplicity circles are akin to communist party "cells." She intends them to be places where people who seek to swim against the current can find the personal strength and intellectual focus necessary to making lasting change in the world.

Reprinted a host of times and now available on audiocassette, Andrews's book has done more than any other recent publication to amplify and focus America's inquiry into simplicity and downshifting. With support from Seeds of Simplicity (a program of the Center for Religion, Ethics & Social Policy at Cornell University), Andrews recently launched the "Simplicity Circles Project," a Web-based study-group database that outlines the dimensions of study circles, maintains online discussion boards, tracks the formation of study groups and circles, and facilitates the formation of new study groups. Some 650 "simplicity circles" are affiliated with the Simplicity Circles Project, and more are added every day. The project is among the most striking examples of the movement's use of the Internet to disseminate its message and draw mainstream America into a culture of frugality.[54]

As a VSM insider, what makes Andrews remarkable is her keen awareness of the troubling "self-help," apolitical quality of the simplicity movement, and her persistent call for the movement to "take the next step." At the 1998 simplicity conference she spoke passionately about the need for simplifiers to come together to create alternative workplace possibilities for *all* workers. The day before, in an extended interview, she expressed hope and optimism that the simplicity movement would coalesce around something like Hochschild's "time movement." This hopeful optimism finds expression in her book (in a closing chapter titled "Transforming Work"), in which she writes that "there are two main problems we need to deal with":

> We need to help people who have been laid off and also help people who remain employed. Even though more and more people are being laid off, there are many more who remain in the paid workforce, often in low-paying, low-skilled jobs that work them long hours. We need to reform the workplace. Ultimately, we must push for flexible working hours, shorter working hours, a flattened hierarchy, small, decentralized work groups, and increased worker participation in decisions. We must raise low wages and limit extremely high wages. Management must find ways to reduce competitiveness and fear, find ways to share profits, create alternative work schedules, give employees more time and opportunity for learning new skills, and do whatever it can to reduce layoffs.
>
> Some of these solutions could be developed by enlightened management, but many of them will have to come from workers finding ways to pressure employers to change. This means finding ways of bringing people together so that they will begin talking and acting collectively to reclaim some of their power. It means supporting unions, joining professional organizations, and starting study circles.[55]

Tom Vanderbilit of *The Nation*, and others like him, appear genuinely unaware that VSM insiders like Cecile Andrews are calling for a confrontation with prevailing workplace dynamics. The problem, however—that is, if you are cheering the VSM toward a higher level of social and political salience—is that the troops do not seem to be listening. Many downshifters, drawing on their social capital of education and connections, have immersed themselves in an *individual* pursuit of a whole host of special workplace arrangements—flextime, job sharing, benefits for part-time work, small-team management structures, and decentralized work groups responsible for getting jobs done, not logging in hours. But they seem too busy cutting their own deals for an exception-to-the-rule, 30-hour-a-week job (with benefits!) to think much about collective action to make such jobs the rule rather than the exception.

In fairness to these downshifters, one could easily imagine them supporting Hochschild's "time movement" if it existed. But none does in the United States, and Andrews and others offer few clues to downshifters about how they might go about building such a movement. The VSM may be as close as we now come in the United States to a time movement, which is not nearly close enough to have any lasting effect on the organization of work for the millions upon millions of workers for whom flexible working arrangements and "voluntary simplicity" remain impractical, risky, or even impossible.

And so downshifters are left to talk among themselves about how to work both sides of the simplicity equation: reducing expenses (so as to buy time and flexibility in the search for rewarding, but limited, work) on the one end, and withdrawing from or narrowly reshaping the workplace on the other. It is an unexpectedly vibrant and creative subculture of formidable size. Established "thriftzines" like *The Tightwad Gazette* or the *Skinflint News*, for example, offer advice on shaving expenses while improving the quality of life, while up-and-comer publications like *Thrift Score* and *Beer Frame* target underemployed young people.[56] Web pages like "Frugal Living" and electronic newsletters like *Frugal Living News* serve as a clearinghouse for everything from how to triple the value of grocers' coupons to how one might approach the boss for flextime.[57]

For those who embrace the disorder of unmoderated electronic discussions, an Internet town hall of sorts thrives at the USENET discussion list ⟨misc.consumer.frugal-living⟩, which consistently ranks among the 25 most active lists (out of almost *20,000* such lists open to the global cyberpublic) in the world.[58] A more structured and comprehensive electronic experience is available through the Web pages of the Simple Living Network (SLN), which since 1994 has provided a wealth of print, video, and online resources to the downshifting community. Indeed, if there is one "place" that best reflects the diverse strands of the movement while capturing its major voices, it is the SLN. In addition to publishing a quarterly electronic newsletter (*The Simple Living Newsletter*) and posting on its Web site a bevy of feature articles, notices of coming events, and essays by regular columnists of prominence, the SLN also connects its members to one another (via simplicity-circle databases and other interactive networking tools) and to six other nationally prominent groups that focus on issues of consumption, simplicity, and sustainability—all this to weave "the web of simplicity."[59] It currently boasts over 170,000 regular readers and supporters (without any significant advertising or self-promotion) and some 7,000 "hits" a day on its Web page.[60]

Study circles, USENET lists, SLN and the web of simplicity, electronic mailing lists, Internet newsletters, interlinked Web sites ... the list goes on. Simplifiers of today, both active and potential, have to be one of the most tightly networked and actively conversational groups in the United States—and their agenda, as Paul Ekin (a British economist who focuses on sustainability) notes, is really quite subversive: frugality, after all, challenges the "near dictatorial power over public policy" of "the notion that human happiness derives exclusively, or even mainly, from the consumption of goods and services."[61] Yet simplifiers insist on holding at arm's length the broader social and political ramifications of their struggle for simplicity. A naive arithmetic informs their politics, or perhaps it is just simple hope: the sum of small, outwardly propagating changes in individual behavior, they believe, will be sweeping social change in how we produce and consume, and govern and judge. They wistfully imagine that fundamental social change—change that will inflict pain on the powerful—can come without any direct confrontation with power, forgetting that the last major successful struggle over work—the late-nineteenth-century campaign for an 8-hour workday/40-hour workweek—succeeded not because workers were "networking" and scrimping, but because they organized around a set of coherent demands and made their collective presence felt.

It would be a mistake to conclude, however, that the VSM is yet another manifestation of a corrosive civic retreat that is eating away at the democratic vitality of the United States. At a time in American history when citizens are withdrawing from almost all significant forms of civic engagement, downshifters are bucking the trend. All available evidence points to their increasing involvement in neighborhood and community; rather than withdrawing and "going off the grid" (Vanderbilt's term), they are using their newfound freedom to reengage the community around them. They volunteer at libraries, help out at schools, tend community gardens, and organize neighborhood book clubs. They are more involved in existing local civic clubs and recreational organizations, and they are more active than most in holding local political leaders accountable.

This is remarkable when one pauses to consider how powerful and ubiquitous the dominant trend toward civic disengagement has become. In his book *Bowling Alone: The Collapse and Revival of American Community*,[62] political scientist Robert Putnam sketches the dimensions of the decline in all its gruesome detail, from the freefalling decline of citizen participation in political campaigns and elections to the steady, almost shocking attrition of active membership in local civic groups and social clubs. Everywhere one looks, Americans are too busy, too tired, too cynical, or too distracted to put themselves in the way of collective enterprises with fellow citizens with whom they might not normally interact, enterprises that simultaneously foster and reward the "arts of association." The consequence, in political science speak, is a waning stock of "social capital"—the degradation of our ability, in other words, to work together and get things done. And as social capital declines, the ability of democratic processes to deliver legitimate and just political outcomes that fire citizens' imaginations and expand their capacity for good judgment comes under siege.[63]

Downshifters will not single-handedly reverse this erosion of social capital. Still, their ability to extricate themselves from a work-and-spend routine that leaves most unable or unwilling to realize the promise of *participatory* democracy is remarkable. They remind those around them that the obligations of citizenship extend beyond the occasional trip to the voting booth or a reluctant willingness to pay taxes, and that the personal and collective benefits of active citizenship are real and lasting. Despite their expansive civic virtue, though, downshifters nevertheless resist seizing on the more broadly political character of simplicity—it is as if the *micro*politics of negotiating new individual workplace rules or trimming long-established household extravagances crowds out any consideration of the *macro*politics of fostering a time movement. On the question of the escapist and even antidemocratic qualities of the VSM, then, the answer remains intriguingly ambiguous.

Enduring?

Just a few years ago, analysts were arguing over whether the VSM could possibly persist. For example, Alan Durning expressed the sense of many when he suggested that the late-1980s rush to simplicity would be short-lived, perhaps merely "another rotation in the binge-and-purge cycle that moral fashion follows as the economy booms and busts."[64] "While simplicity fads have swept the continent periodically," he observed, "most have ended up in consumption binges that more than made up for past atonement."[65]

A decade later, the VSM has proved more enduring and expansive than any might have imagined, due in no small part to the paradoxical economic forces of the time (a booming economy, rising expectations, stagnant wages) and the rising intensity of "luxury fever." Yet the threat to the VSM is not its own episodic history of boom and bust but rather ubiquitous forces of contemporary commodification that have fixed "simplicity" squarely within their sights. As journalist and culture critic Thomas Frank reminds us in *The Conquest of Cool*, countercultural movements in the United States are not so much beaten back as co-opted, commodified, and sold back to would-be protestors for a tidy profit.[66] Why should the simplicity movement, now that it is building up momentum, be any different?

Commodification is no stranger to the VSM. Books on simplicity and downsizing are regular visitors to the *New York Times* list of best-sellers, and the decision by HarperCollins to release Andrews's *The Circle of Simplicity* as a two-cassette book-on-tape (so that workaholic commuters trapped in rush-hour traffic might easily consume the book?) is one more example of capitalism's nimble colonization of every possible market niche. It is hard not to crack a smile or a cynical grin, moreover, when one reads of Bullfrog Films' excitement over the explosive sales of *Affluenza* and *Escape from Affluenza*, or sees, when surfing the Web, that Linda Pierce's *Choosing Simplicity* has broken into Amazon.com's list of its top 5,000 sellers.

But all this is small potatoes compared to the conflict between true simplifiers and "haute simple" or "faux simplicity" looming on the horizon. The opening shots were fired in the mid-1990s, when simplicity rhetoric began appearing in a variety of advertising campaigns. Today, simplicity is used to sell everything from Hondas to closet-organizing systems, leather-bound day planners, elaborate time-scheduling computer software, self-contained Internet-ready personal computers, expensive but easy-to-coordinate business attire, and one-button cell phones. Taking stock of this phenomenon, the *U.S. News and World Report* worries that "simplification has become an elastic marketing concept, a philosophy of frugality transformed into an excuse for yet another shopping spree" and reports that VSM leaders "maintain that the movement is in danger of being co-opted by the very consumer culture it criticizes." But it also quotes Kalle Lasn, founder of Adbusters, who argues that this "simplicity hucksterism" is a welcome indication that the movement is maturing, and growing more threatening to the dominant consumer culture.[67]

The latest, most intense round yet in the commodification wars began in the spring of 2000. That is when AOL Time Warner launched *Real-Simple*, a self-described magazine for busy women looking to simplify, and Christine Carville and Danielle Chang, both in their late twenties, began *Simplicity* magazine (with a parallel Web site), a glossy monthly published with backing from investment bank Goldman Sachs. Both magazines are advertising intensive, to say the least. One issue of *Real-Simple*, for example, begins with 20 pages of advertising before the table of contents, and 116 pages out of 216 total are devoted solely to ads. It suggests to readers that they may wish to use towels more than once, or limit themselves to one pair of socks a day. In a similar vein, *Simplicity* magazine boasts that it will feature comfortable, attractive, and indulgent items geared toward fostering a "quiet simplicity" for professional women aged 25 to 40. Perhaps, suggests *Escape for Affluenza* producer John de Graaf (somewhat tongue-in-cheek), these ventures signal the birth of a "voluntary luxury" movement.[68]

Other heavyweights in the VSM are less amused. Vickie Robin of the New Road Foundation and coauthor of *Your Money or Your Life* angrily characterizes these developments as "slow, deliberate co-opting." Dave Wampler of the Simple Living Network sees both publications as "screwing with people's heads and wallets, and probably turning a tidy profit in the process." "I don't know anyone that is taken seriously in this movement," he writes, "who would claim that buying a certain dress or changing the way you apply makeup or making 'Pineapple Shrimp with Scallion Rice' will make your life more rewarding and meaningful. Come on!" Wampler is not alone. Even a quick visit to the simplicity chatrooms of the Simple Living Network or a glance at the USENET list makes it crystal clear that masses of downshifters are similarly outraged over this new brand of "faux simplicity."

Having assiduously avoided anything formally "political," the VSM is now being forced to acknowledge and do battle with elements of corporate capitalism it has previously sought to escape. What will come of all of this, both for magazine entrepreneurs and the "purity" of the movement, is difficult to say, but from the standpoint of those hoping for a wave of consumptive resistance the outcome could be heartening. "Simplicity," exclaims Vickie Robin, "is not for sale." But as it inevitably becomes for sale, the movement will have little choice but to become more conventionally political, and to devise structures that allow it to speak with one voice against the forces that spread a luxury fever across the land. Such developments bear close watching.

Stepping Back: The Consumption Angle and Consumptive Resistance

The "consumption angle" as described in this book is at once an analysis of our currently unsustainable ways and a call for a new kind of struggle for sustainability. Production, the consumption angle reminds us, is also consumption. Consumption, it insists, occurs not just at the end point of consumer demand, but up and down a chain of material provisioning and resource use. Nodes of production and consumption along this chain come with their own structures, dynamics, and power relations, which interact with one another to cement into place patterns of production and consumption fundamentally at odds with life processes of the planet. Political and economic power—typically concentrated, often highly so, and frequently concealed from easy view—scaffolds the chains and obscures the telling costs of rampant consumerism. The prevailing dogma of consumer sovereignty, which fixes all responsibility for "overconsumption" on individual consumers ostensibly making "free" marketplace choices, blunts efforts to hold the powerful accountable for their central role in the upscaling of desire. A new environmentalism that recognizes these myriad factors and confronts the engines of consumption has never been more necessary.

Held up against the consumption angle, the VSM remains an uncertain and ambiguous thing. To their credit, downshifters and simplifiers, and those who imagine themselves joining their ranks, understand more than most the extent to which *production is consumption*—consumption of time, creativity, control over work, and a sustaining personal sense of fulfillment and accomplishment. Get past the politeness and apolitical sensibilities of the VSM, and what you have is a clear call for production systems that consume far less in terms of human and environmental capital. Of course, since "work" is a politically charged topic, and since most simplifiers come from households for whom ideas of workplace organizing are unfamiliar or taboo, simplifiers displace their anxieties about work onto consumption, thinking that by reducing expenses one can eventually resolve workplace problems. But make no mistake: for a large swath of simplifiers, the core issue is the nature of production and the organization of the workplace, and the dehumanizing and frustrating changes in both over the past two decades. The growing ranks of simplifiers are thus a potentially potent ally in any effort to bring a consumption perspective to bear on our current economic and environmental woes.

For the VSM, *the chain of material provisioning and resource use* is more problematic. Because it dislikes talking comprehensively about work, simplicity becomes a zealous conversation about what one buys and why, to the exclusion of almost everything else. As it grows in mass appeal, downshifting will almost certainly act as a drag on any systematic "up and down the chain" thinking that remains critical to the effective analysis and resolution of the consumption problem—and the winners will be those most adept at manipulating the chain of material provisioning and resource use to their own benefit, at longer-term cost to the environment and human well-being.

The VSM's current trajectory, in other words, spells trouble for those who would facilitate a deeper public understanding of the consumption problem. Rather than continuing to ignore the VSM or imagining it to be a fringe phenomenon, it would be far better to set to work widening existing cracks in the VSM's single-minded preoccupation with the end point of the consumption chain. One chink is the VSM's deep-seated suspicion of corporate power, which is displaced onto consumption and onto a sometimes-obsessive preoccupation with frugality. Another is its recently discovered and currently expanding disdain for processes of commodification and co-optation driven by the explosion of glossy magazines and slick ads extolling the virtues of simplicity. A third is its nascent environmental sensibilities; while not environmentalist per se, many simplifiers recognize that living more frugally also means living more "ecologically." These three elements, and others as well, continuously prod the movement to glance "upstream" at production, packaging, employment, and marketing practices that drive and frame final consumer behavior. Now is the time to accentuate and strengthen these prodding forces through dialogue and coalition building with the primary voices and major networking tools of the movement. The standing ovation showered on Kalle Lasn and his call at the 1998 simplicity meetings for a "pincer movement" that would mesh personal frugality with collective action suggest a large, untapped reservoir of support within the downshifting community for such efforts.

On the question of *power*, finally, the VSM remains on life support. It has not yet developed consistently coherent ways of talking about the distribution and exercise of consumption-amplifying power in any but the most unsophisticated ways. Locked into a rhetoric of the individualization of responsibility, it propagates the all-too-familiar "plant a tree, save the world" environmental mentality. This mentality imagines consumers to be immune to the marketer's ability to tap into environmental concern to sell a host of environmentally *un*friendly products, and draws attention away from inequalities in power and responsibility that occupy the center of the environmental crisis.[69] Ironically, the VSM risks aiding and abetting the very cultural and political forces it philosophically opposes.

But all this could change in a hurry. The U.S. economy is driven by consumer demand, which the simplicity movement hopes to throttle. Fulfillment of this hope would lead to production cutbacks, plant closings, and job loss, at especially great cost to the working poor, who have no savings or deep well of job skills from which to draw in a slowing economy. In theory, simplicity sounds attractive; in practice, it would, absent significant policy change, balance a greater measure of middle-class tranquility on the backs of the bottom 20 percent of American households.

Until recently, the full-employment economy made the class-warfare quality of the VSM easy for all to ignore. Times were good and jobs were plentiful. But all indicators point to an end of these economically anomalous times and a return to the more familiar economic terrain of significant unemployment and, perhaps, onerous levels of inflation. At that point, simplifiers, who are easily plunged into bouts of self-recrimination by critiques of VSM's class bias, will no longer be able to discount the macroeconomic dynamics that convert their good intentions into pain and suffering for others. That will make them uncomfortable indeed.

Signs of that discomfort, and a strategy for ameliorating it, are already appearing. One is Juliet Schor's epilogue ("Will Consuming Less Wreck the Economy?") to her book *The Overspent American*.[70] Schor answers her wrecking-the-economy question with a firm no. "A gradual reduction in consumer spending will not cause much unemployment," she argues, because "the trend towards buying less is likely to be associated with a trend towards working less." "Fewer people would want jobs," she continues, "and the hours worked *per* job could fall." The result: lower economic growth, yes, but more free time, less stress, and less useless stuff. It is a future, Schor insists, worth striving for.

But getting to this future requires an evolution of the rules, norms, and practices that govern work in the United States—dropping the hours worked per job in ways that preserve job security and benefits packages will not happen without determined and spirited national conversation. We are back to the centrality of a "time movement," something that Schor acknowledges in her epilogue but curiously sidesteps. Absent such a movement, a few simplifiers will get to work in better, more accommodating, and more flexible jobs, and they will consume less in the process. Meanwhile, many others without such options will pay the price of economic downturn; their lot will be an involuntary simplicity. The high-minded ideals of the simplicity movement will be soiled in the process, sufficiently so, perhaps, to derail its momentum.

As the economy softens, the VSM will face these sad possibilities, and the press, undoubtedly, will fully illuminate them. There will be the inevitable op-ed pieces and magazine columns charging the VSM with elitist insensitivity to the plight of the poor. The VSM will not take this lying down—if its reaction to the appropriation of "simplicity" by glossy magazines tells us anything, it is that adherents of simplicity are a prickly lot; they take their goals and ideals seriously, and they will react strongly to charges that their behavior only amplifies recessionary economic pressures. But what will they say in their defense? Anything beyond a tacit admission of complicity with economic forces that leave millions of households scrambling for cover will require of them an expanded understanding of power in American society and a newfound willingness to engage the nuts-and-bolts of workplace reform. Pressures already are building in the movement to move in just this direction—and if the economy does head south, these pressures will grow. Advocates of a consumption angle who appreciate how a "time movement" might foster the capacity for restraint would be wise to reflect on the role they might play in hurrying along these changes in the VSM.

As a social movement the VSM remains inchoate. Its numbers are large and growing, but its core ideology with respect to the many challenges posed by the consumption angle remains embryonic. As a social phenomenon it suggests that consumptive resistance flows from the dual pressures of workplace stress and the anxiety that accompanies a methodical upscaling of desire. In many ways simplifiers are no different from the average American; they work hard, raise families, live in both rural and urban areas, hold down a variety of jobs, and express a deep desire to escape the rat race. What makes them a deserving object of continued scrutiny is their choice and ability to act on these desires of escape, desires fueled and facilitated by generally high levels of education, which both raise expectations for "fulfilling work" and confer power to reform or withdraw from the workplace.

Simplifiers are like the proverbial canary in the mine, but with a twist. Like the dying canary that points to the presence of poisonous mine gas, downshifters offer an early warning that something serious is amiss. Moreover, in pursuit of their often narrow and self-interested agenda, they are also whittling away at workplace norms while modeling for others ways of divorcing personal success from willing participation in the work-and-spend cycle that structures much of our lives. Despite its reluctance to think about broader institutional dynamics, the VSM points toward a more coherent time movement—a movement that, by restructuring workplace expectations and structures, could allow millions more to discover the joys of cultivating a capacity for restraint.

10

Jamming Culture: Adbusters' Hip Media Campaign against Consumerism

Marilyn Bordwell

Imagine you're driving to work. As rush-hour traffic grinds to a halt, you sigh, switch on the radio, and reach for a smoke. You survey the familiar urban landscape; a billboard to the left catches your eye. It resembles a familiar cigarette ad, but there's something odd about the image. You look again, more closely this time: it is a colorful cartoon drawing of a thin, pale camel lying on a hospital bed hooked up to intravenous drips. His name, according to the caption, is "Joe Chemo."

Or imagine that you're home, settling in for a night of serious digesting in front of the television, having just gorged yourself at Thanksgiving dinner. You're looking forward to the long weekend, and plan to make your annual pilgrimage to the megamall tomorrow with the kids to get a jump on holiday shopping. As you surf the channels, an odd image appears on the screen: the front end of a smiling claymation pig growing out of a map of North America. The pudgy pink pig wiggles and snorts with glee as an ominous voice-over states: "The average North American consumes five times more than a Mexican, ten times more than a Chinese person, and thirty times more than a person from India.... Give it a rest, America. Tomorrow is Buy Nothing Day." The pig rips a hearty belch; you suddenly feel queasy.

Joe Chemo and Buy Nothing Day are creations of the Adbusters Media Foundation, publisher of *Adbusters: Journal of the Mental Environment* and persistent gadfly to the advertising-saturated, corporate-ruled consumer culture. Based in Vancouver, BC, this small but plucky band of writers, graphic designers, and volunteers coordinates a loosely organized grassroots activism known as "culture jamming" via their award-winning magazine, a snazzy Web site,[1] and an extensive series of advocacy and parody ads: print "subvertisements" and television "uncommercials." Culture jamming aims to liberate the mental environment from the powerful grip of market-structured consciousness by reclaiming airwaves and public spaces to propagate ideas instead of plugging products.[2] Adbusters has declared several annual holidays, including "TV Turn-Off Week" and "World Car-Free Day," and also encourages spontaneous acts of public art and mischief. Adbusters solicits creative contributions from readers and holds annual Creative Resistance Contests to reward the best new subvertisements and jamming ideas.

Adbusters wages guerilla war against the very language of consumer culture: advertising. While ads play the seemingly benign, informative role of alerting us to the variety of products available for purchase, they also shape desires, structure consciousness, and clutter the landscapes of daily life. As graphic design scholar Rick Poynor argues, "Advertising's right to colonise the physical environment of the street and act as primary shaper of the mental environment is taken for granted and there is no officially sanctioned public competition for the thoughts, beliefs, imagination and desires of the passer-by. Apart from other ads."[3] Advertising does more than inspire us to purchase discrete products—this car, that watch, the other cologne; it creates an entire environment of consumerism. Advertising teaches us to consume, but it also consumes resources. It consumes a large part of our public space and increasing portions of our private space; it takes up media airwaves; it exhausts our time, our passions, our energy, and even our mental health; it displaces civic engagement and intimate interaction.

James MacKinnon, *Adbusters* senior editor, states that "advertising is the most important form of communications in the world today."[4] To spread critical awareness about advertising's power, Adbusters fights fire with fire. Rather than launch highbrow missives or scholarly analyses, Adbusters critiques marketing by co-opting the slick, visually appealing techniques of marketing itself. This signature rhetorical strategy lampoons specific name brands—Camel, Marlboro, Benetton, Calvin Klein—but also, and more important, throws into question the entire advertising industry.

Adbusters thus battles the commoditization of culture, the phenomenon described by Jack Manno in chapter 4 of this book. It confronts consumption by targeting the purveyors of excess consumption and challenging them on their own turf. It promotes access to the media as engaged citizens, not passive consumers (see chapter 3). In this chapter, then, I document the ubiquity of advertising and lack of democratic access to mass-media channels, describe the guerilla tactics employed by culture jammers, including the anticonsumption holiday "Buy Nothing Day," and reflect on the possibilities Adbusters holds for sparking a new kind of social movement.

Madison Avenue Behemoth

Advertising is ubiquitous in North American culture—there remain precious few spaces free from brand-name logos and appeals to consume. Advertisements appear on buses and taxicabs, in cornfields, on rooftops, in textbooks.[5] They are inscribed in the sands of public beaches, they adorn Olympic athletes, they grace high school halls and restroom stalls. Experts estimate North Americans encounter as many as 3,000 of them every day;[6] they calculate that the average American child will be exposed to 360,000 of them by the age of 18.[7]

"So what?" you may ask. "Ads don't really affect me. I barely even notice them at all." But perhaps the fact that people barely notice—and hence fail to critically reflect on—these omnipresent advertisements makes them all the more insidious. Indeed, it seems that ads pique our consciousness only when they violate the standard, mundane message of "Buy this," as do the Joe Chemo billboard and the Buy Nothing Day pig spot.

Ad clutter is pricey as well as pervasive. Annual spending on advertising in the United States has been steadily increasing: from approximately $6.5 billion in 1950, to $56 billion in 1980, to about $170 billion today.[8] The price to air a single 30-second spot during the 2001 Superbowl was a record $2.3 million, which works out to $76,666 per second.[9] In recent years, industry growth has been explosive, thanks in part to the advent of Internet advertising and increased marketing efforts by telecommunications and dotcom companies.[10] As more dollars and public space have been devoted to (or devoured by) advertising, companies have had to work ever harder to distinguish themselves from competitors. They have often used sex, violence, or shock techniques that push the envelope, like Calvin Klein's use of youthful nudity or

Benetton's appropriation of politically charged images. Examples include a dying AIDS patient, an overflowing ship of Haitian refugees, real death-row inmates—images that have no connection to sweaters or cologne.

Advertising, according to many critics, is largely responsible for turning whatever civic culture might have once flourished in America into a thriving, voracious, eminently "successful" consumer culture. One thing Americans have been consuming more and more of in recent years is mass-media fare. The average American adult spends 11.8 hours consuming media each day according to a 1998 study by Fairfield Research, Inc.—that is roughly 70 percent of waking hours. Gary Gabelhouse, Fairfield CEO, remarked that "the sheer volume of media products and messages consumed by the average American adult is staggering and growing."[11] And not only are Americans consuming more media, those media are becoming more and more saturated with appeals to consume—advertisements. Just a decade or two ago, films shown on the silver screen and videos viewed at home were free of advertising, save previews for upcoming releases. Now both include ads. Internet service providers and search engines are increasingly cluttered with blinking, animated, and pop-up advertisements. In national newspapers, the ratio of advertising to editorial content is roughly 60–40. Major network television now runs approximately ten minutes of advertising out of every hour. An extreme example of networks milking a show's popularity to cram in extra commercials is the August 29, 1999, episode of the ABC series "Who Wants to Be a Millionaire?," during which there was one ten-minute stretch that included only two minutes of actual show time.[12] That is four minutes of advertising per minute of programming.

Advertisers are also spending an increasing amount of money and energy roping in children. Television is the best venue for reaching kids—seven-year-olds watch 27 hours of TV a week—and commercial TV for children "may well have been the most rapid growing and lucrative sector of the U.S. industry, with 1998 ad revenues pegged at approximately $1 billion."[13] Such efforts to turn young people into world-class consumers and shape their tastes before they can even drive a car (much less purchase one) reap worthwhile benefits: youth garner and dispense an incredible $24 billion in disposable income.[14] Roughly speaking, advertisers on kids' TV networks are spending annually approximately $25 per child on advertising, reaping a reward of about $615 of purchases per child each year.

There is also a clever synergy between advertising and consumption, especially when so many products themselves serve as advertisements. Take a children's story. Perhaps you've purchased the book, received it as a gift, or borrowed a library copy. Chances are, Disney has made an animated film of the story, which your child wants to see in the theater after watching a trailer on TV. Soon thereafter, a spate of toys and related products hit the stores: lunch boxes, T-shirts, toothbrushes, backpacks, stuffed dolls. The story then becomes a hit Broadway musical; you can also purchase the soundtrack or the video for repeated at-home viewing. Finally, a fast food restaurant sponsors the story-turned-popular-culture-fad so that your kid can get action figures with his or her cheeseburger. In the protracted act of consuming this story and its related products, people become walking, living advertisements.

Advertising and consumption have become so deeply ingrained in North American culture that people themselves are becoming branded. Tiny shopping carts available for children at Wal-Mart stores with flags reading "Wal-Mart Shopper in Training" exemplify this process: shoppers no longer think, "I get the things I need at Wal-Mart." Instead, they are encouraged to identify wholly with the store and the act of consuming: "I *am* a Wal-Mart shopper." Clothing catalogs like American Eagle sell an entire lifestyle—suggesting cool books, CDs, road trips, and the like—in addition to logo-adorned sweatshirts. Direct-marketing research, subscriber lists, and various consumption-tracking devices enable companies to peg households as belonging to one of forty-some lifestyle clusters. In advertisers' and marketers' eyes, people are not residents of a neighborhood, members of a community, or citizens of a nation; we are Boomers, DINKs, Yuppies, Generation X, Y, or Z. Who we are turns on what we consume. We are Pottery Barn, Abercrombie, FUBU, Sears.

The Adbusters Media Foundation protests this commoditization of individuals with subvertisements. A powerful print image features a close-up photograph of half of a young toddler's face with the phrase "She's Got Your Eyes" superimposed over her cheek and a television with a big blue eye on the screen in the place where the girl's real eye should be. An uncommercial shows a thirtysomething man slumped on a sofa before a flickering TV screen, his eyes glazed and face expressionless. As the camera pans slowly from his face around to the back of his head, the voice-over says, "The living room is the factory. The product is ... YOU." At that point, we see a UPC symbol tattooed to the man's neck. The voice says, "Snap out of it, America."

Purchasing in response to advertising has a sense of the manufactured, the inauthentic and mind-numbing. Advertising promises a better life—fuller, happier, sexier, easier, more convenient, more fun—and yet always holds out that promise in the as-yet-unrealized future. Studies suggest, though, that North Americans' voracious consumption practices are not making us any happier, as advertising would have us believe.[15] Kalle Lasn, cofounder and editor-in-chief of *Adbusters*, explains that "the whole culture in some sense has been kidnapped by the $162 billion-dollar-a-year advertising industry and the corporations behind it who are giving us this dream of never-ending material progress. They're conditioning us to a false reality."[16] To create an urgent desire for products, advertisers must cultivate a sense of lack, of feelings of inadequacy in the potential consumer. In America, for instance, the malady "halitosis" was invented by Listerine in the 1920s. The company's infamous 1925 ad, "Always a Bridesmaid But Never a Bride," presented bad breath as a personal failing so grave that it would ultimately lead a pleasant and pretty young woman to—horror of horrors!—spinsterhood.

Few ads today make such bald claims, but still they manage to make us feel deficient in many ways. Lasn sees an important correlation between advertising and mental health: "Mental disease has become the number two health problem in the world ... if people start making the connections between advertising and their own mental health ... [they'll realize] that hey, maybe this ad culture is actually part of my stress problem, or part of that lousy way I feel every Monday morning, or part of my depression. Then I think we start looking at advertising a different way."[17] Advertising, in short, consumes us as we consume it.

Battle for the Airwaves

Adbusters grew out of an environmentalist David-and-Goliath confrontation. In 1989, friends and filmmakers Kalle Lasn and Bill Schmalz grew incensed by a multimillion-dollar public relations campaign launched by the British Columbian logging industry. The campaign, called "Forests

Forever," glowingly reassured Canadians that their woodlands were being well managed. Lasn and Schmalz wanted to debunk that corporate rhetoric and tell the environmentalist side of the story, so they created their own ad in response. When they tried to buy airtime for their spot, however, television stations refused to accept their money or show the ad. Outraged, Lasn and Schmalz went to great lengths to publicize the lack of democratic access to the airwaves. Pressure mounted on the Canadian Broadcasting Corporation (CBC) as news coverage of the incident spread: "hundreds of British Columbians phoned the CBC's head office demanding to know why environmentalists couldn't buy airtime whereas the forest industry could."[18] Eventually, the CBC pulled the "Forests Forever" campaign. Energized by their victory, Lasn and Schmalz decided to continue producing campaigns about important social issues—thus Adbusters was born.

During the following decade, Schmalz eventually moved on to other pursuits, but Lasn stayed and presided over the growth of Adbusters and PowerShift, a small ad agency that does occasional work for activist, nonprofit clients.[19] While the magazine, Web site, and Lasn's book *Culture Jam: The Uncooling of America™* broach a wide variety of interconnected issues—politics, environmental problems, corporate domination, overconsumption, suburban malaise, cultural addictions to work, shopping, television, and antidepressants—the one thread connecting them all is the battle for access to mass-media channels. Lasn believes that a major reason people are approaching cultural, psychological, and environmental crisis is the overwhelming lack of democratic communication: the mass media, as Lasn and Schmalz discovered in 1989, are not only driven by corporate money, they are also beholden to corporate ideology. "It's outrageous, really," Lasn remarked, "that a citizen can't walk into his or her local TV station, put the money on the table, and say, 'Hey. Give me thirty seconds of air time. I've got something to say.'"[20]

A primary Adbusters mission, then, is wresting a small bit of media space away from the corporate powers-that-be. Lasn does not intend to cram a particular politics down readers' and viewers' throats; he seems to trust that, given the mental space to clear their heads and the communicative space to share ideas, most people would be more politically aware and engaged, and *doing* something about the problems that threaten the planet. "We need to have a constitutional amendment which basically guarantees peoples' right to communicate, and that doesn't just mean freedom of speech," Lasn says. "We have to go one step beyond freedom of speech to what I call 'the right to access.'"[21] Lasn's call for a "new human right to the airwaves" is actually not nearly as radical as it could be. "Whether the people have some right to free access to TV, I won't argue that for the moment," he remarks. "I won't even argue that maybe two minutes of every hour should be given back to the people. But I'll certainly argue that I, and you, and anybody else should have the same right as Philip Morris, or Monsanto, or General Motors, to walk into a TV station and buy thirty-second spots."[22] In fact, as senior editor James MacKinnon points out, under current media practices, large corporations are ironically the dominant or only voices speaking out on environmental and political issues: "The Chevron Gas company is allowed to talk about the need for clean air, but Adbusters is not allowed to. The Saturn car company is allowed to talk about the need for stronger communities, but Adbusters is not."[23]

One of the dominant myths of the current media system is that people do have access to communication—more now than ever before. It is important to note, though, that while people may have more channels and sheer information at their fingertips now than a decade ago, they enjoy this increased access primarily *as consumers*. Control over the production and dissemination of information has become increasingly centralized as various media corporations merge into ever-larger titans. As media historian Robert McChesney states, "the vast majority of the dominant firms in each of the major media sectors are owned outright or in part by a small handful of conglomerates."[24] That small handful, thanks to recent mergers, has been getting even smaller: six megafirms dominate American mass media. They are AOL Time Warner, Disney, Viacom, News Corp, Bertelsmann, and General Electric. The largest of these firms, AOL Time Warner, is valued at $350 billion; annual revenues are estimated to be $40 billion.[25]

Because of this extraordinary centralization of media ownership, McChesney argues, more communication has not necessarily created a more engaged citizenry or greater democracy. In fact, in recent years, voter turnout rates have fallen, general apathy has increased, and a fog of general resignation seems to have settled over the population. Few people stop to consider whether heavily commercialized mass media could function any other way. "Although people may have once been critical of hypercommercialism, perhaps they are becoming inured to it," muses McChesney. "In a political culture where commercialism appears to be a force of nature rather than something subject to political control, that would be a rational response over time."[26] People have ceded authority for decision making about the media diet to private interests, and then come to see the current setup as somehow natural or inevitable. The capitalist ethos runs so deep that "the notion of public service—that there should be some motive for media other than profit—is in rapid retreat if not total collapse. The public is regarded not as a democratic polity but simply as a mass of consumers," McChesney laments. "Public debate over the future of media and communication has been effectively eliminated by powerful and arrogant corporate media, which metaphorically floss their teeth with politicians' underpants."[27] Kalle Lasn, like McChesney, yearns to reenergize this public debate, and urges citizens to seize back the reins from corporations and take a more active role in governing the vast communication networks.

In the ongoing struggle to secure airtime for Adbusters uncommercials, Lasn has collected a string of rejections—some rude, others rationalizing. In Canada, networks said, "Oh, we have Article Three of our advertising standards guidebook which says we can't air controversial messages—advocacy messages are out of bounds."[28] Some Adbusters spots featuring anticonsumption or antiautomobile messages were rejected because they would hurt stations' business by contradicting all the other advertisements that sell products. From the U.S. front, Lasn collected the following rebuffs: "There's no law that says we have to air anything—we'll decide what we want to air or not" (from ABC New York station manager Art Moore); "We don't sell airtime for issue ads because that would allow the people with the financial resources to control public policy" (from CBS public affairs manager Donald Lowery); and "This commercial ["Buy Nothing Day"] ... is in opposition to the current economic policy in the United States" (from CBS network's Robert L. Lowary).[29] Despite this multitude of rejections, Lasn has persisted, succeeding in placing a "Buy Nothing Day" uncommercial on CNN several years running, paying $10,000 for the single slot.[30]

Lasn has also waged battle in the courts, but with little success to date. In 1995, the Supreme Court of British Columbia dismissed Adbusters' attempt to force the CBC to accept its ads, ruling that the Charter of Rights and Freedoms—roughly equivalent to the U.S. Bill of Rights—does not apply to programming or advertising decisions made by the CBC. According to Lasn, the Canadian Supreme Court threw out Adbusters' appeal without giving an explicit reason, thus exhausting legal options in Canada. The next step would be to take the case to the World Court under Article 19 of the UN Charter of Rights, which declares that people should have the right of access to their own airwaves. But Lasn indicated that this kind of action is difficult for an individual or small organization to launch. Another possibility would be to go directly to the UN Human Rights Commission. The third and perhaps most likely possibility is to initiate a First Amendment suit in the United States. "That could prove to be a lot more interesting than the one we had in Canada, because it's against the big three [networks]," said Lasn. "It makes for good press coverage. Who knows what will finally happen with the legality of it all, but at least it will be out there. In this age when all the media mergers are happening at an exponential rate, people will say, 'Hold on a second! These media mergers are going on and they're saying it's good for us, but I as an individual don't have the right to walk into my local TV station and buy time? What's going on here?' That may be one of the things to tip the balance and launch this Environmental Movement of the Mind that I think will happen."[31]

Guerilla Tactics

Adbusters is part of the creative vanguard for this Environmental Movement of the Mind. It tackles the consumption problem from a popular culture angle, attempting to "uncool" advertising hype and reclaim at least some small slice of public media channels for use by the people rather than by corporate interests. The Media Foundation, started in 1989, published the first issue of *Adbusters* in 1991, and has been steadily gaining momentum. Throughout the 1990s, circulation of the quarterly magazine grew; the Winter 2000 issue saw 100,000 copies printed with six issues published in 2001 and six more issues planned for 2002. *Adbusters* has won numerous awards, including Magazine of the Year (Canadian National Magazine Awards, 1999), and it is Canada's best-selling export magazine, with subscribers in 25 countries.[32] The magazine—which obviously receives no revenue from advertisers—is almost entirely supported by subscriptions and newsstand sales; only about 8 to 10 percent of the Media Foundation's budget is funded by grants.[33]

The Adbusters Web site was launched in 1995, and the Culture Jammers list-serve currently has about 20,000 subscribers around the world. About 10,000 copies of the Culture Jammer's Video—a collection of Adbusters' TV uncommercials and a 12-minute educational spot—circulate in universities, high schools, and at public access TV stations.[34] In 1999, Lasn published *Culture Jam: The Uncooling of America™*, a book that brings together in a colorful, impassioned, occasionally raucous fashion a decade's worth of reflection and activism.

Adbusters' rhetorical style is witty, slick, and often irreverent. The cover of the January/February 2001 issue, for instance, features an aerial view of Jesus crucified amidst endless heaps of garbage; the caption wishes readers a "Merry Christmas & Happy New Year." Inside, a multipage spread on consumption presents a shocking image of two juxtaposed bellies: one dark-skinned (presumably African?), grossly distended by malnourishment, the other white, hairy, middle-aged, and clearly overweight. Below the copy begins, "No Connection. One billion people are dying of starvation. Another billion are dying of excess."

The signature Adbusters parody ads also employ guerilla visual tactics: they closely resemble corporate advertisements, but with a slight turn of phrase and twist of image. The friendly, cool cartoon character Joe Camel becomes Joe Chemo—a sad, sickly version laid out on a hospital bed or in an open coffin. Adbusters' take on the infamous Absolut Vodka ads features the tall, clear bottle slouching limply to one side in a puddle of liquid, below which reads the copy, "Absolute Impotence." Such subvertisements provide a kind *of perspective by incongruity*:[35] by co-opting familiar themes and logos, they not only caution readers about potential dangers of smoking and drinking—a function already served by the Surgeon General's warnings on liquor bottles and cigarette ads in the United States—they also jar us into contemplating the damage done by the advertising industry itself. One of Adbusters' television uncommercials, "Obsession Fetish," links fashion advertising to women's poor body image and eating disorders. The 30-second spot features stylized black-and-white shots of nude body parts, clearly reminiscent of Calvin Klein ads. The final sequence shows the bony backside of a Kate Moss–

like model convulsing to the side as we hear her breathe heavily and moan . . . and then vomit into a toilet. The voice-over concludes, "The fashion industry is the beast."

In addition to creating these uncommercials and subvertisements, Adbusters seeks to propagate them to a wide audience. The Media Foundation essentially gives away its designs to anyone willing to buy public space for their display, pledging to send a broadcast-quality copy of uncommercials to those who have purchased airtime on local, public access, or national TV networks. Other culture-jamming activities extend beyond the ad parodies: Adbusters encourages and reports on a multitude of resistant actions, from the renegade rappellers who strung up a "Buy Nothing Day" sign from the five-story-high rafters in the Mall of America to student and parent groups protesting the introduction of Channel One TV in public schools. Jammers searching the Adbusters Web site can download Car-Free Day posters and bogus traffic tickets to stick under unsuspecting windshield wipers. The tickets read, "Vehicle Infraction: This form of transport incurs economic costs on the city which have not been included in the retail price. Your operation of this vehicle makes you personally liable for the following: climate change, ozone depletion, smog-related health problems [the list continues]. . . . Your fine = $17,000. If you dispute the fine, the trial will be held in your own conscience." Other antiauto acts of public mischief have been featured in the magazine and on the Web site. In one instance, some Jammers in faux "public works" suits placed orange pylons in a downtown intersection to square off curbs, slow down motorists, and give pedestrians and cyclists a fighting chance during morning rush hour.

Give It a Rest

One of the most popular and widely publicized of the Adbusters campaigns is Buy Nothing Day, a 24-hour moratorium on shopping. First declared in 1992, the holiday is now celebrated by more than a million people spanning the globe in nations including Nigeria, Korea, Israel, Panama, and Latvia, as well as Canada, the United Kingdom, the United States, and Japan.[36] For the holiday's first few years, Lasn explained, Adbusters just picked a random day in September for Buy Nothing Day, during a time in which not much else seemed to be happening. "For two or three years we did it that way, for no good reason," he said. "But then Vicki Robin, who runs the New Roadmap Foundation [in Seattle, WA, and coauthored *Your Money or Your Life*], became interested in Buy Nothing Day. We were at a conference together in California, and she suggested that it could be more powerful if we do it on that crazy day after Thanksgiving.... It's a good day in general because it's the beginning of the Christmas shopping season all around the world."[37] The official Buy Nothing Day now occurs on the Friday after Thanksgiving—the largest single American shopping day—although Lasn explained that Buy Nothing Day groups in some countries move the date around a bit to make it work best for their respective cultures.

Each year, Adbusters generates a new poster design (available on the Web site) and mails out thousands of postcards to publicize the event. Groups in various locations stage street performances to celebrate the day and encourage holiday shoppers to go home, or to give "priceless gifts of time and love." In 1997, Seattle singing groups, including the Raging Grannies and the Frugalettes, transposed Christmas carols into antishopping tunes, and an "affluenza doctor" issued advice to those infected with the shopping bug. That same year in Christchurch, New Zealand, people dressed as rodents and staged a "rat race."[38] Other Buy Nothing Day activities include encouraging passersby to cut up credit cards on the spot, and distributing holiday gift-exemption vouchers and consumption checklists that ask: "Do I need it? How many do I already have? How much will I use it? Can I do without it? How will I dispose of it when I'm done using it? Are the resources that went into it renewable or nonrenewable?"[39]

Adbusters' TV uncommercial promoting Buy Nothing Day, described at the beginning of this chapter, has aired on CNN several years running. Similar versions of the "Give It a Rest" spot were made featuring gluttonous European and Japanese pigs. However, Buy Nothing Day has not been openly embraced by all who encounter the holiday. Some people react with outrage: "How dare you suggest I not buy what I please with my hard-earned dollars?" Lasn reports that Adbusters typically fields a number of irate phone calls on or around Buy Nothing Day, hearing acerbic remarks like "Why don't you get the hell back to China, you Commie pinkos!"[40] The freedom to consume, it seems, is deeply embedded in the fabric of North American ideology. If some citizens confront consumption, others find it a personal affront, even an unpatriotic act.

So has Buy Nothing Day succeeded? Perhaps it is too early to tell. The popularity of the holiday seems to be steadily growing, and festivities have been covered by the *New York Times*, *Wall Street Journal*, and National Public Radio. Year 2000's gross post-Thanksgiving sales were down from the previous year, though it is doubtful that the decline can be credited as much to Buy Nothing Day culture jamming as to a dip in the economy. Still, if Adbusters and its cohorts around the globe succeed in getting people to pause and think, that is an important first step toward taming the consumption beast.

Neither Left nor Right But Straight Ahead

Adbusters clearly has a political, environmentalist edge, and its various materials are suffused with a revolutionary rhetoric. The "Autosaurus" uncommercial ominously warns: "It's coming, the end of the age of the automobile." The Autumn 1998 issue of *Adbusters* asserts that

> the next revolution—World War III—will be waged inside your head. It will be, as Marshall McLuhan predicted, a guerilla information war fought not in the sky or on the streets ... but in newspapers and magazines, on the radio, TV and in cyberspace. It will be a dirty, no-holds-barred propaganda war of competing worldviews and alternative visions of the future. We culture jammers can win this battle for ourselves and for planet Earth.[41]

Lasn points to the Seattle street protests against the World Trade Organization (WTO)—the so-called battle in Seattle—and other demonstrations against genetically modified crops, environmental destruction, and corporate domination as evidence of pressure building toward an inevitable eruption.

In his book *Culture Jam*, Lasn advocates "meme warfare," a battle for the symbols and catchphrases that control peoples' values and behavior. The new activists must not only take to the streets, they must also devise and propagate new memes that will shift the dominant paradigm. Some of the culture-jamming memes include:

> *True Cost:* In the global marketplace of the future, the price of every product will tell the ecological truth.

Demarketing: The marketing enterprise has now come full circle. The time has come to *unsell* the product and turn the incredible power of marketing against itself.

The Doomsday Meme: The global economy is a doomsday machine that must be stopped and reprogrammed.

No Corporate "I": Corporations are not legal "persons" with constitutional rights and freedoms of their own, but legal fictions that we ourselves created and must therefore control.

Media Carta: Every human being has the "right to communicate"—to receive and impart information through any media.[42]

Such memes serve as potent mind bombs, working to uncool consumption, dethrone corporations, and summon individuals to the impending revolution.

Despite all this politically charged, revolutionary rhetoric, Lasn shies away from hardened ideology. In defining culture jammers, Lasn declares, "We're not slackers, we're not academics, we're not feminists, we're not lefties."[43] At first glance, the Adbusters campaigns seem to be aligned with leftist politics, yet one of their slogans is "Neither left nor right but straight ahead." Why does Lasn disown the Left? "After spending most of my life being a lefty," he explains, "about ten years ago I became really disillusioned with the left. I looked at the political landscape and activism and what everybody was doing, and I became convinced that if we are going to launch a new activist wave, we will have to jump over the dead body of the left. To me, the left is moribund."[44] Other cultural critics have also noted the fracturing of the Left, particularly through the 1980s and 1990s, which has led to widespread disillusionment. Todd Gitlin describes this phenomenon in *Twilight of Common Dreams*,[45] noting how identity politics pits group against group, and forecloses the possibility of a common language. The Left has splintered into a number of competing special-interest groups, and building coalitions among them has been a difficult, tenuous project.

Lasn feels that the Left was more viable back when the debate between socialism and capitalism was alive, prior to the dissolution of the Soviet Union and communist regimes in Eastern Europe. Now, Lasn is not necessarily against capitalism; he just insists that markets need to reflect the true costs of environmental damage inflicted by certain goods and services. He remarks,

> We don't have a level playing field between generations. Our marketplace is doing a great job right now, and the people who are having the brainwaves in the first world are making a lot of money, and maybe that's fine. But still, our own economy is running off the natural capital of the planet. We're still using up the forests, and creating global warming, and doing things at the expense of our own children. This is the big one that I'm worried about. I'm always scared that this old lefty rhetoric about inequality and gender gaps and all that—which is all fine—but ... you're missing the big point. And so while we argue about how much somebody working for Nike in Malaysia should be getting paid, we're forgetting about the fact that the whole planet is going to dissolve.[46]

Culture jamming may indeed be the social movement of the new millennium—in part because it eschews any particular dogma. Culture jamming is passionate but not puritanical.[47] It is a loose assembly of grassroots fragments, rather than a monolithic, single-cause, hierarchically organized movement. Adbusters traces the roots of the culture-jamming spirit to a number of historical "flashpoints"—the Black Panthers, Women's Liberation, the student and worker riots in Paris in 1968, Vietnam War protests, the Zapatista rebels.[48] Arguably, such a genealogy contradicts Lasn's rejection of the political Left, but it does reflect a spirit of resistance, the sense of underdog daring exemplified by such groups and movements.

Adbusters coordinates a steadily growing network of activists in their campaign against overconsumption and misconsumption (chapter 2). And its work has been gathering momentum. Its clever, irreverent, but serious message appears to be striking a chord throughout North America and beyond. Adbusters is one of several pockets of resistance; other similar culture-jamming groups include San Francisco's Billboard Liberation Front, Australia's Billboard Utilizing Graffitists against Unhealthy Promotions (BUG-UP), and a smattering of Web sites and "zines" (alternative magazines). Even Ralph Nader got into the subvertising act with his presidential campaign commercials.

Adbusters is not immune to critique; in fact, the magazine regularly prints several negative letters to the editor each month, ranging from polite critique to scathing diatribe. In her book *No Logo*, fellow Canadian and cultural critic Naomi Klein faults Adbusters for hypocritically marketing the antimarketing message via selling products: videos, calendars, t-shirts, postcards. She writes, "*Adbusters* has simply become too popular to have much cachet for the radicals who once dusted it off in their local secondhand bookstore like a precious stone," pointing out the difficult dilemma of how to get out the message without selling out.[49] Is *Adbusters* no longer radical when you can buy it at Barnes and Noble?

Yes, *Adbusters* has been growing in circulation, influence, and renown. But this is best understood as evidence of growing awareness and appeal for the culture-jamming attitude—not as "selling out." Adbusters manages to strike a balance: it is not about completely junking capitalism, or killing your TV, or swearing off lattes or red meat for life. It does not have an absolute doctrine, a political platform one must adhere to, or a hard-and-fast line to toe. Rather, Adbusters is about clearing the mental environment, thinking for one's self, reclaiming authentic acts, and generally resisting the numbing, draining grind of consumer culture. Adbusters' version of culture jamming is appealing because it builds on some of the most deeply seated democratic beliefs—the right to freedom of communication and association, the privileging of individual liberty and self-determination—but also frames those individual rights within a larger context of global ecology and corporate domination. Adbusters recognizes that moral superiority talk, even responsibility talk and alarmist rhetoric, will not succeed. Lasn and his colleagues do succeed, if nothing else, in getting people to think outside the lines, to confront their own consumption and the very institution that promotes it—advertising.

11

Think Globally, Transact Locally: The Local-Currency Movement and Green Political Economy

Eric Helleiner

During the 1980s and 1990s, "neoliberalism" emerged as the dominant economic ideology across the world. Like nineteenth century economic liberals, contemporary neoliberals advocate global economic integration through the liberalization of trade and investment flows. They also seek to narrow the possibilities for various kinds of political intervention in the domestic economy through deregulatory initiatives and "neo-constitutionalist" measures such as the creation of independent central banks.[1] As this ideology has gained influence, it has become increasingly common to view individuals primarily as private consumers rather than as public citizens when economic issues are discussed.

In this new context, it is perhaps not surprising that those who oppose neoliberalism have begun to focus on the potential role of consumption as a political tool. If votes at the ballot box have been unable to redirect economic policy, perhaps people's "votes" in the economy—that is, their consumption choices—can promote different values in the economic realm than neoliberal ones. One example of such a consumption-based oppositional movement is the fair trade movement, which seeks to link Northern consumers to "progressive" producers in the South. A second includes the various consumer boycott campaigns targeting products made by firms or countries that are violating human rights or environmental standards. A third is the voluntary simplicity movement, which seeks to encourage the more affluent of the world to reduce their levels of consumption and reject the materialistic values of the emerging transnational consumer culture (see chapter 9).

In this chapter, I focus on a different consumption-based oppositional movement: the local currency movement. Beginning in the early 1980s, citizens in countries across the world, from Australia and Japan to Canada and Britain, have created hundreds of local currencies. These currencies serve as a means of exchange within a clearly defined local community network and are not convertible into the national currency or any other currency. In some instances—such as the well–known "Ithaca HOURS" from Ithaca, New York—local currencies are issued in a physical form such as a paper note. In many other instances—such as the popular Local Exchange Trading Systems (LETS)—the currency exists just as a bookkeeping entry usable only by the local network of people (ranging from half a dozen to several thousand[2]) who have become members through the payment of a small membership fee. When transactions are conducted between LETS members, they are reported to a central accountant who credits or debits the respective accounts. There are often ceilings set for a maximum debit or credit in the system, but no interest is charged on debit accounts.

These local currencies are not an entirely new phenomenon in history. Similar monetary structures have existed in the past, most recently and extensively during the early 1930s.[3] But previous local currencies have existed only temporarily, usually in response to brief episodes when money was scarce. Supporters of local currencies today, by contrast, are part of a more sustained transnational movement that aims to use monetary structure as a tool for permanent social change, as a means of confronting the neoliberal order and its reduction of citizen to consumer.

Although local currencies have attracted considerable media attention, academics have been slower to analyze their emergence. Much of the detailed writing about them has been authored instead by "practitioners" who are involved in the movement itself and whose writing is aimed primarily at attracting supporters.[4] Some important academic analyses have begun to appear in the last few years. These are largely empirical, describing participants and the impacts of these currencies on local economies.[5] In this chapter, I take a different tack, focusing on mostly conceptual questions: (1) how do local-currency advocates use consumption behavior as a political tool to challenge global neoliberalism? and (2) what alternative political-economic order are they seeking to construct through this tool? My objective, in other words, is to analyze political strategy rather than evaluate empirically its effectiveness. I do this by arguing that local currencies are an innovative way of steering consumption behavior to challenge three goals of neoliberals and their construction of "the consumer." These goals are: (1) promotion of an ever-expanding scale of economic life, (2) "depoliticization" of the economy and its management, and (3) advocacy of individualistic identities. In place of these objectives, local currencies are meant to promote a more localized sense of economic space, an enhanced capacity of local communities to manage money to serve political goals, and a more communitarian sense of identity. These objectives stem not from the traditional opponents of economic liberalism—Marxists, economic nationalists, or social democrats. Rather, they stem from political-economic thought associated with the "green" movement.[6]

Challenging the Expanding Scale of Economic Life

A key objective of neoliberals in the contemporary age has been to promote international free trade and the "globalization" of economic life.[7] The expansion of the scale of economic life on an increasingly global basis encourages worldwide competition and an efficient division of labor. This spatial objective of neoliberals is the prime target of local-currency advocates. A recurrent theme in their writings is the value of localized, small-scale economies. They create local currencies explicitly to foster this more decentralized sense of economic space by altering transaction costs to encourage participants to change their consumption patterns. The primary change is to "buy locally."

Since local currencies are inconvertible outside the local network of participants, a holder of this form of money can only spend it on local goods or services. In Margrit Kennedy's words: "Because green dollars [a common term for local currencies] cannot leave the local area to buy Japanese cars or dresses from Hong Kong, every commercial transaction encourages the development of local resources."[8] This kind of "intra-local" trade thus encourages local firms to expand and acts as a form of protectionism against "imports."

The transaction costs faced by a buyer in searching out a local seller are reduced by the publishing of regular newsletters for the local-currency members. In their pages are listed all the various goods and services that members are willing to offer for purchase with the local currency. The founder of the Ithaca HOURS system, Paul Glover, notes that the system's newsletter rivals the local Yellow Pages and is "a portrait of our community's capability, bringing into the marketplace time and skills not employed by the conventional market."[9]

Glover's comments also highlight how this new, locally-oriented consumption behaviour encourages members of local currency networks to make products or perform services that they might not have otherwise made or performed. Since every purchase eventually must be settled with an offsetting sale within the network, participants in local currencies often comment on how this monetary system has prompted them to rediscover skills that can be used to create products or services for sale to other local participants. These are often skills associated with "informal" or "household" sector activities that these individuals had abandoned in favor of purchases from the conventional market. Local economic skills and economic activities are thus fostered in ways that they are not in the conventional economic system. In Guy Dauncey's words, a local currency "awakens the natural creativity of initiative which we all possess, but which many people lose touch with when they receive a regular salary or welfare-cheque."[10]

Local-currency advocates are not the first to recognize that currency reforms can alter transaction costs to promote new conceptions of economic space. The same was true of nation-builders in the nineteenth and twentieth centuries who saw the creation of an exclusive and homogeneous national currency as a key part of their project to construct a national economy. Prenational monetary systems were quite heterogeneous, with various kinds of currency circulating simultaneously within the territory of a state. The creation of national currencies was explicitly designed to eliminate the transaction costs associated with making nation-wide economic transactions. Policymakers also sometimes used the new national currency structures as a tool to discourage economic transactions with the outside world by making them inconvertible.[11]

In a similar way, neoliberals in the contemporary era have promoted monetary reforms that alter transaction costs to foster economic integration between countries. They have, for example, pushed countries to remove national exchange controls and make their currencies fully convertible. Similarly, in the European context, a common argument in the creation of the Euro was its role in reducing transaction costs associated with cross-border economic transactions in that region. Earlier in the post-war period, leaders were keen to see the growth of "eurodollar" markets; that is, markets in which U.S. dollars were used in offshore locations. This new kind of "transnational" form of money played an important role, especially in the early years of its growth, in fostering a global economy, enabling transnational private financiers and firms to avoid national capital controls in the post-1945 era.[12]

Local-currency advocates are thus not unique in using monetary reform to alter transaction costs to foster different kinds of economic space. What is unique, however, is their goal of fostering *localism* in the economic realm. To understand the source of this preference, it is important to recognize their connection to the broader "green" movement. The greens have emerged over the last several decades as an important social and intellectual movement as well as a key political force in many countries. They are sometimes associated with the environmental movement, but the two should not be confused. The greens are concerned not just with environmental issues, but with a broader political-economic project that I describe below. Moreover, many environmentalists are not "greens"; they may approach environmental issues from a liberal, Marxist, or social democratic perspective rather than a "green" one.

What is particularly distinctive about the greens' approach to political-economic issues is their enthusiasm for the decentralization of social life. One reason is environmental, the other more social. Greens argue that environmental problems have been exacerbated by the large-scale nature of economic life in industrial societies. Long-distance trade produces more pollution and draws on more renewable and non-renewable resources than more localized economic activity. In addition, intense competitive pressures engendered by large-scale economic integration encourage downward harmonisation of environmental standards and practices. The spatial extension of economic practices frequently transforms isolated and diverse local ecosystems into more integrated, monoculture-based ones, a transformation that can increase the fragility and vulnerability of ecosystems and undermine traditional patterns of human life that have co-evolved with local ecosystems.

Furthermore, the greens worry that people's environmental awareness and sense of responsibility is undermined as economic processes are spread out over greater distances. When unsustainable harvesting of resources or the dumping of waste takes place in locations far away, individuals are not aware—as they are in more local economies—of the environmental consequences of their consumption behavior. Indeed, as the influence of distant events and decisions increases, lifestyles and worldviews that were deeply embedded in local settings and tradition are increasingly marginalized, according to green thought. To use Vandana Shiva's phrasing, the "local" is increasingly "globalized" in ways that prioritize decontextualized, short-term perspectives and cosmologies over more ecological, long-term conceptions of social life, conceptions that value the future and the past as much as the present.[13]

Although the greens are best known for their environmental concerns, equally important is their social critique of large-scale economies. They worry, for example, about forms of concentrated wealth and corporate power found in the large national and international markets. They also express concern about the coercive power of distant and powerful bureaucratic agencies that usually emerge to regulate and stabilise large-scale markets. Some of these agencies are associated with the nation-state, which, as commonly portrayed in green literature, is a political community in which individuality is submerged by mass ideologies such as nationalism and in which coercive power is concentrated in large-scale, difficult-to-control state authorities. As economic activity has become increasingly global in the post-1945 period, greens also worry about the growth of powerful international bureaucracies such as the WTO and IMF.

In large-scale economies, the greens also argue that individuals experience a sense of alienation; large and anonymous economic forces seem to control their destiny in hard-to-understand ways. People's alienation is reinforced when their participation in the economy no longer seems closely tied to meaningful local values and social relationships. Thus, while a larger and more extensive market-based division of labor may increase economic efficiency and the overall production of material goods, the greens argue that it does not necessarily translate into a higher "standard of living" in a broader sense. Indeed, the greens are keen to advocate a less materialistic lifestyle, partly for environmental reasons but also because modern consumerism is seen as devoid of deeper spiritual content and meaningful community context.

As a result of these concerns, the greens are active supporters of initiatives to counter large-scale economic spaces, both at the global and national level. They promote small-scale markets and economic activities that draw on local factors of production and cultivate a simpler way of living as well as a self-reliant, democratic, and environmentally friendly ethic. Because of their hostility to coercive power, the greens place particular emphasis on the role of voluntaristic community-based and individual initiatives in promoting this "turn to the local." They are hostile to ideas involving centralised planning or external top-down initiatives. Green initiatives, by contrast, include such things as the creation of local producer and consumer cooperatives, community-based micro-credit schemes, and community-supported organic agriculture projects. The creation of local currencies also fits within this vision since membership in the local-currency network is on a voluntary basis, open to anyone in the local community.

It is important to recognize that the greens are not seeking to create an exclusively localized economic space with local currencies. It is unusual for members of such networks to transact a majority of their economic life within the network. Even individual payments by members are frequently made partly in local currencies and partly in national currencies (local currencies are usually denominated in units that correspond one-to-one to the national currencies to facilitate this process). These practices partly reflect the practical difficulties of totally "delinking" from today's modern large-scale economy. But they also reflect a conception of economic space that is different from the exclusive and homogeneous sense of national space that dominated economies in the age of the nation-state. Greens often invoked as a model what Karl Polanyi described as the parallel existence of local and transnational economies in the preindustrial age. This is the kind of economic world the greens seek to cultivate with local currencies.[14] They also cite Fernand Braudel's description of the preindustrial economy in which three levels of economic activity coexisted: that of "material life" (economic activity existing outside of the market), the market economy (transparent small and medium-scale markets), and capitalism (large oligopolistic firms operating in transnational and global markets).[15]

Both Polanyi and Braudel highlighted how each of these parallel economies in the past were served by distinct parallel currencies. The objective of today's local-currency advocates is to recreate this world of "currency pluralism" that existed before modern homogeneous and exclusively national currencies were created.[16] Their strategy, in other words, may not directly confront the global economy but it does seek to create parallel spaces alongside the global economy, spaces in which alternative ways of living can flourish. In pushing for local currencies, the greens are thus not calling for a total rejection of national currencies. Rather, they envision what James Robertson calls a "multi-level currency system."[17] In Dauncey's words, a local currency "is not meant to replace ordinary money—it is meant to be used alongside it.... Green money will be preferred for local trading, and ordinary money saved for the import-export side of the community economy."[18]

A Tool for Active Macroeconomic Policy: "Repoliticizing" Money and Its Management

Local-currency advocates seek to challenge not only the globalization of economic life but also the neoliberal goal of "depoliticizing" the economy and its management. Neoliberals worry that when the economy is actively managed by politicians, the result is often economic disaster: short-term political pressures override concerns about economic efficiency. Even without such short-term political pressures, many neoliberals question whether the politicians can ever have enough information to effectively manage a modern complex economy. Politicians' efforts may result not just in economic chaos but also in a loss of individual freedom as the preferences of "planners" take precedence over those of individuals. In place of the "visible hand" of politicians, neoliberals prefer the "invisible hand" of the marketplace. And to the extent that the economy must be managed by a central authority, they prefer this management to take place in an automatic and rule-bound fashion, one that is insulated from the short-term, ill-informed, and often arbitrary interference of politicians.[19]

In the monetary realm, this neoliberal position has led to initiatives that reduce the ability of states to pursue discretionary monetary policies. The abolition of capital controls, for example, has been driven partly by a desire to see global financial markets exert a healthy "discipline" on those governments employing undue discretionary Keynesian macroeconomic measures.[20] Neoliberals have also pushed governments to make their central banks more independent and to establish clear "monetary rules" for the conduct of monetary policy. Many neoliberals have gone further to call for the abolition of central banks altogether. Some favor the introduction of currency boards in which discretionary monetary management becomes impossible, while others call for private actors to take over the issuing of money as a way to prevent governments from being able to manipulate money.[21]

These initiatives to both deregulate money and "depoliticize" its management worry local currency advocates, who argue that communities should manage money collectively and do so precisely in a discretionary way. Some highlight environmental reasons, arguing that the creation of money must be strictly controlled by local communities to achieve a more sustainable economy, one that is not so focused on endless growth and accumulation of capital.[22] More common in the local currency literature, however, is a concern with the social costs of deregulated and "depoliticized" money. One such concern is that money can no longer be managed in a discretionary way to solve economic problems such as unemployment. Another concern is the vulnerability of local communities around the world to the volatility and crises that characterize deregulated global financial markets today. Modern money is portrayed as a risky form of money "disembedded" from community values and preferences.

However much local-currency advocates criticize neoliberal thought, they do not want to return to the past when money was managed by nation-states in more active ways. In some cases, this sentiment reflects their belief that the nation-state is no longer capable of performing this role in the age of globalization. But it also derives from a green critique of the nation-state and national currencies. As discussed already, the greens believe that nation-states are too large a community to be genuinely democratic and responsive to local needs. To them, national currencies were never infused with the kind of organic sense of social embeddedness they seek. Rather, these monetary structures are often seen as having been managed to serve the interests and power of distant state bureaucrats and elite interests.[23] In advancing this criticism of centralized state economic planning, the greens in fact sometimes sound quite similar to neoliberals themselves.

Even if state officials and politicians were genuinely committed to managing the national currency in a discretionary way to serve community interests, many greens also believe that this monetary tool is too large to be responsive to local needs. Local-currency advocates often argue, for example, that a key macroeconomic problem facing poor local communities is the outflow of money to large, wealthier metropoles within a national currency zone. Local communities are also buffeted by macroeconomic developments in far away parts of the national currency zone in ways that do not always serve their interests.[24] Only by creating smaller subnational currencies, many local-currency advocates believe, can macroeconomic problems of local communities be addressed.

In what precise ways does a local currency become a tool for local communities to manage their economies in a discretionary way? To begin with, it enables them to influence the overall supply of money in the community. From an environmental standpoint, the supply of local currencies in LETS networks grows only when it is matched by an exchange of real goods and services. In this way, greens argue that local money is less inclined to encourage exponential growth disconnected from the ecological constraints faced by the real economy.[25] Local currency thus deals directly with issues of overconsumption (chapter 2) and in a way that national currency cannot.

From a social standpoint, advocates highlight how local currencies can provide a primitive macroeconomic stimulus to communities that are starved of the national currency. Local currencies do this simply by providing a medium of exchange to facilitate economic transactions that were previously inhibited by monetary scarcity. A stimulus may also be created by the fact that local-currency networks can provide the poor or unemployed with access to interest-free credit (since no interest is charged on debits in the system).[26] Similarly, wealthier members of the network who are accumulating money will be encouraged to spend rather than hoard it because it does not earn any interest in the system. In Kennedy's words, local currencies thus "punish those who sit on their surplus money."[27] Indeed, many local-currency advocates draw on the ideas of Silvio Gesell, who argued that interest-bearing qualities of money were a central cause of boom and bust cycles in the macroeconomy.[28]

While the money supply expands endogenously in line with the volume of exchanges in a LETS network, local-currency systems employing physical notes must manage the supply of money in a different way. In the Ithaca HOURS system, for example, a twice-monthly meeting open to all members—the "Barter Potluck" or "Municipal Reserve Board" meeting—makes decisions on the amount of new notes to be issued.

When they decide an increase in the note supply is warranted, a decision is also often made to combine the overall macroeconomic goal of expanding the money supply with an equity objective: some of the new notes are issued in the form of grants and loans to community-based organizations.[29]

Local-currency advocates also highlight the macroeconomic role that the local currency can play as a primitive form of capital control between the local community and external monetary system. Because the money earned is inconvertible into external currencies, local-currency systems encourage money to remain within the community instead of flowing out. As Glover puts it: "We printed our own money because we watched the Federal dollars come to town, shake a few hands, then leave to buy a rainforest and fight wars. Ithaca's HOURS, by contrast, stay in our region to help us hire each other."[30] In a similar way, local currencies can reduce the vulnerability of the participants to upheavals in the national and global monetary systems. In Thomas Greco's words, "Just as a break-water protects a harbor from the open sea, a local currency protects the local economy from the effects of the global market and the manipulations of centralized banking and finance."[31]

Local currencies can also be used to promote certain shared principles held by the community concerning the value of different goods and services. Whereas national currencies value goods according to their market price, many local currencies promote alternative measures of value based on some common community value. Systems that price all goods and services according to a labor system of value are, for example, quite common; that is, prices are determined by the time involved in making or performing the good or service. Underlying this choice is an egalitarian ethic; as one LETS participant in the United Kingdom put it, "an hour is an hour ... no job is more important than any other."[32]

Similarly, many local currency networks actively encourage the kinds of work that are often unpaid in the national economy—subsistence activities, "informal" sector activities, "household" work. These are included in the network and assigned a monetary value. This new monetary instrument thus gives legitimacy and recognition to activities in a way the mainstream economy does not. Since many of these activities have traditionally been "women's work," some advocates claim that local currency fosters greater gender equity.[33] Others highlight how local currency helps people break their dependence on the modern consumerist market economy while recognizing the social and environmental benefits of a simpler lifestyle.

In sum, by choosing to transact within a local-currency network, individuals not only foster a more localized sense of economic space but also create a macroeconomic tool that can be actively used to promote shared community values. Advocates of local currencies are, in other words, seeking to restore a sense of collective social purpose or "embeddedness" to the functioning of money, a characteristic that national currencies and other forms of money appear to have increasingly lost in this age of globalized financial markets and "depoliticized" money. In Andrew Leyshon and Nigel Thrift's words, advocates are promoting an "alternative morality" of money to that which has become dominant in today's world.[34]

Reorienting Identities: From Individualism to "Persons-in-Community"

In addition to reconfiguring the geography of economic life and "repoliticizing" the management of money, local currencies challenge a further aspect of neoliberalism: its radical individualism. As explained already, neoliberal thought is based on a conception of society as a collection of isolated individuals pursuing their own self-interest.[35] By contrast, green thought is guided by a more communitarian worldview that sees humans as "persons-in-community."[36] Greens argue that people realize their full identity and meaning only in the context of the broad social values and experiences of their community. Supporters of local currencies hope that this form of money can strengthen this realization process. Indeed, this motivation appears to play a significant role in explaining people's interest in local currencies. Thorne, for example, reports that the prime motivation many British LETS members gave for joining a local currency network was to gain a "sense of belonging" or to be involved in "community building."[37]

The idea that a form of money can promote a communitarian ethic may at first seem odd. After all, some of the most famous analysts of the sociology of money have come to the opposite conclusion. Both George Simmel and Karl Marx, for example, commented on the power of money to undermine traditional social relations by transforming social and personal ties into impersonal and instrumental economic calculations. Money's ability to assign value in a standardized way to diverse items was seen to dissolve the concrete relationships of traditional societies and replace them with abstract and impersonal social relations. For this reason, Marx referred to money as a kind of "radical leveler" that "does away with all distinctions" associated with traditional social relations.[38] Likewise, Simmel talked of the "colorlessness" of money, its "uncompromising objectivity" and its indifference to "particular interests, origins, or relations," features that derived from its ability to "become a denominator for all values."[39]

But this view of the social meaning of money is misleading. Viviana Zelizer has argued convincingly that earlier sociologists of money ignored how money has always been profoundly embedded in various cultural and social structures and thus invested with social meaning.[40] She demonstrates this point by highlighting the pervasive practice of earmarking national currencies and the creation of special forms of localized currency during the very historical period that Marx and Simmel were analyzing, the nineteenth century.

During this same period, I have elsewhere shown that nation-builders were keenly aware of the potential for national currencies to be used to promote new collective national identities.[41] At the symbolic level, policymakers recognized that coins and notes—being "among the most mass-produced objects in the world"[42]—could act as important carriers of nationalist imagery aimed at constructing a sense of collective tradition and memory. By reducing transaction costs within the nation, a common homogeneous national currency was also seen to be similar to a national language; it would bring members of the nation together by facilitating "communication." As tools for national macroeconomic management, national currencies were also linked to the sense of "popular sovereignty" that was crucial to the new nationalist sense of collective identity. Similarly, policymakers hoped that a collective experience of monetary events through a common national currency would bolster the feeling of being a member of a national community with a shared fate. Finally, because trust plays such a large role in the use and acceptance of modern forms of money, national currencies were seen as something that might encourage identification with the nation-state at a deeper psychological level. If the value of the national currency remained stable over time, trust might be fostered in the national community that issued and managed that money.

Supporters of local currencies share similar notions about how money can foster a sense of community belonging. Creators of local currencies that use physical notes, for example, have fully exploited the abilities of these notes to carry messages symbolizing localism. For instance, Glover describes the symbols that Ithaca advocates chose for their currency: "The design of Ithaca HOURS honors local features we respect, like native flowers, powerful waterfalls, crafts, farms, and our children. The designs of Federal dollars, on the other hand, honor slave holders (Washington, Jefferson, Hamilton, Jackson) and monuments of corporate governance."[43] The LETS networks, which do not employ physical forms of money, also attempt to exploit the symbolic potential of money through the names assigned to them; in Britain, for example, Greenich uses "anchors," Canterbury employs "tales," Totnes's currency is the acorn.[44]

At a more concrete level, local currencies are seen as fostering a sense of community by bringing people together through trade. As a flyer for LETS Toronto puts it, "by trading in the community, people get to know each other."[45] Trading fairs and newsletters also foster community sentiment through this common currency. In these ways, consumption behaviour is used not to reinforce a sense of possessive individualism, but rather a sense of communitarian values that often rejects materialism as a value. As one Canadian member of LETS put it, "Just about every time I trade through the LETS system I get to meet someone personally. I've got to know an extra 100–150 people in this way. To me, that wealth of relationships in the community is synonymous with economic wellbeing."[46]

Some local-currency advocates argue that the collective experience of membership in the network fosters a sense of shared identity. In Paul Glover's words: "We encounter each other as fellow Ithacans, rather than as winners and losers scrambling for dollars."[47] The way local currencies provide a common macroeconomic tool may foster a sense of local pride and sovereignty, too. Local-currency advocates also draw the link between trust and collective identities that nationalists drew. Indeed, local currencies rely even more on trust for their acceptance than national currencies because such currencies have no market value; that is, they are not bought and sold in foreign exchange markets. Their value depends entirely on the willingness of members to accept them. This in turn may encourage a sense of collective identity among users. Across the front of Ithaca HOURS bills, for example, is written "In Ithaca We Trust." In LETS networks, there is an added dimension of trust required for proper functioning: members are trusted not to abuse the ability to run up large sums in their debit accounts that risk never being repaid (and on which no interest is charged). Some LETS networks try to reduce this potential for abuse by setting maximum limits on debit balances. Most also use a form of peer pressure: the account balances of all members are available for public scrutiny. Trust is still important, though, in part to highlight the positive role debts can play in encouraging a sense of commitment to the community. As literature produced by a LETS network in Winnipeg noted: "With barter credit we are never in debt to the person we trade with. Our commitment is to the community to return value for value received and to keep trade moving."[48] Williams also notes that "'debt' is positively encouraged, being seen as a commitment to put energy back into the system at a later date."[49] In this way, supporters of LETS hope that this form of money builds an economy based more on what Polanyi referred to as a sense of "reciprocity" than pure "exchange."[50] The originator of the LETS idea, Michael Linton, says that LETS is best defined not as a barter system, but as "a promise by people in the community to people in the community."[51]

In sum, local currencies are explicitly designed to foster a sense of collective identity that challenges the radical individualism promoted by neoliberals. Indeed, whereas the act of consumption in mainstream society is often seen as a selfish and individualistic act, in a local-currency network it is seen as a community-building act. "The LETSystem restores the quality of gift economy to the modern trading economy," says Dauncey. "People in Courtney who trade in Green Dollars say that they experience an uncommon 'lightness of being' in their trading relationships.... While ordinary money is inherently privatized, and individualistic, Green Dollars are a community money, carrying an inherently social message."[52]

One additional point must be noted here about the kind of identity that local currencies promote. I have highlighted the emphasis greens place on seeing individuals as members of local communities. But they also are keen to stress that these localist identities not be seen as parochial and inward-looking, but rather as identities located in a global context. In the words of a famous green slogan, the greens push people to "think globally" while acting locally.

The admonition to "think globally" is partly designed to encourage people to retain strong links to ideas and groups beyond one's local community, to avoid the danger of localist frames of reference becoming stagnant, and to acquire knowledge about the urgency and seriousness of the global economic and environmental predicament, a knowledge they hope will prevent complacency at the local level. "Thinking globally" is thus aimed at cultivating a sense of membership within a community of shared fate at the planetary level, challenging nationalist ideas and encouraging a desire to promote change as part of a transnational movement. But the phrase is also one that is meant to remind green activists to be skeptical of transnational solutions to the world's problems dreamt up by distant bureaucrats or advocates. To "think globally" is to recognize the diversity and complexity of local environments and peoples around the world, to see the need for initiatives informed by local knowledge and local community initiatives. In this way, "thinking globally" reinforces the urgency of the need to "act locally."[53]

The commitment to these values is clear in local-currency networks, where newsletters make frequent reference to the link between local action and global environmental and economic issues. It is also apparent in the transnational nature of the local-currency movement. Participants in local currency experiment are quite conscious that they are part of a global movement and are keen to share ideas. The LETS idea, for example, has been actively promoted through The Other Economic Summit process (which has shadowed the G-7 summits every year since the early 1980s) and Internet sites as well as through the transnational advocacy of people such as Michael Linton. But greens are also very protective of the local autonomy of each local-currency network. They emphasize the need for each community to devise a network suited to its local needs and values.

Conclusion

The local-currency movement is a political movement that challenges the priorities of global neoliberalism in the contemporary age by altering the consumption patterns of individuals. It does so in a unique way: by creating a new form of money at the subnational level. I have attempted to show that this monetary instrument is designed to alter people's consumption in ways that challenge three parts of the neoliberal project. In place of the neoliberal promotion of an expanding scale of economic life, local currencies foster a more decentralised and local economic geography. While neoliberals hope to "depoliticize" the economy and its management, local currencies are designed to restore communities' ability to manage economic life actively, to serve their collective values and priorities. Finally, the radical individualism promoted by neoliberals is challenged by local currencies meant to encourage a more communitarian sense of identity.

The challenge that local currencies pose to the neoliberal project is also interesting because its intellectual roots stem not from the theoretical camps typically studied by political economists, the traditional challengers of neoliberalism: Marxism, economic nationalism, or social democracy. Instead, its inspiration comes from green theory. As I have outlined, a distinctive feature of the greens' perspective on issues of political economy is their focus on the *scale* of social life. While the traditional schools debate the respective virtues of states versus markets, the greens ask a different question: What is the appropriate scale of the political or economic institution? The significance of scale to the greens is that they believe, in E. F. Schumacher's memorable phrase, that "Small Is Beautiful"; much of their political activity is devoted to implementing a localist vision oriented to appropriate scale.[54]

Part of the appeal of the greens today is indeed their focus on the politics of scale. When the nation-state dominated the political arena, the scale of human community generated less controversy; it was assumed, not debated. But as the nation-state's future is increasingly called into question, many of the key political debates in the contemporary age—for example, those surrounding globalization and regional integration—have come to revolve around issues relating to the scale of human community. This has provided an important political opening for the greens to gain an audience for localism. This is particularly true in the monetary sphere, where the growth of eurocurrency activity, the creation of the Euro, and an increasingly pervasive "dollarization" in many parts of the world have all combined to encourage people to question national currencies as a "natural" or "inevitable" means of organizing commerce.[55]

A second interesting feature of the green political project is its somewhat ambiguous relationship to neoliberalism. In this chapter, I have emphasized the challenge that green thinking and local currencies pose to neoliberal priorities through a value system that prioritises community values and environmental sensibilities and through a rejection of worldwide economic integration and laissez-faire economics. At the same time, however, the greens' enthusiasm for small-scale market activity and voluntary initiatives in a local community context, as well as their frequently strong anti-statism, can sometimes dovetail with liberal goals of scaling back the role of government and delegating social welfare functions to local civil-society groups. As a result, while greens and neoliberals clash strongly in their views on global economic issues, the two can sometimes find common ground at the local level.

The local-currency movement is an important example of this point. The Australian government, which has shifted in a neoliberal direction since the mid-1980s, has supported the creation of local currencies as a means for unemployed people to reduce their dependence on the social security payments from the state.[56] Similarly, local-currency initiatives have occasionally received support from neoliberals who see these monetary structures as an important initiative challenging the state's monopoly over currency.[57] Indeed, while attacking the contemporary global economy, some local-currency advocates themselves have cited favourably Hayek's call for the privatization of money production (what he calls the "denationalisation" of money) and his critique of national currencies as monetary structures designed to centralise state coercive power.[58]

The local-currency movement thus has a paradoxical relationship with neoliberal economic priorities in this age of globalization. To its advocates, the movement is definitely working against the global neoliberal agenda and its associated values of individualism, mass consumerism, globalization, and laissez-faire economics. And in some ways it does pose a radical challenge to this agenda. By demonstrating in a concrete way how a different kind of economy can function, it challenges the neoliberal "TINA" thesis—"there is no alternative" to neoliberal economic policies. But neoliberals themselves do not see local currencies as a significant challenge to their interests. In part, this is because the amount of economic activity and numbers of people actually involved in local-currency networks are tiny to date.[59] But it is also because local currencies promote change through a mechanism that neoliberals themselves endorse: the voluntaristic behaviour of individuals acting as consumers in the economy. In this way, the local-currency movement highlights a limitation of the use of consumption as a political tool: it seeks to promote social change by playing within the means and many of the rules and terms that neoliberals have established as the new terrain of politics.

12

Caveat Certificatum: The Case of Forest Certification

Fred Gale

Consumption is a condition of life on earth. Plants combine light, carbon dioxide, water, and minerals to survive. Animals eat plants and other animals. Unlike plants and other animals, however, humans have a measure of choice about what and how much they consume—a choice related to their class and their society's level of economic development. Rich, well-organized people have more choices than do poor, unorganized people. Research on consumption and activism have tended to ignore this "consumption problem," though. One reason, as the introduction to this book notes, is that pointing the finger at any specific activity—car racing, boxing, gambling—invites the charge of hypocrisy. What about my car, the international conferences I attend, and the vacations I enjoy? A far larger part of the problem, however, is the degree to which the environmental and social impacts of final consumption decisions are systematically shaded (chapter 5) and hidden from consumers, who would be more sensitive to consumption issues and more cautious in their consuming if presented with credible, easy-to-use information on the impact of their consumption. The real problem, in other words, is that despite commonplace assumptions about the power of consumer choice to shape production, about the "sovereignty" of consumers (see chapter 14), consumer choice is often manipulated to serve well-organized interests. This is done through carefully crafted advertising campaigns, the lobbying of governments, or the restriction of information to consumers about production practices and attendant social and environmental impacts.

Ecocertification and labeling schemes (ECLs) are one way to narrow the information gap between production and consumption, and reclaim for consumers some power and some degree of responsibility in their decision making. This is especially important now, as distancing increases. The spatial rift in national and global economies separates the harvesting and processing of raw materials from that final purchase at Wal-Mart. Those who would span this divide with ECLs assume that if consumers learn (1) that the values inhering in a product conflict with their broader values (for example, equity, environmental, feminist, communal values), and (2) that a substitute product exists that conflicts significantly less with those values, then (3) they will choose the second product over the first.

To implement ECLs, however, many theoretical and practical questions arise. How, for example, do purchasers determine the values that inhere in a commodity? How reasonable is it to expect consumers to analyze each and every product bought? Can they rely on certifying organizations and buy only those labeled goods? Can the company that simultaneously makes the product and certifies it (what is known as *first-party certification*) be trusted? Or are second parties (industry groups and governments) and third parties (independent auditors) more trustworthy? What role should the state play in the development and implementation of ECLs? Are ECLs "economic instruments" of public policy, or are they forms of trade protectionism?

In this chapter, I address these questions by examining ECLs' potential for, on the one hand, restructuring commercial timber relations, that is, simultaneously promoting "cautious consuming" and sustainable forestry practices; and, on the other hand, avoiding "improvements" that do little more than reinforce existing industrial and governmental practices. To minimize the danger of such "greenwashing," I argue that consumers need a simple maxim or rule of thumb to help them sort through the claims being made by proponents of different approaches. I identify five criteria to analyze the legitimacy of competing ECLs and use them to evaluate three popular schemes in the forest-products industry. I conclude that only one of the current schemes—that established by the Forest Stewardship Council—ranks high across all five criteria.

Ecocertification and Ecolabeling: A Brief History

A neoclassical economic rationale exists for promoting ECLs. According to the theory of perfect competition, buyers and sellers are assumed to have "perfect information." In practical terms, this means that if you are buying a secondhand car, you know the price, quality, servicing costs, and delivery time of all secondhand cars on the market. In fact, however, consumers operate in an environment in which information is far less than perfect. They confront claims and counterclaims across all relevant variables. In such circumstances, it is especially difficult and costly to obtain good, never mind perfect, information to screen out the best cars from the lemons. Most consumers "satisfice," therefore, purchasing products that are "good enough" rather than those that are "best."

In the secondhand car market, where consumers need protection against fraud, good reputations and effective guarantees matter. In other areas of production and consumption, concern about fraud gives way to concern about safety. If products are toxic or hazardous, governments normally regulate their availability. Regulations include warning labels on products designed for use at home and work, prescriptions for drugs to ensure that only a small amount is available at a time, licenses that entitle certain individuals to use specific products, and age limits for certain classes of goods. It is notable that regulation via prescriptions, licenses, and age limits contradicts a pure theory of consumer preferences since some purchasers are prevented from buying the products they desire. Government and industry have found nothing wrong up to now in using labels to inform the public about the products being purchased.

There should be no controversy, then, in using labels to provide "information" to consumers on the products they buy. Business has long used labels to market products, to make them stand out from those sold by competitors, and to develop "brand" loyalty. Government, too, has long required that some products be labeled dangerous, hazardous, age restricted, or toxic. The practice has been unremarkable so long as control over information remained in the hands of government and business. With the rise of social movements using certification and labeling as market-based tools to promote progressive social values, the practice has suddenly become contested.

The use of ECLs has expanded in several markets in the past decade, driven largely by health, safety, and environmental concerns about the consequences of industrial approaches to production. Consumers are switching to organic produce as concern grows over the industrial model of agriculture and its reliance on herbicides, pesticides, genetically modified plants, and offal feedstock. Organic production is experiencing growth rates of about 20 percent in the United States, Japan, and France and accounts for about 10 percent of total domestic food production in Austria and Switzerland.[1] Advocates of coffee ECLs hope to reduce industrial coffee-production damage by promoting small-scale, peasant-controlled, ecologically sustainable approaches.[2] In the early 1990s, concern over biodiversity loss led environmentalists to demand an ECL for tropical timber products.[3] This campaign was subsequently extended to temperate and boreal timber forests, as concern grew about timber management practices in developed countries.

The development of ECLs in the forest-products industry is particularly interesting. The story begins in the mid-1980s, when Koy Thompson of Friends of the Earth, United Kingdom, and Tony Synott, of the Oxford Forestry Institute, developed a proposal to study the feasibility of ecocertification and labeling in the tropical timber industry. They submitted their proposal to the British Government, which approved and forwarded it to the International Tropical Timber Organization (ITTO). The proposal was considered at the ITTO's November 1989 Council meeting, where it met a brick wall. Chip Fay of the Environmental Policy Institute noted at the time that "no pre-project proposal has ever stirred up more debate.... Malaysia, Indonesia and the Cameroon led the opposition and with the notable exception of Brazil the pros and cons fell largely along producer and consumer lines."[4]

Producing-country objections resulted in the proposal's "reformulation" to examine the broader questions of "incentives" for sustainable forest management. While this reformulation did not preclude the study of certification and labeling, it was a project of much broader scope. The subsequent history of certification and labeling at the ITTO confirmed its inauspicious start. Although numerous studies were carried out, producing countries and the tropical timber industry opposed all schemes as potential nontariff barriers to tropical timber imports.[5]

If certification and labeling of timber products had remained in the hands of the responsible international body—the ITTO—developing-country and industry opposition combined with developed-country indifference would have blocked its serious consideration and further development. The vehemence of the opposition at the ITTO meetings convinced some environmental groups that ECLs might prove effective in promoting better management of forests. Moreover, the inability of the ITTO to actively pursue the matter created a vacuum into which several environmental organizations moved, notably the World Wide Fund for Nature (WWF). Building on the experience of the organic food movement, several meetings took place after 1989 that culminated in the founding meeting of the Forest Stewardship Council (Toronto, Canada, October 1993).[6] Subsequently, FSC has become a global player in forest certification and labeling and the bête noire of the global forest industry.

The founding of the FSC in 1993 forced businesses and governments to take ECLs seriously. The immediate response in many countries was the establishment of alternative forest certification and labeling schemes. In Canada, for example, the Canadian Pulp and Paper Association (CPPA) paid one million Canadian dollars to the Canadian Standards Association (CSA) to develop an ECL for the Canadian forest industry. In the United States, the American Forest and Paper Association (AFPA) initiated its Sustainable Forestry Initiative (SFI). The African Timber Organization (ATO) began to develop an ECL to cover its members, while Indonesia initiated discussions that eventually led to the establishment of a certification body, the Lembaga Ekolabel Indonesia (LEI). Most recently, small forest operators in Europe established the Pan European Forest Certification (PEFC).

By the end of 2000, forest managers could certify their operations under one of several competing schemes that varied significantly in standards, processes, and rigor. The proliferation of ECLs is the central difficulty confronting consumers, who need a rule of thumb or maxim to enable them to choose the best schemes and to avoid deceptive claims. To develop such a maxim, the next section presents a framework to analyze the legitimacy of competing approaches.

Criteria for Legitimacy

Enabling interested consumers to make choices that advance their broader social values by purchasing products with socially and/or environmentally damaging consequences requires the development of criteria to evaluate the legitimacy of different ECL approaches. Building on a Canadian Parliamentary report,[7] five criteria—scientificity, representativity, accountability, transparency, and equality—can be identified that together provide a broad evaluative framework. These generic criteria, used frequently in various combinations by social scientists to judge the legitimacy of alternative social arrangements, enable a comparison to be undertaken of the strengths and weaknesses of different ECL approaches.

Scientificity, the first criterion, is crucial from an environmental perspective because the processes by which products are made are as important as the products themselves. Consumers require assurance not only about a product's ecological soundness, but also about the soundness of the processes involved in its production. In general, processes that are less energy-, chemical-, and resource-intensive generate products that are more environmentally sustainable, but complexities arise when tradeoffs must be made between one category and another. Is a product that uses fewer chemicals but more energy more sustainable than one that uses more chemicals but less energy? Or are they judged as equals?

Profound disputes involving science arise over approaches to resource management. In fishing, for example, stock management has been based on the concept of maximum sustained yield (MSY). Managers use MSY to calculate the maximum number of fish that can be harvested from a given fishery using estimates of population size, age profile, and birth and mortality rates to calculate mean annual recruitment. The MSY paradigm treats individual fish populations as largely separable from neighboring fish populations and the surrounding environment. It is a reductive paradigm because the inherent embeddedness of fish populations in the wider ecosystem is overlooked.

In recent years the reductive scientific paradigm used in natural-resource management has been challenged by an alternative that emphasizes complexity, interdependence, and chaos.[8] A systemic approach to fisheries management situates fish populations within larger, complex food webs, recognizing that the webs' complete specifications are unknown and unknowable. Within such a paradigm, it makes little sense to manage a single fish population since any act of management ripples through the entire food web.[9] The focus shifts from individual fish populations to ecosystem management, generating demands for a great deal more information and different kinds of information. Managers no longer aim to maximize sustained yield because they recognize that accurate data are not available to make the required calculations. Rather, they consider how best to maintain ecosystem function, composition, and process.

The criterion of scientificity, then, enables one to distinguish one certification approach from another on the basis of which scientific paradigm the ECL validates. ECLs based on systemic science offer stronger guarantees to consumers that broader ecological values are protected because this approach to natural-resource management focuses on ecosystem sustainability rather than maximum yield. ECLs based on reductive science reinforce business-as-usual practices that are part of the problem, not part of the solution.

The second criterion—representativity—draws attention to the array of interests that participate in the development of an ECL. The number of interests represented in a scheme varies from one to many and, as a broad generalization, the more interests that are represented, the better, this because all ECLs require trade-offs to be made across many dimensions—social, economic, cultural, class, and environmental. To ensure that such trade-offs do not systematically advantage some interests at the expense of others, it is imperative that a diversity of interests are actively represented throughout the various phases of a scheme's development and implementation.

The third criterion to judge an ECL's legitimacy is accountability. Accountability refers to the processes put in place to obtain feedback and address criticisms of a scheme's operation. Such processes may include consumer surveys, public consultations, community forums, and formal inquiries. In particular, ECLs should be accountable to (1) indigenous and nonindigenous communities in the forested area being considered for certification, and (2) the growing number of informed consumers who are establishing the market niche for "green" products.

The fourth criterion, transparency, is a major public policy issue today. It reflects public concern over government and business secrecy and the lack of timely information and public debate on key matters of public policy. Transparency refers to the amount of publicly available information on the certification standards applied and the openness of the processes adopted at all stages of the scheme's development and implementation. The more transparent an ECL, the better.

The final criterion for assessing the legitimacy of ECLs is equality. Schemes should treat producers as equally as possible, within and between countries. In particular, certification and labeling should be available regardless of a producer's size or location. If small producers experience higher costs than large producers, they will be squeezed out of the certified-products market, unless a remedy is established. If the ECL operates only in a single country, it could disadvantage producers in other countries.

To evaluate the legitimacy of different schemes, each ECL can be ranked as high, medium, or low across each of the five criteria. Each criterion is equally weighted. Schemes that average a high rank across all five criteria are legitimate and provide a strong degree of assurance to interested consumers that their purchases validate their broader ecological values. Schemes that average a low rank across the five criteria on the other hand are most likely examples of "greenwashing." Cautious consumers must investigate all ECL claims, since purchasing goods labeled by schemes with a low legitimacy ranking does not achieve their broader environmental and social objectives. Worse, if interested consumers are unable to distinguish between ECLs of high and low legitimacy, the overall objective of ecocertification and labeling will be defeated.

ECLs, Legitimacy, and the Forest-Products Industry

The criteria just developed can be used to evaluate the legitimacy of ECL schemes in the forest-products industry. Currently, there are more than a dozen such schemes in operation, each of which claims to guarantee to consumers that their purchases validate their broader environmental values. Most forest ECLs have been developed for national markets. Included here are the American Forest and Paper Association's Sustainable Forestry Initiative (SFI), the Canadian Standards Association's Sustainable Forest Management System (CSA), the Lembaga Ekolabel Indonesia (Indonesian Ecolabel Scheme), the Finnish Forest Certification System (FFCS), the United Kingdom Woodland Assurance Scheme (UKWAS), and Malaysian National Timber Certification (MNTC). At the regional level, the Pan-European Forest Certification Scheme (PEFC) was launched by small, private forest interests in Europe, while the African Timber Organization's scheme covers members from many African exporting countries. At the international level, two ECLs are used to certify forest management, one run by the Forest Stewardship Council (FSC) and the other by the International Organization for Standardization (ISO).

The proliferation of ECLs in the past decade testifies to certification's increasing importance. The forest-products industry is not monolithic, however, and attitudes to certification and labeling vary. Support for ECLs tends to be greater downstream among retailers of forest products than upstream among companies responsible for forest management and lumber production.[10] The emergence of so many schemes generates consumer confusion, however, over the claims and counterclaims of different certification bodies. One way of overcoming this confusion is to analyze each claim systematically across the five criteria developed earlier. In the following paragraphs, I present a preliminary analysis of three schemes that are in common use in North America—SFI, CSA, and FSC.

The American Forest and Paper Association's Sustainable Forestry Initiative

In 1993, the American Forest and Paper Association initiated discussions for an ECL scheme now known as the Sustainable Forestry Initiative (SFI). SFI was a reaction to the pending establishment of the Forest Stewardship Council by environmental organizations. It commenced operation in 1995 after two years of intraindustry discussions under the guiding principle that "environmental conservation and responsible ecological practices make good business sense."[11] Central to the SFI approach is the SFI Standard, which sets out the principles and objectives to which members must adhere to qualify for an ecocertificate. SFI's Principles and Objectives are kept under constant review and have been modified frequently.[12]

SFI is grounded in an industrial approach to forest management. As previously discussed, the dominant forest-management concept is maximum sustained yield, designed to achieve high levels of timber production from stands of natural and planted forests using clear-cutting, genetically modified seedlings, and herbicides and pesticides. The industrial forest-management perspective is embraced in several paragraphs of the SFI Standard. For example, paragraph 4.1.5.1.2 permits the use of 120 hectare-average clear-cuts; paragraph 4.1.2.1.6, genetically modified seeds; and paragraph 4.1.2.1.3, herbicides and pesticides.

SFI provides weak protection for biodiversity and wildlife (paragraphs 4.1.6 and 4.1.1) and consistently privileges economic over other forest values (i.e., paragraphs 2.2.1 and 4.1.2). Objective 2 is to "ensure long-term forest productivity and conservation of forest resources through prompt reforestation, soil conservation, afforestation, and other measures," an objective that makes no distinction between natural and planted forests.[13] Given its promotion of industrial forest management, the SFI Standard ranks low on the criteria of scientificity.

On the criterion of representativity, SFI also ranks low. The scheme was developed exclusively by members of the American Forest and Paper Association, which remains its chief sponsor. AFPA members include some of the largest U.S. forest companies, such as Weyerheuser, Rayonier, Georgia-Pacific, Boise Cascade, and Hancock.[14] AFPA claims on its Web page that its expert panel comprises "independent experts representing conservation, environmental, professional, academic, and public organizations." Notably absent from the panel, however, are nominees of mainstream environmental organizations. Several panel members are government forestry officials, and there are delegates from the National Wild Turkey Federation, the Ruffed Grouse Society, and the International Paper Company. Such environmental under-representation and industry and government over-representation renders the "independence" of SFI's expert panel questionable.

AFPA recognizes it has a problem with representativity. W. Henson Moore, AFPA president and CEO, has acknowledged the need "to look for more ways to broaden multi-stakeholder involvement in both the management and verification process of the SFI program if we expect to increase its credibility."[16] Of the external "environmental" organizations supporting SFI, many are bird-hunting groups such as Ducks Unlimited, Quail Unlimited, and the Ruffed Grouse Society.[17] Such groups are criticized for their narrow, instrumental approach to environmental protection. Their purpose is to ensure the continued supply of birds for their members' sport hunting, not ecosystem management per se. Consequently, such groups favor ecosystem manipulation to maximize the production of a single species, an approach that is the wildlife counterpart of industrial forest management. Several other AFPA members promote plantation forestry; these organizations include the American Tree Farm System, the National Association of State Foresters, the Society of American Foresters, and the Pulp and Paper Workers Resource Council.

How does the SFI Standard rank against the third criterion of accountability? Section 4.3.1 of the Standard, referred to as Objective 9, requires AFPA to "publicly report" progress of program participants in fulfilling their commitment to sustainable forestry.[18] To accomplish this, SFI participants submit annual reports on their progress to AFPA, which compiles and presents the aggregate data. Objective 10 requires SFI members to "provide opportunities for the public and the forestry community to participate in the commitment to sustainable forestry" (section 4.3.2). Performance against this criterion is measured by the use of "mechanisms for public outreach, such as: [toll-free] 800 numbers, environmental education, and/or private and public sector technical assistance programs."[19] These paragraphs present evidence of a concern with accountability within the SFI Standard, although such accountability is almost exclusively one-way. That is, members adopting SFI are expected to provide information to the public at large, but few opportunities exist for consumer and public input into standards development, implementation, and evaluation. In 1999, however, AFPA experienced difficulty with one specific user group: loggers. Loggers complained that "there was no established procedure for reporting perceived violations of the new sustainable forestry practices" and, when these concerns were not addressed by AFPA, they "withdrew from the semi-annual SFI National Forum meeting in the summer of 1999" in an effort to "force attention to their concerns and frustrations over inconsistencies within the SFI program."[20]

As a consequence of the loggers' protests, AFPA established a joint Loggers Task Group and developed an additional system of accountability through which perceived violations of the SFI Standard can be reported. Pending state development of appropriate procedures and systems, the AFPA has put in place an Interim Inconsistent Practices Report Protocol (IIPRP), which "allows for the filing of anonymous complaints of possible SFI program infractions, which are then investigated and reviewed by a representative of the SFI Expert Review Panel."[21]

SFI has become more accountable. However, the accountability mechanisms are modest and skewed toward program participants. Consumers who are pushing green consumption forward are largely ignored, as are major environmental and indigenous peoples' organizations. Taking into account AFPA's recent response to loggers' protests and the establishment of an inconsistent practices protocol, SFI ranks medium in terms of accountability.

A major environmental criticism of the SFI Standard relates to the fourth criterion. It is that there has been a lack of transparency, both in the scheme's initial development and in its subsequent implementation. The SFI Standard was developed exclusively by AFPA members without public information on the role of large corporations in standards development. Notably, there was little information available on the range of standards initially considered and on how they were whittled down to 11 "Objectives." Such gaps in public information would be reduced by the timely release of SFI Program committee documents and the inclusion of relevant interests (including environmental, consumer, and indigenous peoples) within the SFI Program committees. AFPA did establish an Expert Review Panel, but panel membership is neither reported in its annual report nor available on its Web site. While the panel is clearly improving the implementation of the SFI Standard, the lack of information about panel members and working procedures is a concern. Also, AFPA's process to investigate complaints of "inconsistent practices" is secretive. According to AFPA's annual report, "all complaints will be thoroughly investigated" by the Expert Review Panel and those unwilling to reform their practices will be "dropped from the program."[22] However, the AFPA's annual report does not provide any details of past investigations, nor does it indicate an intention to make public in the future the names of companies dropped from the program. In sum, AFPA's inherent secrecy merits a low rank on the criterion of transparency.

The final criterion for evaluating the legitimacy of the AFPA scheme is equality. What is the distribution of costs and benefits of SFI for various types of producers in different countries? The question focuses attention on the direct and indirect costs of becoming an SFI member. According to its recent annual report, AFPA has achieved considerable success in recruiting members into SFI. The report notes that "90 percent of the industrial forestland in America is controlled by our membership, and all of it is being managed under a new, environmentally conscious, sustainable forestry ethic."[23] Given that the vast majority of those engaged in the management of industrial forestland are also members of the SFI program, the scheme is clearly equitable for AFPA members. It should be noted, however, that AFPA's success is due to the modest direct and indirect costs of AFPA membership. The costs of SFI include the time required to complete an application form and the payment of a modest annual licensing fee. With direct costs so small, it is not surprising to find a large number of companies joining, especially when the program helps them maintain their market share. Moreover, because the SFI scheme is based on a North American industrial forest-management paradigm, the indirect costs of membership are also marginal since few, if any, changes are required in forest-management practices and procedures.

From an equity perspective, it is notable that SFI reflects forest-industry standards in the United States. Stakeholders in other countries were excluded from participation at the scheme's inception. The scheme may benefit large U.S. forest companies such as Bowater, Domtar, Rayonier, and Weyerheuser at the expense of foreign companies in Brazil, Indonesia, Malaysia, Chile, and elsewhere. This national approach to certification adds legitimacy to developing-country arguments that "voluntary" certification will prove to be a nontariff barrier to Third World imports. Membership in SFI has been extended to non-AFPA members, which partly offsets the allegation of national bias. In the longer run, any inequality is to be overcome by harmonizing SFI with other national schemes such as Canada's CSA and Europe's PEFC. The objective, according to SFI's annual report, is to "establish a consistent, credible system for international mutual recognition of a sustainable forestry certification program."[24] SFI can be ranked medium on the equality criterion, therefore, reflecting its broad membership based on weak standards and low direct and indirect costs.

In summary, AFPA's SFI ranks low across three criteria (scientificity, representativity, and transparency), and medium across two criteria (accountability and equality). Its overall legitimacy ranking is low to medium, providing interested consumers with few guarantees that their purchases validate their broader ecological values.

The Canadian Standards Association Sustainable Forest Management System

In Canada, as in the United States, industry has played a leading role in the development of a national ecocertification scheme in the forest-products sector. Unlike AFPA, however, the Canadian forest-products industry worked through governmental and nongovernmental channels, specifically the Standards Council of Canada (SCC) and the Canadian Standards Association (CSA, now CSA International). In 1993, the Canadian Pulp and Paper Association (CPPA) requested that the CSA—an industry-dominated standards development "nongovernmental organization" accredited by the SCC—develop an ECL as part of a strategy to guarantee access to foreign markets for its members' exports. Two developments motivated the CPPA: increasing environmental activism in Europe to boycott Canadian forest products and the impending inaugural conference of the Forest Stewardship Council. In response, CSA established a technical committee to develop an ECL modeled on an Environmental Management System (EMS) approach similar to that used by the International Organization for Standardization (ISO). The Canadian Sustainable Forestry Certification Coalition (CSFCC), a broad coalition of Canada's major forest-products companies, promotes the resulting technical document[25] as a "made-in-Canada" solution to its certification and labeling problems.

An analysis of CSA across the five legitimacy criteria reveals a number of similarities to AFPA's SFI scheme. While it could be argued that CSA is scientifically neutral on the first criterion of scientificity because it adopts an "environmental management systems" (EMS) approach, a closer analysis reveals otherwise. In general, EMS approaches are agnostic with respect to practices and normally do not establish performance guidelines. Rather, they focus on the organizational processes and procedures put into place to achieve stated objectives. The CSA is not a pure EMS, either, because it endorses a set of forest-practices standards that emerged from the Montreal Process and were consolidated in a 1994 report by the Canadian Council of Forest Ministers (CCFM).[26] The set of criteria and associated critical elements of the CCFM remain firmly embedded in an industrial approach to forest management and are thus not enough, in themselves, to establish sustainable forestry. In addition, the CCFM standards include a statement of principles that locates forest management in a broader ecosystem framework and a set of norms that enable one to assess whether a company is transgressing them or not. In the absence of ecosystem principles and agreed norms, the CCFM's criteria and indicators permit business-as-usual forestry practices.

Under the CSA scheme, a company meets its certification requirements if it demonstrates due diligence in implementing its EMS and conforms to the broad criteria and indicators set out in the CCFM document. Unacceptable practices from within an ecosystem forest-management paradigm—practices that cause substantial environmental damage such as large-scale clear-cutting, planting of genetically modified seeds, or extensive and intensive chemical use—are viewed as legitimate under the CSA/CCFM approach. Consequently, the CSA ECL ranks low in terms of scientificity.

To this low ranking for scientificity must be added a low to medium ranking for representativity, despite claims about the diversity-of-interest representation on its technical committees. Notably, bona fide environmental groups were not represented. Chris Elliott, in a detailed study, observed that

> NGO participation in the TC [CSA technical committee] was more apparent than real for several reasons. First, although the chief executives of two of the major Canadian NGOs working on national forest policy issues ... were listed as members of the technical committee (CSA 1994, Appendix 2), neither had agreed to this.... Second, the NGO category of membership was actually characterized as "Environmental and General Interest" and at various meetings included a diverse range of representatives.[27]

While CSA attempted later to draw more representative stakeholder input into its scheme, it failed to attract the necessary support from critical groups. Its technical committees lacked representation from indigenous peoples and experts in ecosystem-based forest management. The representativity of the CSA scheme ranks as low to medium, therefore—a ranking that recognizes its greater but ultimately incomplete effort to obtain broader stakeholder participation than AFPA.

How does CSA fare on the criterion of accountability? Its approach mandates a substantial input from the public and other stakeholders during the certification process, when a company is seeking to register a designated forest area. Companies seeking certification must consult with stakeholders affected by their plan, including Canada's indigenous and nonindigenous communities. Furthermore, under the CSA, determining whether a designated forest area qualifies for certification is the responsibility of a third-party auditor. Unlike AFPA, it is not possible for a company to certify itself. Thus, under the CSA system, a company is accountable publicly and professionally to outsiders for achieving its forest-management objectives.

On the other hand, there is a notable lack of procedural mechanisms for *ongoing* public input into CSA's EMS. No standing committee has been established with widespread stakeholder representation to review the scheme's implementation or to deal with criticism and disputes. The major organization endorsing CSA, the CFSCC, is a consortium of Canadian forest-products associations with a narrow mandate. Moreover, the criteria and indicators of the CCFM were developed through an intergovernmental process that lacked input from ecosystem-based foresters, ecologists, environmentalists, and indigenous peoples. Taking the above into consideration, the CSA scheme ranks medium in accountability.

How much information does CSA provide and how open are its processes? The CSA provided high-quality information about its proposed system at the outset and made its key technical documents (Z-808 and Z-809) available to interested parties. The commercialization of the CSA in the late 1990s meant, however, that the organization now charges fees for these documents. Unlike AFPA, a free, online version is not available on CSA's Web site. Thus, the availability of information actually *declined* in the past two years. Further, because CSA's mandate includes only the development of an ECL, no central body exists to consider the scheme's shortcomings and further development. Responsibility for implementing and improving the scheme rests with individual corporations. Members of the public and consumers with complaints about processes and practices can bring these to the attention of the company and comment during an audit of a company's performance. Such input may be difficult to give in practice, though, since the company sets the agenda, determines the timetable, and pays for the audit. Consequently, CSA ranks low to medium in terms of transparency.

On the final criterion, equality, it is notable that the initiative to develop CSA was taken by Canada's largest forest corporations. Their overriding concern was to secure continued access to export markets in Europe, Japan, and the United States. ECLs based on EMS approaches can be costly to implement. In the absence of specific arrangements for small- and medium-sized enterprises (SMEs), such schemes favor larger corporations. The costs of implementing CSA's ECL include expenditures on developing the environmental management system, on initiating and maintaining a public consultation process, on altering and implementing a modified forest-management regime, and on engaging the services of an accredited auditor. As a rule, only the larger corporations can afford these costs, a statement confirmed by a PriceWaterhouse-Coopers report.[28] The report estimates the cost of CSA certification for BC's Small Business Forest Enterprise Program (SBFEP) at more than C$200,000 per operation, well beyond the means of most SMEs.

In addition to its possible discriminatory impact on small forest operators, CSA's ECL, like the AFPA's SFI, was developed exclusively in one country with input from nationally based industry. International representation was not solicited and was viewed as unnecessary, since it would always be possible to "harmonize" the scheme with other national schemes later. As Jennifer Clapp notes in relation to ISO, this develop-first-and-harmonize-after approach discriminates against developing countries, because the standards established focus "primarily on the concerns of industrialized countries rather than those of developing countries."[29] CSA ranks low in terms of equality, reflecting its high cost of implementation, the lack of special arrangements for SMEs, and its discriminatory potential against Third World producers.

In summary, CSA's ECL ranks low on the criteria of scientificity and equality, low to medium on representativity and transparency, and medium on the criterion of accountability. While the scheme has greater legitimacy than AFPA'a SFI, its overall low-to-medium legitimacy ranking indicates that the environmental values of interested consumers remain inadequately addressed.

The Forest Stewardship Council's Scheme

The Forest Stewardship Council's scheme grew out of frustration on the part of environmental organizations at the combined failure of the forest industry and governments to implement environmentally sustainable forest policies and practices. Since its founding in 1993, FSC has grown in size and influence. Headquartered in Oaxaca, Mexico, it has national offices in countries around the world, including the United States, Canada, the United Kingdom, the Netherlands, Sweden, and Brazil.

In evaluating the overall legitimacy of the FSC, a notable contrast to CSA and AFPA is its approach to scientificity, which is grounded in a systemic scientific paradigm. FSC Principle 6, for example, sets out a strong requirement for the protection of biodiversity and other forests values: "Forest management shall conserve biological diversity and its associated values, water resources, soils, and unique and fragile ecosystems and landscapes, and, by so doing, maintain the ecological functions and the integrity of the forest."[30]

The application of Principle 6 is elaborated in several subparagraphs and requires forest managers to protect the full range of forest values. While the FSC Principles do not ban clear-cutting as a management tool, Principle 6 and its several further provisions significantly constrain its use as a forest-management option. In forests where clear-cutting is permitted, the size of the clear-cut is far smaller than current industry norms. The systemic scientific paradigm of FSC is reflected also in Section 6.8, which bans the use of genetically modified organisms (GMOs). Interested consumers are thus guaranteed that their purchases do not contribute to the spread of largely untested biotechnology. The scientificity of FSC ranks high, therefore, because it embraces an ecosystem approach to forest management.

On the criterion of representativity, too, FSC ranks high. FSC has established a unique institutional structure governed by a General Assembly composed of three equally weighted chambers representing environmental, social, and economic interests.[31] The diversity of interests represented at the General Assembly encompasses a large number of environmental groups active in forest-related issues throughout the world (such as the Rainforest Action Network, Greenpeace, and World Wide Fund for Nature, for instance). Important economic stakeholders include forest-products retailers such as Home Depot, Boots the Chemist, B&Q, and Sainsbury's Homebase, as well as forest-management companies like Northland Forest Products Inc. and Sweden's AssiDoman Skog & Tra. Social interests are represented by such groups as the Arkansas Public Policy Panel, the Mountain Association for Community Development, CUSO, the National Aboriginal Forestry Association, and the Taskforce on the Churches and Corporate Responsibility.

Especially important in terms of representativity is the inclusion of indigenous peoples' groups. Principle 2 states that "the legal and customary rights of indigenous peoples to own, use and manage their lands, territories, and resources shall be recognized and respected." Moreover, it is this proposition that has made the FSC Principles and Criteria unpopular with governments around the world. Principle 4 also provides that "forest management operations shall maintain or enhance the long-term social and economic well-being of forest workers and local communities,"[32] which strengthens indigenous peoples' and local community's claims to benefit from their local forest resources. FSC ranks high on representativity, therefore, in terms of the diversity of interests involved in the development and implementation of its scheme.

On the criterion of accountability, FSC is the more open and democratic of the three schemes studied here. As a membership organization, FSC welcomes companies, organizations, and individuals that endorse its Principles and Criteria. Requirements for membership include the submission of a membership application, a statement of support for the FSC, organizational details if the application is on behalf of an organization, evidence of charitable status for nonprofit organizations, and a statement of support from two other FSC members.[33] The business of the FSC is carried on by a board of nine directors: two (one each from the North and the South) represent economic interests, and the remaining seven represent environmental and social interests. Although economic interests are represented, the environmental and social values of forests are protected institutionally and given weight.

The board of directors is accountable to the membership at annual General Assembly meetings, which are open to outside observers. The proceedings of General Assembly meetings are posted on the FSC Web site, and a review of past proceedings provides strong evidence of active membership participation and concern.[34] The ongoing participation of consumer groups, environmentalists, and industry at FSC's General Assembly contrasts markedly with the closed processes that characterize the AFPA and CSA schemes. FSC ranks high, therefore, in terms of accountability.

FSC ranks high too on the criterion of transparency. Its Principles and Criteria were developed through an open process that involved negotiations with a large number of interested parties. Regional forest-management standards now being developed in a number of countries —including the United States, Canada, Brazil, and Peru—involve broad-based participation from each of FSC's chambers. A controversy has erupted in Eastern Canada about the process for developing Maritime Regional Standards. The development of these forest-management standards began in April 1996, with 120 individuals representing nine diverse groups, including indigenous peoples, environmentalists, foresters, small business, woodlot owners, community groups, youth, and government.[35] Large industry was represented by the chief forester of a major Canadian East Coast company, J. D. Irving Limited, and a representative of the Nova Scotia Forest Products Association. Following two years of negotiations and consultations, disagreement remained with large industry over the draft Maritime Regional Standards, especially the use of biocides, exotic species, and the size of protected areas.[36] Following industry claims of bias in the representation of interests in the committee, FSC Canada formed a Dispute Resolution Committee in September 1999 to investigate. Although the committee found that there had been "adequate and balanced representation," large industry was still not satisfied and pressed its case at the international level with FSC International. Subsequently, in February 2000, FSC International launched its own commission of inquiry "to examine matters surrounding the preparation and endorsement of the FSC Maritime Regional Standards in Canada."[37] The point is that there was a great deal of transparency in the process, even in the case of a dispute involving a major stakeholder in the development of forest standards in Maritime Canada. Not only was the issue publicly discussed and debated, but there was also a process of inquiry at the national and international levels that enabled the dispute to be adjudicated. Such openness is notably lacking in the other AFPA and CSA schemes. Under the AFPA scheme, a secretive expert review panel receives confidential complaints from foresters and adjudicates these without public scrutiny. Under the CSA process, there is no central body that can engage in such adjudication arrangements. Members of the public with complaints about standard setting and forest practices in a Defined Forest Area may make their concerns known to the auditor, who, being paid by the company, may find it convenient to overlook or and downplay their importance. The FSC scheme thus ranks high on the criterion of transparency.

FSC fares least well on the criterion of equality. The organization's early focus was to establish a performance-based system of ecocertification that provided strong guarantees to consumers that their purchases validated their broader environmental values. FSC paid less attention then it might have to the cost of obtaining such a guarantee and to the impact on SMEs. As a consequence, only the more dedicated small producers have sought FSC certification, and many stridently opposed the scheme. Opposition has been especially active in Europe among small private-forest landowners, who formed the Pan European Forest Certification Scheme (PEFC). PEFC members embrace a set of industrial forest-management standards similar to those set out in the Montreal Process and the CSA (although they were developed under a different process, the Helsinki Process). Opposition to FSC by members of PEFC is due less to the direct costs of FSC's ECL and more to the indirect costs involved in shifting from industrial to ecosystem forest management.

In response to growing protest from small producers, FSC instituted a system of group certification. If small holders in a region desire FSC certification, they can now establish an organization under FSC's guidelines.[38] This approach enables independent, third-party auditing companies to certify groups of producers rather than individual members, thus spreading the costs of certification across a larger number. Group certification therefore helps small producers participate in its ecocertification scheme. In addition, and unlike AFPA and CSA, FSC developed its approach with substantial transnational input.

The overriding purpose of FSC is to improve forest management within states, not to enable some states to obtain a competitive advantage over others. The development of FSC's group certification and, for that matter, its organizational principles as a whole have been international exercises balancing group representation within countries (among economic, environmental, and social groups) and between countries (between North and South). This balancing is an ongoing institutional feature of the FSC, not only through its three-chamber structure, but also through the voting arrangements within each chamber. FSC's by-laws state: "To achieve a balance between Northern and Southern perspectives within each chamber, Northern and Southern organizations and individuals shall have 50% of the voting power. In effect, there shall be 'Northern' and 'Southern' chambers within each of the three chambers."[39] Given these institutional features and its recent efforts to accommodate the concerns of small producers, FSC ranks as medium to high on the criterion of equality.

In summary, the FSC scheme ranks high across the criteria of scientificity, representativity, accountability, and transparency. It ranks medium to high on the criterion of equality. The overall legitimacy of FSC is thus high. Interested consumers buying products with an FSC logo can be confident that their purchase validates their broader environmental and social values.

Conclusion

Table 12.1 summarizes the results of my comparison of the three certification and labeling schemes. The FSC scheme is superior to AFPA and CSA. Of the three schemes reviewed, only FSC's can be considered legitimate, providing a credible guarantee to consumers.

To the extent that these findings can be generalized, a simple rule of thumb emerges, one that interested consumers can employ to increase the likelihood that their purchases validate their broader ecological and social values: Purchase goods certified by schemes endorsed by reputable environmental organizations, and be skeptical of industry- and government-sponsored logos. Industry and government schemes are likely to safeguard large-scale industrial systems of production and the competitive advantage of individual states. They are likely to be exercises in greenwashing, not in fundamental change of inherently unsustainable practices.

As is the case with most rules of thumb, following this one does not guarantee correct decisions. Certainly consumer decision making would improve to the extent that consumers went the next step and, say, employed the five criteria used here—scientificity, representativity, accountability, transparency, and equality. But given the effects of commoditization (chapter 4) and distancing (chapter 5), even this simple rule

Table 12.1
Comparison of legitimacy of three ecocertification and ecolabeling schemes

Scheme	Scientificity	Representativity	Accountability	Transparency	Equality	Legitimacy ranking
AFPA	low	low	medium	low	medium	low to medium
CSA	low	low to medium	low to medium	low to medium	low	low to medium
FSC	high	high	high	high	medium to high	high

of thumb is likely to improve consumers' decision making because, unlike with industry and government practices, it highlights noncommercial values and helps build ecological and social feedback into the end-use decisions of purchasers.

Of course, the above maxim raises the question of how consumers can know whether a particular ecocertification scheme is sponsored by a "reputable" environmental organization. This task has been made more difficult over the past 20 years as industry has set up organizations with environmentally friendly names, organizations that amount to little more than fronts for corporate interests. A reputable environmental organization is one that has a broad membership base, one that actively pursues campaigns and projects to protect the environment in the relevant sector, and one that has "brand" recognition. It is an organization that is recognized by its peers as a senior member of the local, national, or international environmental community. Consumers in doubt about a specific ECL, or the organizations endorsing it, can consult their local and national environmental NGO networks to locate such organizations.

The proposed rule of thumb reinforces several themes of "cautious consuming" (chapters 1 and 14). First, consumers need to be literally cautious in making any assumptions about the validity of ECLs. *Caveat emptor*, buyer beware, has its counterpart here in *Caveat certificatum*, beware of a certificate. Most schemes do not validate the broader social and environmental values of committed consumers. With commoditization and distancing, purchasers must remain skeptical of all certification and labeling claims. In the short to medium term, the onus remains on consumers to inform themselves of the relative merits of different schemes, to, in effect, decommoditize and reduce distancing.

Second, ECL schemes highlight the potential power of cautious consuming, thus engendering better producing throughout the commodity chain. It is notable that, with FSC's early success in organizing members of the forest industry and environmental groups, governments and industry rushed to create their own schemes, however marginal their proposed changes in industrial practices may have been. Why did they respond so quickly and with such determination? The answer is that FSC has threatened industry's market share, profitability, competitiveness, jobs, and investment. If ECLs were insignificant, industry and government response would have been insignificant. Governments and industry, in effect, fear that cautious consuming will catch on. They are now doing their level best to muddy the certification waters, to dupe unsuspecting consumers into believing they are making environmentally sensible purchasing decisions when they buy goods with industry- and government-sponsored ecolabels.

If this judgment appears unduly harsh, consider the emerging politics at the international level. Governments and industry are now organizing to outlaw schemes like FSC's, using intergovernmental organizations such as the World Trade Organization and the International Standards Organization. The Canadian government, for example, is pushing to have voluntary certification and labeling included in the WTO's Technical Barriers to Trade Agreement (TBT). Included in the TBT at Annex III is a Code of Conduct for standards development. The code provides for standards to be developed by *a single body within a national jurisdiction* and for *minimum overlap between national, regional, and international standards*. If the Canadian strategy succeeds, it is possible in the future that governments could ban the use of the FSC ecolabel on the basis that it (1) was not developed by an appropriate national standards body, (2) does not conform to ISO standards, and (3) constitutes an unnecessary barrier to international trade. Certification by reputable environmental organizations and its implicit cautious consuming are, indeed, a threat.

Finally, ECLs exploit the unequal power relations that exist between consumers and producers through the forest-products commodity chain. As Conca notes (chapter 6), when profits and power reside downstream in the retail sector, upstream producers who mine and harvest the primary resource become vulnerable. In the present context, it is precisely the downstreaming that enables environmental organizations to put pressure on the end of the chain where power lies—big retailers like Home Depot, Boots the Chemist, B&Q—knowing that these companies will channel pressure upstream to extraction and processing companies like Weyerheuser, Rayonier, Georgia-Pacific, Boise Cascade, Abitibi-Consolidated, and Hancock. Information pickets outside large retail outlets make consumers aware of the alternatives. Meanwhile, WWF-sponsored "buyers' groups"—loose associations of large corporations operating in the downstream retail sector of the forest industry—contribute by encouraging retailers to commit to purchasing only FSC-certified timber. Pressure from both sources flows upstream to the large timber companies at the top of the chain.

Cautious consumers can effect dramatic change through their purchasing power. Indeed, cautious consumers demanding FSC-certified timber have had a profound effect already on global timber markets, out of all proportion to their numbers. There is tremendous scope, therefore, for the FSC model to be extended to other sectors and to other products, the most strategic being coffee and tea. Environmentally committed consumers seek these products already, buying organic produce at local community markets and alternative trading stores. What ECLs can do, however, is to make the same choice available to the interested-but-not-fully-committed consumers who want to "do the right thing" but not at great personal inconvenience. This larger segment of consumers constitutes the mass of shallow ecological purchasers that might yet be enlisted to achieve deeper ecological objectives.

13

Citizens or Consumers: The Home Power Movement as a New Practice of Technology

Jesse Tatum

In the United States today, there are an estimated quarter of a million "home power" residences[1] whose occupants produce their own renewable electricity. Most of the energy comes from photovoltaic (PV) panels, otherwise known as solar cells, but more and more comes from small wind and microhydroelectric systems. In a world accustomed to top-down decision making where consumers are expected to wait for tax incentives before even reducing their consumption, this is an astonishing fact. What is more, the home power movement[2] does not stem from government programs, confounding all official assumptions about how renewable electricity might eventually be adopted in residences. In fact, there has been almost no contact between power authorities and home power's developers. This is not a movement in which adopters mention the Department of Energy or seek public funding for their activities.

Home power thus flies in the face of conventional conceptions of consumer behavior. Its participants develop much of the technology themselves (except the actual photovoltaic cells) and willingly pay up to twice the price of utility power to put their systems in place. As a departure from routine construction and energy-supply patterns, home power entails substantial investments of time on the part of homeowners; they must learn about electricity use and home power equipment alternatives, as well as design systems unique to their energy requirements and to local resource availability. Once installed, home power systems require some ongoing involvement in operation and maintenance and may entail continuous adjustment of one's energy-use patterns according to weather-related energy availability. As a result, home power promotes and sustains an altered sense of function and consequence at every flip of an electrical switch drawing from the energy of the sun.

Home power may not be for everyone. On the other hand, end-use efficiency measures that have become an integral part of the movement mean that only about one-tenth of a typical home's roof area is required for energy supply. This plus the fact that home power systems can be applied almost anywhere means that such sources could ultimately supply the bulk of home electricity needs even in densely populated urban areas. As an increasingly well-developed and well-demonstrated alternative, home power could conceivably take wing on a grand scale, profoundly transforming major segments of the economy and reducing the environmental footprint of both the developed and developing worlds.[3]

The promise may be great, but the practice of home power is far from what most residents know in their everyday patterns of energy use. Even the most environmentally committed people would probably find themselves confounded either by the technical aspects or by the seeming impossibility of squaring the demands of the technologies with the demands of busy schedules and fast-paced lives. Most significantly, perhaps, the typical energy user in North America would experience a disconnect between accustomed patterns of consumption and the profound reorienting of daily life home power adopters experience.

In the contemporary political economy where energy is cheap and convenient, home power is certainly not for everyone. Yet it is in many respects a mainstream middle-class phenomenon that includes high school teachers, architects, builders, medical technicians, lawyers, and doctors. It seems, in fact, to have a draw unlike anything else that I, a long-time student of energy use, have encountered. For example, an illustrated talk I give shows home power homes around the country, describing the technologies and the owners' experiences. The audience, whether comprised of college students or the most jaded of Department of Energy bureaucrats, is quickly drawn in. In fact, I sense that while government officials cannot conceive of a connection to their professional functions, they are just as engaged at a personal level as others. It is as if members of the audience have been searching for something like this. Whether they actually make the choice is, in the contemporary energy situation, less important than the fact that home power is so appealing. One reason may be the nature of the home power choice.

When individuals choose home power they are making more than a conventional consumer choice, more than a decision to consume one kind of kilowatt hour, even a "green" one. Instead, they are choosing to step away from conventionally constructed acts of consumption. The choice is not an expression of preference, as one might choose, say, between a Porsche and a Ford, but a rejection of consumerist choice as a whole, a rejection of the automobile itself and, at the same time, an embrace of new modes of transportation. Ultimately, the significance of home power in this country does not lie in how many energy systems are installed, or even in how much energy they produce. Rather, its significance lies in the movement's fundamental departure from traditional consumer behavior, a departure that renders it invisible on ordinary radar screens and essentially unintelligible among government officials, journalists, and others who operate with the dominant market images of "production" and "consumption."

Indeed, in the two decades or more since photovoltaic panels first became available outside NASA's space programs (i.e., with ARCO Solar's panels in 1979), it has become increasingly clear that home power is not about "production" or "consumption" in the usual sense of these terms. Drawing from extensive ethnographic exposure to the home power movement over a period of more than ten years,[4] I see this movement as a response to widespread neglect of core human needs, needs that are more substantial and more universal than those embodied in the dominant images of "the consumer." As a part of that response, the technical innovation and social engagement of the movement point to an untapped reservoir of technological and behavioral alternatives that could lead to significant and more widespread reductions in environmental damage—and all without a sense of sacrifice. I elaborate these ideas in the last section of this chapter after briefly describing some of the movement's history, its technical achievements, and the views and commitments typical among home power people.

A Brief Technical History

Technologically, home power systems have arisen from an elaborate participatory research effort[5] that began with fairly primitive, isolated, and yet parallel design efforts by people all over the country. Now there is a highly sophisticated range of technologies existing in a coordinated, though not centrally controlled, network.

Beginning in the "back to the land" efforts of the 1960s, the very first "home power" systems innovators used automobile batteries to power simple lighting systems, radios, and tape players in tents, vans, and other primitive shelters in rural areas from California and Washington State to Arkansas and Vermont. Batteries were removed from cars to power these primitive home energy systems, then replaced and recharged during the occasional drive to town. Simple generators for directly charging batteries were cobbled together from lawnmower engines and old car alternators. In the 1970s, before photovoltaic panels and small wind machines became readily available,[6] significant effort was devoted to refurbishing old wind machines as a power source—machines often originally intended for remote agricultural use. In the absence of efficient, reliable, and affordable inverters for converting direct current stored in batteries to household alternating current, most home power systems remained very small, powering 12-volt lights, converted blenders, and other small appliances.

In the 1980s, however, photovoltaic panels, commercial spin-offs from NASA's space programs, surpassed old wind machines on the basis of simplicity and reliability. While the panels, like wind machines, do not always produce energy (i.e., when the sun is not shining), they have no moving parts to maintain and do not involve the continuing headaches of the mounting towers required for wind machines. High-efficiency, high-reliability inverters became available such that new home power homes would routinely include inverters and wiring for 110-volt outlets and appliances. By the 1990s, very small wind machines (some that do not require mounting towers and are designed to be modular in the way that photovoltaic panels are) and very small hydroelectric generators have become increasingly common complements to photovoltaic arrays. From a traditional economic perspective where cost and the convenience of the consumer are paramount, one would not install any of these three systems (PV, wind, hydro) unless the resource regime (sun, wind, water) could provide energy continuously, all day and year round. But in the home power context, local weather and seasonal variations in resource availability often allow the different systems to complement each other. Adopters thus tend to include more than one system even if the utilization and economic return on each is reduced.

From the days of the earliest home power pioneers, it has been clear that energy end use, as well as energy supply, would have to be carefully redesigned to fit the new ways of living. Electricity use for low-temperature heat has been eliminated and replaced by super insulation or passive solar space heating (with wood heat as backup). Solar water heaters are backed up by gas-fired "flash" heaters, devices that heat water as it flows without a hot-water storage tank and plumbed in series with the storage tank of a solar water-heating system. These heaters allow efficient use of the solar heater and storage tank while ensuring delivery at the tap of water at the expected hot-water temperature. Fluorescent lighting has largely replaced incandescent, with roughly a four-to-one efficiency improvement. And in very recent years, new light-emitting-diode (LED) options have appeared with a further efficiency improvement of nearly a factor of ten. Refrigerators, one of the largest electricity components in the typical home, have been radically transformed as well. Gas-fired refrigerators have reemerged in homes with very small electricity supply systems. And electric compression refrigerators with vastly increased insulation and other efficiency measures have reduced electricity use by a factor of about six compared with traditionally available units for use with full-sized home power systems.

These technical developments, some simple, some sophisticated, have emerged from an expanding cottage industry and a network of participatory researchers. The energy-efficient refrigerator, for example, is made in a small plant in Arcata, California, staffed in significant part by Humboldt State University students. Many or most of these moves on the end-use side are no more cost effective by traditional economic standards than are the renewable energy supply systems. For instance, a single white LED lightbulb can now cost in the neighborhood of $80 or more, retail. Thus neither supply nor end-use technologies are cost competitive when compared with traditional alternatives. When combined, however, they reduce overall system costs in absolute terms, and the patterns that are attractive to home power people become more affordable.[7]

In sum, more than three decades of self-financed participatory research has vastly improved the technology of home power and greatly reduced the barriers to implementation once faced by the movement's earliest pioneers.

Home Power Practice and People

Most of the leading figures in the home power movement today began with work on their own homes back in the 1960s or 1970s. Working initially just to develop their own systems (sometimes their neighbors', too), they have in many cases unexpectedly found themselves with small home power businesses that have since grown either locally (e.g., in the relatively dense home power markets of certain parts of California or the Southwest) or as mail-order enterprises. Early pioneers Richard and Karen Perez played an especially significant role with their *Home Power Magazine* connecting the initially diffuse and isolated efforts of home power enthusiasts. They regularly feature home owner design and installation stories and make the technical aspects of home power systems highly accessible. The magazine's advertising has contributed significantly to the development of a nationwide equipment and installation infrastructure. A section on letters from readers provides a forum to exchange experiences and suggest ways to improve equipment. Richard and Karen Perez first published *Home Power* on newsprint and with modest advertising support, distributing it free by leaving copies in laundromats, knowing that "back to the land" folks without power did not have their own laundry facilities. Now printed in color on slick recycled paper, it has a print run for subscribers and magazine distributors of about 24,000, with another 48,000 readers online.[8]

Visiting home power homes across the country, it is not uncommon to find roof-mounted PV panels of different size and appearance. Purchased two or more at a time, these panels attest to a history of system expansions and improvements. Remotely sited in most cases, home power homes without conventional power supplies generally have not been eligible for mortgage financing. So their owners have developed them progressively as funds become available. One often finds small and relatively primitive structures near more substantial homes, the former having served as an affordable home for a period of years as the latter got built. One Arkansas family, for example, now uses its former home, a single-room, dome-shaped structure, as a workshop (with its own small PV electricity supply), having finished an attractive log home nearby from timbers taken from their own land. Even newcomers, like a young couple in Kentucky—one with a new bachelor's degree (mechanical engineering) and master's degree (science and technology studies) from Rensselaer Polytechnic Institute, the other a primary school teacher—choose to live in small and relatively primitive temporary quarters (in this case a 12- by 15-foot "house" of their own construction) while building their larger, more permanent home. Good fortune or, as in the case of an architect and his family near Clarkdale, Arizona, careful planning may also result in integration of what were early temporary quarters into a finished home. Notably, finished homes generally remain modest and retain something of the backpacker's style or the "voluntary simplicity" of their owner's origins and continuing pattern of life.

Owner construction has been quite common, as has owner design and implementation of home power systems. Local materials also are often employed. A landscape painter and his family in Kentucky, for example, used the undried lumber available from a local mill to side the house they built. In other cases, a portable sawmill is purchased or run by a neighbor, or logs or stone are used directly from a person's own land. Cooperative construction, while not as common, is also widespread, both in multihome community settings where participants may purchase a large parcel of land together, and in smaller groups of two or three neighbors or friends.

Home Power in Amherst, Wisconsin

Cooperative home power efforts have coalesced in increasingly activist communities in a number of states. The home power community in and around the small town of Amherst, in central Wisconsin, serves as a model or, perhaps, as a prototype.

Among Amherst's early pioneers were several people who moved to the area in their college years in the late 1960s, settling initially in communal living situations on old farmsteads, sometimes without electricity. Only about 12 miles from Stevens Point and its branch of the University of Wisconsin (UWSP), Amherst is a small community whose downtown boasts only one intersection busy enough to justify a four-way stop sign. The land, disturbed by glacial action, was, and remains, reasonably inexpensive, with small and financially marginal farmsteads the predominant pattern.

One of these early pioneers—I will use the pseudonym *Roy*—completed his bachelor's degree in the environmental studies program at

UWSP and began selling wood stoves from the barn on his farmstead. With the energy crisis of 1973, Roy got into the business of fabrication and installation of air-heating solar collectors, using models one of his professors had shown him years earlier. Moving his business with a partner to a former hardware store building in Amherst, he then continued on his own when the loss of federal solar tax credits led his partner to leave the business. Adding a line of home power equipment, he gradually expanded the business, eventually selling it to a home power firm marketing nationwide, and continuing as an employee with that firm. Roy, his wife, and two children ended up staying in the farmhouse as all but one of the original friends left the area. They now heat the house with some of his solar collectors and a wood stove. Roy has installed a PV power supply and a separate PV array to charge an electric car that he bought secondhand. The one other remaining person from Roy's original group, still a close friend, lives with his own family on adjacent land in a separate home-powered home.

Another early pioneer in the area also began in a communal living arrangement, arriving after one year of college studying sculpture. He is now a partner in a small construction firm that specializes in construction of solar-heated homes that combine active water-collector systems with heat distribution in a cement slab floor, and passive solar features. He also builds masonry wood-stove systems for backup heat—systems that produce heat from a single fire over many hours, avoiding the need to keep a traditional wood stove burning continuously. He and his family live in a large home he has built on the same land he first came to, with continuing energy experimentation ongoing in his house. A 1-kW wind machine is the most recent addition to the primary PV system for electricity. He and two other local residents were the first in the area to purchase PV panels, doing so just as they became available. They not only got a price break on the group purchase but have cooperated both in home construction and in sharing their home power–development efforts. Early patterns of cooperation continue, both in the construction partnership's practice of including their customers in some element of each home's construction, and in their policy of not requiring any particular commitment of hours from their employees. They communicate and cooperate, but do not have fixed work requirements.

About a dozen home power adopters and enthusiasts, including the two described above, got together in 1990 to launch a renewable energy fair. Under the auspices of their Midwest Renewable Energy Association, it has become an annual event, attracting some 100 volunteers and 10,000 attendees, all on an annual budget of about $100,000. Held at the local county fairgrounds, this three-day event now includes over 100 workshops on every conceivable home power topic, as well as equipment displays by over 80 indoor and outdoor vendors. Attracting volunteer participation and attendance from across the country and internationally, this fair has become a major focus for home power efforts and has become a model for similar fairs in about five other states.

Among more recent arrivals in the community are Tehri and Roak Parker, who bought a formerly Amish house and added a PV array purchased at the area's energy fair. After working on the fair for several years, Tehri has become the director of the Midwest Renewable Energy Association, and her husband, with a law degree in environmental law, now works as a local public defender. The annual energy fair includes tours of several home power homes, with two of the three described here among the half dozen or so routinely made available.[9]

Home Power as Divergent Choice

On the face of it, home power is just another way of consuming. Rather than buying electicity from the utility company, paying whatever price is set by the market and by regulators, home power people buy the equipment and produce their own. If they are "smart shoppers," if they purchase the best equipment for the dollar and use it right, they might get their electricity cheaper than their neighbors. What's more, as "early adopters," home power users might get ahead of their neighbors, positioning themselves to take advantage of new products as they come on the market.

To the extent that the mainstream press covers home power, this is the typical slant. It is, however, a serious misreading of the motivations and sensibilities driving the home power movement. It may also be a serious misreading of the significance of the movement to society at large. From extensive ethnographic study over a period of more than ten years, I have come to the conclusion that home power is less "alternative consuming" and more "alternative living." It is a search for a new "form of life,"[10] one driven by concerns that cross the ideological spectrum and occur all along the income scale. Home power adopters attempt to carve out meaningful relations with technology, natural resources, place, and community. If they are rejectionist, it is largely with respect to the dominant patterns of consumption—that is, they reject the prevailing belief that happiness and fulfillment come from that which is purchased, from having a wide range of consumer choices, shopping wisely, and, inevitably, consuming more and more.

With respect to technology, home power folks are designers, pioneers of self-provisioning in a world that puts comfort and convenience among its highest values. Most home power adopters are not "technophiles" in the ordinary sense of the word. While many are technically inclined, especially those who become equipment suppliers, many others are not. Taken as a whole, the movement is not an energy technology equivalent of computer hackers or model rocket builders. The technology of home power is less instrumental and more an expression of a particular sense of one's place in the world—how one relates to others and to the environment.

The home power choice has a strong resource and environmental resonance. Home power people are generally highly motivated to shift from fossil fuels to renewable energy sources, as well as to reduce overall consumption and increase end-use efficiency. They see home power technologies as a means of expressing a different kind of relationship with the natural environment, one that embraces, for example, close monitoring of weather conditions to determine energy availability and hence the activities that draw on that energy.

Environmental values are also implicit in the siting of homes in relatively remote and attractive natural settings, a choice that may compel more travel and, thus, greater reliance on fossil fuels. But home power people are not, as a whole, commuters. Rather, their choice of place is both for environmental reasons (being "closer to nature," say) and for reasons of community. In fact, home power people often choose place and community before addressing the question of how to make a living. Place tends to remain firm even in the face of material—that is, financial—difficulties. Home power people tend to "put down roots," whether they live in an entire community of home power adopters or in a more conventional community setting where they are relatively isolated in terms of lifestyle. Either way, work patterns tend toward a diverse mix of activities, often including unpaid work like raising one's own food, maintaining one's own residence and vehicles, and procuring backup fuel by cutting firewood (often collectively with neighbors). In home power families, often only one parent has conventional paid employment and one takes on the home schooling of children.

Thus, a central theme in the home power movement is to reject dependence in one's relations with natural resources, technology, and work—patterns of dependency that are inherent in a contemporary consumerist lifestyle. The choices of home power people are largely incommensurate with conventional patterns, even including mainstream energy-saving programs. A solar energy program sponsored and organized by a utility company might appear to satisfy the needs of home power advocates. California's Sacramento Municipal Utility District is a case in point, where the utility designs photovoltaic power systems and installs them on individual homeowners' roofs. But the utility company provides what home power provides only in a strictly technical sense—electricity from the sun and the wind. What such programs miss is the development of key relationships with nature, with technology, and with each other—considerations that are at least as important for home power people.

The difference lies in the fact that the utility-company programs create relationships that are remote, impersonal, and contractual. The utilities are typically large and impersonal organizations, the only human contact being that between meter readers and home owners. While a utility "green energy" program would certainly ease adoption of certain technologies in a consumption context, it would fail to afford the kinds of engagement with the world that are central to the home power movement—that is, a more meaningful connection with both the material and the human. A utility company's "green" choice, while still better than the relationship implied in the conventional consumer's flip of a switch, is not the same as an individual's choice for independence from such distant, contractual relations. For those seeking such independence, even a drastic reduction in photovoltaic panel prices, leading to a boom in commercially installed systems, would not substitute for the satisfaction of designing and knowing one's own system and having a clearer sense of the resource and environmental consequences of one's own choices. Reducing distancing (chapter 5, this volume), or the extreme separation of production and consumption, is a prime motivation among home power users.

Adopters' Interests and a New Practice of Technology

One way of interpreting the home power movement is to distinguish "citizens" from "consumers." As designers they do not simply select from a predetermined menu of commodities available in the marketplace but actively participate in the shaping of the very technologies they rely on for essential services. Activated as agents of choice in fields far broader than those normally accessed by ordinary consumers, they have taken a kind of *ownership* of problems[11] beyond even the immediate questions of energy-resource depletion and environmental degradation. Acting directly and indirectly to reinvigorate "community" in directions reminiscent of seemingly outdated notions of citizenship, they help create "community" by joining with others to willingly "accept certain responsibilities for the common."[12]

With such nonconsumerist choice, such deliberate design and engagement, citizenship therefore emerges in conjunction with technique. Technologies "become woven into the texture of everyday existence." As Langdon Winner has written, "The devices, techniques, and systems we adopt shed their tool-like qualities to become part of our very humanity. In an important sense we become the beings who work on assembly lines, who talk on telephones, who do our figuring on pocket calculators, who eat processed foods, who clean our homes with powerful chemicals."[13] This "becoming" derives in part from how people procure the necessities and niceties of life, and, therefore, has significance beyond the energy itself. Technologies express and instantiate certain relationships among people and between people and the natural world.

In this sense, home power adopters are embracing a different and, for them, more attractive form of life. Their reward is *intrinsic* to participation in the design process, including what would normally be regarded as the added *burdens* of problem formulation, consideration of means and ends, and the physical challenges of actual implementation. The reward does not lie simply in the "consumption" of energy as a commodity. Home power, unlike consumerist choice, is not an attempt to "disburden," not a way of relieving the chooser of the demand for thought and effort. Rather, home power cultivates skill; it has an "orienting" effect in a person's life beyond its value as a mere commodity. Affording a "wealth of engagement" rather than an "affluence of consumption," it comes without a sense of sacrifice that would otherwise be implied in a conventional consumerist perspective,[14] where consumption, stimulated by marketing, seems to promise a desirable experience for every new acquisition. As that promise eludes fulfillment, a sense of longing results, creating yet more individual vulnerability to new acquisitions thrown up by the marketeers.[15] Through active design and hands-on engagement, home power participants step off the consumption treadmill, rejecting marketing-led design[16] and going beyond notions of "mimetic" desire in which people take their cues as to what is valuable by observing others' possessions.[17] Although working initially in relative isolation and without well-articulated models, home power adopters aim for enduring satisfaction, not the fleeting pleasures of continuous purchasing.

Ordinary consumption, it must be noted, flows from a set of market alternatives that are themselves shaped by the institutions of the marketplace, *not* simply or entirely by the prospective interests of citizens or consumers. Always implicated in the shaping of a product is the need, for example, of the producer to make a living through the mechanisms of the marketplace. The distortions that result range from things like delays in the first introduction of more efficient fluorescent lighting ascribable to electric power producers concerned about implied reductions in power use,[18] to the more flagrant abuse of consumers perpetrated over the years by the tobacco industry.[19] The accumulated effects of this sort of distortion may have carried us far from the real interests of ordinary people as advertising and other mechanisms have been employed in the service of particular patterns of consumption. What might otherwise be viewed as a wide range of consumption alternatives may actually suffer from a radically impoverished "one dimensionality."[20] As home power enthusiasts have taken direct control over elements of design or "technology shaping" ordinarily reserved for other market players, they have emerged with an ensemble of interrelated technological alternatives that suggest, in the context of such speculation, that market processes may simply have gotten, for them, intolerably far from interests or commitments they hold at a very fundamental level.

In this light, home power might be seen as a maturation of the 1960s response—one made possible over the span of decades of dogged effort only by the depth and indelibility of the alienation of that time. Home power people have not, in this view, been behaving in any recognizable way as mere "consumers." Working to recover a sense of agency and connectedness, and a sense of independent capacity, they are trying to recover essentials in the direct actions of living, essentials that had been lost through the delegations inherent in the act of consuming. If ordinary consumption exploits an ultimately illusory desire for control, appearing to offer "whatever the consumer wants," home power has been more concerned with recovering a more basic and perhaps more essential sense of being "equal to" the world: a sense that one can function effectively, on one's own behalf, in the world.[21]

Conclusion

It may be tempting to dismiss home power as a fringe movement, as a small and insignificant collection of people tinkering with collectors and windmills and generally rejecting the mainstream of modern life. But if the mainstream is associated with overconsumption, with patterns of production and consumption that threaten life-support systems, then it may be precisely in such movements that insights—maybe even models—can be found.

Lewis Mumford argued that some of the greatest epochal transformations of the past have been more a product of "a steady withdrawal of interest from the goods and practices of 'civilization,'" than of political shifts within a constant and universal pursuit of such goods and practices.[22] Writing more than three decades ago, Mumford noted that "though desertions and dropouts are still insignificant in quantity, something like a large-scale withdrawal and reversal [could] actually be in the offing."[23] Home power reaffirms the notion that people's needs and desires cannot always be satisfied on the consumption treadmill. It also tells us that our expectations and our institutions can in fact change to meet those needs and desires.

Hard questions remain about the accessibility of home power to larger numbers of people. On the one hand, the movement has seen dramatic growth. Several decades ago, the pioneers had to make an abrupt leap from the conventional and familiar (just order the power and flip a switch) to the unknown (how else can I provide the basic energy services I need?). To the extent the technologies were available, there were few precedents, let alone guidebooks and magazines, for home adoption.

Many of these early movers invested considerable effort and money only to fail. But the fact that so many pioneers have continued and others have followed is testament to the strength of their convictions within the movement. Today, the way has been cleared and opened to a significant degree as the home power alternative has been explored, tested, and developed. On the other hand, there is in home power both a rejection of modern patterns of production and consumption and an embrace of alternatives—a position that leaves little middle ground either institutionally or technologically.

The significance of home power is not measured primarily in the numbers of people in its ranks. In the final analysis, the home power movement is indicative of deeply seated and widely felt needs that are not currently satisfied in the marketplace. The movement may, in fact, be in some measure incommensurable with a framing of the world in terms of "market" concepts. There is cause for hope in the experience of the home power movement that material practices responsive to contemporary environmental concerns are possible and *without* a great sense of sacrifice. We may be able to forgo the dream of two cars (now, a van and a sport utility vehicle) in every garage without a sense of deprivation. To the degree that contemporary consumption patterns undermine community, environmental sustainability, situatedness, and a sense of orientation—indeed *meaning* in one's life—we may urgently need to ask what further opportunities, in addition to home power, may be latent in material and sociocultural organization. What prospect is there in the near term, for example, for alternative transportation or food systems analogous to home power? And how might such prospects be combined in new and yet more attractive alternatives for the future?

14

Conclusion: To Confront Consumption

Thomas Princen, Michael Maniates, and Ken Conca

We began this book with a conundrum. On the one hand, concern about consumption and its effects—ecological, social, and otherwise—is bubbling up all around. On the other, discussion of these problems in policy and academic circles is surprisingly impoverished. At risk, we have asserted, are the integrity of biophysical systems vital to all life and the habits of citizenship essential to democratic forms of governance. We have suggested that confronting consumption must begin with a willingness to ask uncomfortable questions, to look for answers via new categories of thought and argument. The environmental discourse of "sustainability" that has been so popular over the past couple of decades has served a useful purpose by building bridges between conservation and development and ecology and economics. But it encourages modes of thinking that neither question nor respond to the underlying forces driving the escalation of needs and desires around the world.

At one level, then, this book has been about encouraging alternative questions and ideas. Among them have been the social embeddedness of consumption, the chainlike character of resource-use decisions and power relations shaping consumption practices, the hidden forms of consuming embedded in these linked stages of economic activity, the dangers of distancing, individualization, and commoditization. At another level, though, this book has been about insisting that confronting consumption can be, and must become, a driving focus of contemporary environmental scholarship and activism—and that a different kind of environmental politics is needed, one that is at once individual and collective. This *politics of consumption* would necessarily engage people in a variety of roles: *scholars*, as builders of concepts for understanding consumption and its discontents, and as chroniclers of social experiments emerging in response to overconsumption and misconsumption; *activists*, as political catalysts pushing for change in public opinion and policy; *educators*, as communicators of scholarly and activist discoveries, and as agents of critical questioning of material growth and consumerism in industrial society; and *everyday citizens*, who, it should be clear, have the most to lose by failing to confront the consumption juggernaut and the most to gain by creatively responding to the consumption challenge. Our hope is that scholars, activists, educators, and everyday citizens will find the combination of theory and practice in this book intellectually productive and, perhaps, pragmatically inspirational.

As teachers, we often end our classroom lectures with "take-home" messages—core ideas or enduring lessons we want our students to remember long after the details of our lectures, much to our chagrin, have been forgotten. We advance here three such "take-home" messages, derived in part from the preceding chapters and in part from our ongoing efforts to push the thinking ahead. One is the overriding need to pose critical, "out-of-the-box" questions. The second is the pernicious idea of consumer sovereignty and its impact on policymaking and public perception. And third is the centrality of cautious consuming as an emergent organizing principle for economic decision making and everyday life.

Questioning "Outside the Box"

Confronting consumption begins with posing questions outside the mainstream, questions that make explicit what is being consumed and by whom. It is to ask about the centers of decision making that dominate resource extraction and waste-sink filling. It is to put "value subtracted"—the hidden costs incurred when impacts are diffuse or displaced in time and space—alongside "value added"—the tangible benefits that are so easily highlighted for short-term gain. It is to ask whether the dichotomies of supply/demand and production/consumption are the best lenses through which to view economic life. It is to ask what is lost when creeping commercialism treats more and more realms of life as if they were items to be marketed and used up.

Asking such questions calls attention to the seemingly natural processes of individualization, distancing, and commoditization that structure people's lives. Individualization is almost second nature to those of us reared in the cultural context of North American individualism. Children learn from an early age that it is what one does as individuals that counts most; by extension, if something goes wrong, the individual is to blame. To pin environmental degradation on individual choice rather than structural constraints seems perfectly sensible. Similarly, distancing is natural for generations of North Americans, accustomed as they are to long-distance transport of food and manufactured goods. Americans are largely ignorant of the historical processes—often violently coercive toward people as well as nature—that accompanied the making of integrated national markets as early as the nineteenth century.[1] To be sure, the costs were huge—the great cut-overs of temperate forests and the elimination of bison rangeland, for example—but they were always out of sight and out of mind. So when North Americans now think about their cornucopia of goods, it is to marvel at the variety and availability, not to question the sources and conditions under which they are produced. Without historical knowledge, and with the emergence of global production patterns of the sort dissected in chapter 6, abundance appears even more sensible, more "natural," and more cost-free.

Confronting consumption thus means raising thorny questions about this abundance rather than celebrating it or working on the margins to uncover the odd pollutant. It is to demystify widespread degradation in the midst of such abundance and pollution. It is to connect seemingly individual acts with collective structures, to juxtapose choice with constraint, abundance with scarcity, cleanliness with degradation, and citizenship with consumerism. It is to see that individualization and distancing follow logically from a social construction of commerce that celebrates the individual as consumer yet neglects the collective as rule maker, that celebrates the efficient and the growing yet finds it difficult to entertain the possibility of too much consumption, too much degradation, or too much cost displacement.

To confront consumption is also to put *consumerism* in its place—and to keep it there. Facile applications of consumerist reasoning invade ever more realms of seemingly nonmarket, noncommercial activity. For example, consumerist language has become a favorite of U.S. electoral politics, with negative impacts that are well known. Market analysts (political consultants) poll consumers (potential voters) to ascertain tastes so that the producers (political parties and consultants) can use the most advanced marketing techniques to sell their product (the candidate). Like the top-selling breakfast cereal or athletic shoe, the product that best responds to consumers' expressed preferences wins. Investors (political action committees, wealthy individuals, corporations, labor unions) invest (donate) scarce resources (large sums of money) in the market research and product placement so as to earn a return (legislative and administrative favors) later. With billions of dollars pouring into U.S. elections, gaining office becomes a contest of fundraising and advertising. It appears democratic because people vote. But real influence, both during and after the election, resides with those with the resources and information flows to shape "the message" (candidate plus script) and circumscribe the consumers' (voters') choices at the polls.

Another case of misplaced consumerist reasoning is the new language of the university and of education in general (an idiom the authors of this book know all too well). Production (teaching) must be streamlined (made efficient) with larger classes, advanced electronic services, and more flexible instructors (temporary, part-time, and low-paid), so as to maximize value (job skills) for the customer (student). To compete with other producers, courses must be packaged attractively (eye-catching titles, lots of visuals, cheery instructors) and be convenient (not too much writing or critical reflection; few, if any, low grades). As it turns out, such language does more than emulate the commercial world or allow administrators to impersonate corporate CEOs. Under the guise of producers responding to consumers, university life is fast shedding its highest ideals of liberal education and philosophical inquiry for business's lowest ideals of bottom-line obsession and short-term profit maximization. But there is little resistance because, as faculty, students, and staff routinely are told, budgets must be balanced and, after all, it's what the "consumers" of higher education desire.[2]

Ultimately, to confront consumption is to take a "consumption angle" on the challenges of overconsumption, an approach that dares to ask why people consume and whether they consume too much, that eschews the production angle, which, with new technologies and practices, might at best slow the pace of environmental degradation and social erosion. Working through the dimensions of the "consumption angle" (part I), an important source of rethinking is the numerous experiments taking place nearly everywhere, many small-scale and grassroots. We have chosen a few from North America not because they are representative, nor because we see them as "the answer." Rather, we selected these cases because they engage the consumption questions in provocative ways. These cases, together with the conceptual tools of earlier chapters, help one resist the temptation to interpret the consumption problem as "bad consuming," as misinformed decision making or greedy self-aggrandizing among a pampered elite. The concepts of individualization, distancing, and commoditization bring a structural dimension to the problem of overconsumption that deserve a more privileged place in contemporary thought, policymaking, and activism. What's more, these cases and concepts point toward a path of cautious consuming, a path that is at once respectful of limits and, yet, as the on-the-ground cases show, involve considerable innovation and personal risk taking.

Unmasking the Sovereign Consumer

Ideas matter in a society because they often go beyond the descriptive and analytical to the political. As they become the language of public discourse, they favor some interests and agendas while suppressing others; they simultaneously justify and absolve.

Consumer sovereignty is one such idea. It is central to mainstream understandings of the industrial political economy; it tends to justify processes that sustain the consumption treadmill and policymakers' obsession with growth—what turns out to be material growth, not necessarily growth in life quality. And it absolves nearly everyone of responsibility.[3] Conceived as an essential building block for an elegant theory (neoclassical economics) and highly useful for predicting price and output changes, consumer sovereignty has been appropriated by select commercial interests and their governmental representatives to justify all manner of resource exploitation.

The reasoning is seductively straightforward: industry only responds to consumers' wants and needs. A firm would be happy to produce fuel-efficient automobiles or wood from sustainably managed forests or electricity from windmills, but the demand isn't there. If the public really wanted cleaner production, more efficient use, and better management of natural resources, preferences would shift, and the marketplace would respond. Moreover, say the appropriators of consumer sovereignty, if individual consumer preferences become collectively destructive—if exploding preferences for SUVs and electronic gadgets and foreign travel, for instance, lead to undesirable levels of dependence on oil imports or increased risk of climate change—the problem is ethical, educational, and political, not commercial. Preferences among the mass of consumers, the argument goes, may be askew, but the corrections should occur in one's place of worship, in the school, or in the legislature, not in the factory or bank. To suggest that industry should make such corrections is to violate both private choice and public choice, two pillars of an open society and an efficient economy, indeed, of democracy itself.

Such reasoning might make sense if two core assumptions of the economic model—namely, complete information and insignificant externalities—were reasonably approximated. Yet several of the chapters in this book point to the woefully inadequate, information-impoverished, decontextualized character of consumer choice in the face of commoditization (chapter 4), distancing (chapter 5), trends in global distribution (chapter 6), political-economic patterns of past and present commodity chains (chapters 6 and 7), and underlying processes of individualization that divert attention away from the structural (chapter 3). From the perspective of long-term ecological and social sustainability, the model's assumptions are not even close. Even so, such reasoning might make practical sense if industry did indeed merely "respond" to consumer demand. A $170 billion annual bill for advertising in the United States alone (chapter 10), not to mention massive lobbying of governments at all levels, suggest otherwise.

Confronting consumption is thus to reject the "consumer knows best" construction intimately associated with the idea of consumer sovereignty and so popular among boosters of commercial expansion. It is to challenge those who would escape the responsibility of their production and deny their complicity in the massive manipulations of public perception and policymaking. It is to shine light on processes of taxation and subsidy that distort the prices that consumers see and limit their choice in the marketplace. This construction is only one rhetorical device to advance the interests of industrial expansion and dodge issues of impact, limits, and responsibility, however.

Another is the efficiency claim. This too has a long and respectable theoretical pedigree: the market—that combination of consumers choosing from a basket of goods to maximize personal utility and producers optimally mixing factors of production to maximize gains—is the most efficient means of allocating society's scarce material and human resources. What is not much talked about, however, are the highly subjective value choices underlying this ubiquitous efficiency claim. When introductory economics textbooks argue for free markets and consumer sovereignty in the service of societal efficiency, they have a very specific definition of efficiency in mind: a material outcome in which the existing distribution of income and power is accepted as given, and where everyone is made "at least as well off as before" without any redistribution of assets, power, or influence. This is an inherently conservative conceptual construct: it pushes decision makers toward growth mania (if redistribution is taboo, the only way to lift the poor from their plight is to "grow" the economy), it justifies a skewed distribution of income and power, and it too easily assumes crucial enabling conditions such as perfect consumer information and unlimited resources.

What, we wonder, if efficiency were understood differently—say, as a condition in which the economy generates the most human happiness over the long run while maintaining the resilience of environmental systems and the integrity of close, noncommercial, community networks essential to participatory democracy? Clearly, a set of policies might emerge that question prevailing distributions of power, privilege, and prestige. Instead, murky, conservative, timid, and mainstream ideas about a particular kind of efficiency quietly militate against such possibilities, sheltered by the unexamined aura of objectivity.

A third rhetorical device within consumer sovereignty is the mantra of low consumer prices. Political and business leaders frequently point out that Americans "demand" cheap gasoline for their cars and low-priced electricity for their homes. Some imply that American consumers *deserve* low prices. Few rationales are as effective in justifying a public expenditure or opposing a regulation. California's energy crisis of 2001, for example, was preceded by partial deregulation where wholesale prices would be competitive but end-use prices would be capped. Consumers, so went the argument, could not bear the burden of market reorganization, even if real costs increased. When those costs did increase, the state (that is, California taxpayers) spent billions of dollars to keep electricity prices low for its citizens (that is, for California electricity users). At the same time, the U.S. government rolled back efficiency standards for refrigerators and hot-water heaters because, as the industry complained, the increased prices (short-term appliance *purchase* prices, not long-term energy expenditures) would be a "cruel penalty" on consumers.[4] When President George W. Bush reneged in early 2001 on his campaign promise to reduce carbon emissions as called for in the Kyoto Protocol, the first reason given was that carbon dioxide was not a pollutant. The second reason, the one that people could take seriously, was that reducing emissions would raise consumer prices. When Microsoft, the software manufacturer, was accused of monopolistic practices, it took a position with a long history of success in U.S. antitrust cases: the true beneficiaries of Microsoft practices, however predatory and monopolistic they may seem to some, are the consumers: more choice and lower prices. When garbage is shipped from New York, Illinois, and Ontario to Michigan, state officials and waste-management companies explain that such transfers keep dumping fees low, meaning that roadside pickup of unlimited quantities of household trash remains cheap.

The list is endless but the logic is suspect. Certainly everyone wants to spend as little as possible, but only when there are no implications down the line. Low prices only make sense when all else is equal or when trade-offs are clear. I want lower gasoline prices *if* highway congestion doesn't increase; I want cheap electricity *if* brownouts remain unlikely; I want cheaper satellite TV *if* the further attraction of staying indoors and channel surfing doesn't further degrade the social networks of my neighborhood. That boosters of low prices can avoid spelling out such conditions and avoid being held accountable for such trade-offs is testimony in part to the low level of political discourse in the United States. But the mantra of low prices, along with the "consumer knows best" construction and the efficiency claim, also testify to the power of the idea of consumer sovereignty, an idea that, although a technical term conjured up by economists, suffuses much of contemporary policymaking.

The explanation for such power may lie in the aptly chosen term *sovereign*. Pharaohs, emperors, and kings and queens enjoy luxuries and conveniences that are widely recognized as the appropriate trappings of power. It's their right and their duty. What's more, such entitlements are recognized by everyone, from the exalted ruler to the commoner. It's the proper order of things. Now, it would seem, consumers have theirs: the right to maximize choice and the right to get the lowest possible price. Everyone from political leader to CEO to worker recognizes this essential rightness, however implicit it might be. It's the proper order of things. Political leadership is exercised when consumer entitlements are promoted through a new housing development, shopping center, or trade agreement. Citizenship is expressed when the sovereign buys a fuel-efficient car, sweatshop-free shoes, or ecofriendly cleansers.

Such sovereignty knows no bounds. Its imperialism is not geographic (although much of globalization relies on the idea) but commercial and extractive: the sovereign must have more and more goods at ever lower prices and its agents must scour the globe to find the inputs and waste sites to make it all possible. Of course, this grand entitlement scheme for consumption can persist only as long as its boosters can defer, displace, and obscure true costs and pressing trade-offs; distancing and individualization become essential processes in sustaining the scheme.

In the end, however, the idea of consumer sovereignty simply doesn't hold water. It is a myth convenient for those who would locate responsibility for social and environmental problems squarely on the backs of consumers. It makes the idea of unlimited economic growth appear both natural and inevitable. But consumers do not have perfect information and they are not insulated from the influence of marketers. Their consumption choices seem broad but are in fact tightly constrained (one chooses, in today's marketplace, between a red car and a blue one, not between an automobile-based transport system and a mass-transit based one, and this is no accident[5]). Their "rule" over the investment and production decisions of major corporations (decisions that are made months if not years before a product actually is launched) is limited at best.[6]

Consumer sovereignty as an idea depends on a complacent media, as well as an unquestioning academy and other institutions, to continually assert the idea. The media is a chief purveyor as it fuels supposedly efficient and costless growth. It accepts outright a logic that associates economic "strength" with consumer spending, even excess spending. "Even as the stock market fell and unemployment rose," a front-page article in the American paper-of-record proclaimed in early 2001, "consumers continue to spend more than they earned, increasing their debt levels to buy new homes, cars, and other items." As a result, "the economy showed surprising resilience ... extending the longest American expansion on record." Borrowing money from the bank and resources from future generations, it seems, strengthens the economy; personal liability aggregates to social asset. The only worry, so this popular appropriation of consumer sovereignty goes, is that these debt-ridden consumers won't do it long enough: "Economists and government officials remain concerned that consumers will rein in their spending in the coming months."[7]

To confront consumption is thus to launch a wholesale debunking of the myth of the sovereign consumer. It is to unmask such a beast to find a citizen, a community member, a defender of values that can never be reduced to market prices. It is to question the mainstream press's coverage of the economy, the environment, and, for that matter, the dominant but ultimately narrow and self-defeating idea of progress. It is to challenge the popular understanding of how the economy works and how it affects our everyday lives, an understanding that crosses the political spectrum.

Opportunities for mounting such challenges will expand as the real-world contradictions of a theoretical sovereignty mount as, inevitably, they will. Infinite expansion on a finite planet is impossible biophysically. Catering to the mythical sovereign—the insatiable consumer—will hit ecological and social walls, rendering such expansion impossible and such sovereignty illegitimate. The search for alternatives will intensify. What are now fringe movements and experiments—simple living and local currencies and home power, for instance—will undermine the "sovereign."

Cautious Consuming

Confronting consumption is to take seriously systemic problems of excess, to accept overconsumption as a real outcome in a political economy that can't ask when enough is enough. It is to resurrect seemingly outmoded concepts and norms such as thrift, frugality, self-reliance, simplicity, and stewardship, and put them in the context of ecological and social overshoot. It is to consider "cautious consuming" as an antidote to "exuberant producing."

When a powerhouse economy like California's experiences an energy crisis, to confront consumption is to scrutinize the easy assumption that the problem is a "supply shortage," or even inefficient use. It is to raise the very real possibility that the problem is actually too much consumption: more energy use than atmospheric waste sinks and dammed rivers can withstand, and more purchasing than can be justified by assumptions of individual or social welfare gain. When prices for meat or computers drop dramatically, it is to challenge the notion that "consumer benefits" outweigh the costs of factory hog production or ever more commoditized personal space. When concerned citizens protest forest clear-cuts along scenic highways or a power plant next door or a housing development on prime farmland, it is to pursue the protestations to their logical ends and ask: Is the problem solved when production shifts to distant shores while consumption of wood products and energy and open space proceed apace?

To consume cautiously is to defy commercial relations that always proceed apace. It is to challenge the ethical moorings of a political economy that knows no bounds, that acts as if widespread irreversible degradation and growing inequality can be addressed with yet more economic goods. For some, it is to create a politics of consumption by interjecting "consuming what and by whom" questions into mainstream policymaking, even when such efforts fall on deaf ears. For others it is to experiment with alternative means of material provisioning, with new economic and social relations. It is to downshift, to barter, to buy with local currencies, to grow one's own, all the while remaining acutely aware of and responsive to growing socioeconomic inequity. For still others it is to hold decision makers accountable when their policies make such choices difficult. When, for instance, the U.S. Food and Drug Administration wrote standards for organic food that included genetically modified organisms, irradiation, and fertilizer from municipal sludge, growers, retailers, and consumers rebelled, forcing a rewrite. In the end, the agency excised industrial agriculture's provisions.

Cautious consuming springs from the recognition that patterns of risky behavior permeate the global political economy. Recall, for example, chapters 2 and 5, in which Princen developed a logic of human separation from nature, a severing of negative ecological feedback, and the creating of positive social feedback all combining to heighten the risk of ecological overshoot. In chapter 3, Maniates argued that those with power can deflect responsibility and take risks by individualizing social and environmental problems. In chapter 4, Manno showed how commoditization depreciates social, psychological, and ecological values by reducing them to one value, market price. And in chapter 7, Clapp illustrated the risky character of displacement as a response to pollution where tighter regulations only create new bulges in an already squishy and unmanageable balloon.

Supplanting risky behavior with cautions consuming means neither sacrifice nor trading the good life for austerity and doing without, as part III of this book sought to show. Cautious consuming does mean being thoughtful, simplifying where necessary to achieve balance in the personal sphere, and accepting more responsibility for patterns of work and end use in the larger social sphere (chapters 9 and 11). What's more, as the chapters on movements for forest-products certification, local currency, and home power suggest, cautious consuming need not mean rejecting markets entirely or moving to purely local, zero-distance patterns of production and consumption. It does mean being willing to do several things: apply limits to commoditization, reduce distancing, locate responsibility and consequence among key decision makers, and nurture social and ecological capital. And it means organizing to oppose the forces of imprudent, cost-displacing activity, an example of such opposition being the work of Adbusters (chapter 10).

Ultimately, the challenge is not just to confront consumption but to transform the structures that sustain it. Individualization, commoditization, and distancing follow logically from a social construction of commerce that celebrates the individual as consumer yet neglects the collective as rule maker. Consequently, a central transformational challenge will be to connect seemingly individual acts with collective structures.

Donella Meadows—whose life as a systems analyst, public intellectual, activist, and community builder came to an end as we were completing this book—would say that the transformation begins by asking the tough questions, questions about unending material growth, about the purpose of economic activity, about what is being consumed, about who is benefiting at whose expense. But such questions, we must stress, are only "tough" in the context of a sovereign that knows no bounds, a political economy that worships material growth, an environmental ethic that confuses cleanup and amenity with long-term sustainable resource use. Such questions, we hope to have demonstrated in this book, are easy—or, at least, they become easier—with the consumption angle as an analytic lens and with cautious consuming as a mode of practice.

Notes

Chapter 1

1. The World Commission on Environment and Development, *Our Common Future* (New York: Oxford University Press, 1987) is credited with introducing the term into popular environmental discourse. On the rhetoric of sustainability at the Earth Summit, see Pratap Chatterjee and Matthias Finger, *The Earth Brokers: Power, Politics and World Development* (London: Routledge, 1994).

2. Juliet Schor, *The Overspent American: Upscaling, Downshifting, and the New Consumer* (New York: Basic Books, 1998); Joe Dominguez and Vicki Robin, *Your Money or Your Life: Transforming Your Relationship with Money and Achieving Financial Independence* (New York: Viking Penguin, 1999); John de Graaf and Vivia Boe, *Affluenza*, a coproduction of KCTS/Seattle and Oregon Public Broadcasting (Oley, PA: Bullfrog Films, 1997); Remarq: retrieved in September 2000 from the World Wide Web at http://www.remarq.com

3. Norman Myers, "Consumption: Challenge to Sustainable Development . . ." and Jeffrey R. Vincent and Theodore Panayotou, ". . . or Distraction?", *Science* vol. 276 no. 5309 (April 4, 1997): 53–57.

4. See for example Ted Bernard and Jora Young, *The Ecology of Hope: Communities Collaborate for Sustainability* (Gabriola Island, BC: New Society Publishers, 1997); Daniel Mazmanian and Michael Kraft, eds., *Toward Sustainable Communities: Transition and Transformations in Environmental Policy* (Cambridge, MA: MIT Press, 1999); Thomas Prugh, Robert Costanza, and Herman Daly, *The Local Politics of Global Sustainability* (Washington, DC: Island Press, 2000); William Shutkin, *The Land That Could Be: Environmentalism and Democracy in the Twenty-First Century* (Cambridge, MA: MIT Press, 2000).

5. See Alice Hubbard and Clay Fong, *Community Energy Workbook: A Guide to Building a Sustainable Economy* (Snowmass, CO: Rocky Mountain Institute, 1995); Mark Roseland (with Maureen Cureton and Heather Wornell), *Toward Sustainable Communities: Resources for Citizens and Their Governments* (Stony Creek, NY: New Society Publishers, 1998); Wolfgang Sachs, Reinhard Loske, Manfred Linz, and Timothy Nevill, *Greening the North: A Post-industrial Blueprint for Ecology and Equity* (London: Zed Books, 1998); Michael Shuman,

Going Local: Creating Self-Reliant Communities in a Global Age (New York: Free Press, 1998).

6. See for example Robert Kates, "Population and Consumption: What We Know, What We Need to Know," *Environment* vol. 42 no. 3 (April 2000): 10–19; John Holdren, "Environmental Degradation: Population, Affluence, Technology, and Sociopolitical Factors," *Environment* vol. 42 no. 6 (July/August 2000): 4–5.

7. See for example United Nations Environment Program, *Global Environment Outlook 2000* (London: Earthscan Publications, 1999) p. 2, which argues that "a ten-fold reduction in resource consumption in the industrialized countries is a necessary long-term target if adequate resources are to be released for the needs of developing countries," or United Nations Development Program, *Human Development Report 1998* (New York: Oxford University Press, 1998) back cover, which asserts that long-term prosperity hinges on "forg[ing] consumption patterns that are more environmentally friendly, more socially equitable, that meet basic needs of all and that protect consumer health and safety."

8. International Institute for Sustainable Development, *Developing Ideas*, issue 16 (January/February 1999).

9. See for example Daniel Bell, "The Public Household—On 'Fiscal Sociology' and the Liberal Society," *Public Interest* no. 37 (Fall 1974): 29–68. In this classic work, Bell characterizes growth as a powerful "political solvent."

10. A striking exception is the work of William Ophuls. See William Ophuls and A. Stephen Boyan, Jr., *Ecology and the Politics of Scarcity Revisited: The Unraveling of the American Dream* (New York: Freeman, 1992), which explicitly challenges this disciplinary bias, and William Ophuls, *Requiem for Modern Politics: The Tragedy of the Enlightenment and the Challenge of the New Millennium* (Boulder, CO: Westview Press, 1997). The discipline may be showing renewed interest in raising hard questions about the corrosive effect of consumerism, commoditization, and growth. See for example Robert Putnam, *Bowling Alone: The Collapse and Revival of American Community* (New York: Simon and Schuster, 2000); Michael Sandel, *Democracy's Discontent* (Cambridge, MA: Belknap Press, 1996).

11. See for example Daniel Miller, ed., *Acknowledging Consumption: A Review of New Studies* (London: Routledge, 1995); Alan Warde, "Notes on the Relationship between Production and Consumption," in Roger Burrows and Catherine Marsh, eds., *Consumption and Class: Divisions and Change* (New York: St. Martin's Press, 1992), pp. 15–31; Richard Wilk, "Emulation, Imitation, and Global Consumerism," *Organization and Environment* vol. 11 no. 3 (September 1998): 314–333.

12. Raymond De Young, "Some Psychological Aspects of a Reduced Consumption Lifestyle: The Role of Intrinsic Satisfaction and Competence," *Environment and Behavior* vol. 28 no. 3 (May 1996): 358–409; Guliz Ger, "Human Development and Humane Consumption: Well-Being Beyond the Good Life," *Journal of Public Policy and Marketing*, vol. 16 no. 1 (May 1997): 110–125; Marsha L. Richins, "Valuing Things: The Public and Private Meanings of Possessions,"

Journal of Consumer Research vol. 21 no. 3 (December 1994): 504–521; Tibor Scitovsky, *The Joyless Economy: The Psychology of Human Satisfaction* (New York: Oxford University Press, 1976).

13. A notable exception is Paul Stern, Tom Dietz, Vernon Ruttan, Robert Socolow, and James Sweeney, eds., *Environmentally Significant Consumption: Research Directions* (Washington, DC: National Academy Press, 1997).

14. For an important exception to this tendency, see Fred Gale and R. Michael M'Gonigle, eds., *Nature, Production, Power: Towards an Ecological Political Economy* (Cheltenham, UK: Edward Elgar, 2000).

15. See for example Amitai Etzioni, "Voluntary Simplicity: Characterization, Select Psychological Implications, and Societal Consequences," *Journal of Economic Psychology* vol. 19 no. 5 (October 1998): 619–643; Harwood Group, *Yearning for Balance: Views of Americans on Consumption, Materialism, and the Environment* (Takoma Park, MD: Merck Family Fund, 1995); Roger Rosenblatt, ed., *Consuming Desires: Consumption, Culture, and the Pursuit of Happiness* (Washington, DC: Island Press, 1999); Juliet Schor, *The Overspent American: Upscaling, Downshifting, and the New Consumer* (New York: Basic Books, 1998); Robert Frank, *Luxury Fever: Why Money Fails to Satisfy in an Era of Excess* (New York: Free Press, 1999). For counterarguments, see Stanley Lebergott, *Pursuing Happiness: American Consumers in the Twentieth Century* (Princeton, NJ: Princeton University Press, 1993); James Twitchell, *Lead Us into Temptation* (New York: Columbia University Press, 1999).

Chapter 2

1. Economics, curiously enough, is the only discipline that has spawned efforts (albeit well out of the mainstream discipline) at biophysical grounding. See for example Nicholas Georgescu-Roegen, "Energy and Economic Myths"; Kenneth E. Boulding, "Spaceship Earth"; and Herman E. Daly, "Steady-State Economy"; all in Herman E. Daly and Kenneth N. Townsend, eds., *Valuing the Earth: Economics, Ecology, Ethics* (Cambridge, MA: MIT Press, 1993). For attempts within an institutionalist tradition, see Robert Costanza, Bobbi S. Low, Elinor Ostrom, and James Wilson, eds., *Institutions, Ecosystems, and Sustainability* (Boca Raton, FL: Lewis, 2001); Thomas Princen, "From Property Regime to International Regime: An Ecosystem Perspective," *Global Governance* vol. 4 no. 4 (October–November 1998): 395–413.

2. William Leach, *Land of Desire: Merchants, Power, and the Rise of a New American Culture* (New York: Pantheon Books, 1993).

3. Herman E. Daly, *Beyond Growth: The Economics of Sustainable Development* (Boston: Beacon Press, 1996).

4. John H. Adams, "Oilgate in the Arctic: Message from the President," *Amicus Journal* vol. 20 no. 4 (Winter 1999): 2.

5. Daniel Miller, ed., *Acknowledging Consumption: A Review of New Studies* (London: Routledge, 1995); Richard Wilk, "Emulation, Imitation, and Global

Consumerism," *Organization and Environment* vol. 11 no. 3 (September 1998): 314–333; Leach, *Land of Desire*.

6. This neglect of the externalities of consumption does have its exceptions. Within economics, see Herman E. Daly, *Beyond Growth*, and Juliet Schor, "A New Economic Critique of Consumer Society," in David Crocker and Toby Linden, eds., *Ethics of Consumption: The Good Life, Justice, and Global Stewardship* (Lanham, MD: Rowman & Littlefield, 1997). Within anthropology, see Roy Rappoport, "Disorders of Our Own: A Conclusion," in Shepard Forman, ed., *Diagnosing America: Anthropology and Public Engagement* (Ann Arbor: University of Michigan Press, 1994).

7. For an extended critique of the production perspective including that adopted by many environmentalists, as well as an explanation of why pursuing a research agenda on consumption and environment engenders resistance, see Thomas Princen, "Consumption and Environment: Some Conceptual Issues," *Ecological Economics* vol. 31 no. 3 (December 1999): 347–363.

8. Paul Stern, Thomas Dietz, Vernon W. Ruttan, Robert Socolow, and James L. Sweeney, eds., *Environmentally Significant Consumption: Research Directions* (Washington, DC: National Academy Press, 1997).

9. Bruce Nordman, "Celebrating Consumption." Paper, Lawrence Berkeley Laboratory, 90-4000, Berkeley, CA, 10 pp., 1995. Available from the World Wide Web at http://eetd.LBL.gov/EA/Buildings/BNordman/C/consmain.html

10. Cutler J. Cleveland, Robert Costanza, Charles Hall, and R. Kaufman, "Energy and the U.S. Economy: A Biophysical Perspective," *Science*, vol. 225 no. 4665 (August 31, 1984): 890–897; Lee J. Schipper, "Carbon Emissions from Travel in the OECD Countries," in Stern et al., *Environmentally Significant Consumption*.

11. Klaas Jan Noorman and Ton Schoot Uiterkamp, eds., *Green Households? Domestic Consumers, Environment, and Sustainability* (London: Earthscan, 1998).

12. Greg A. Keoleian and Dan Menerey, "Sustainable Development by Design: Review of Life Cycle Design and Related Approaches," *Journal of the Air and Waste Management Association* vol. 44 no. 5 (1994): 645–668; Thomas E. Graedel and Braden R. Allenby, *Industrial Ecology* (Englewood Cliffs, NJ: Prentice Hall, 1995).

13. Marsha L. Richins, "Valuing Things: The Public and Private Meanings of Possessions," *Journal of Consumer Research* 21 (1994): 504–521; Aaron Ahuvia and Nancy Wong, "Three Types of Materialism: Their Relationship and Origin." Paper, University of Michigan Business School, Ann Arbor, April 13, 1997; Guliz Ger, "Human Development and Humane Consumption: Well-Being Beyond the 'Good Life,'" *Journal of Public Policy and Marketing* vol. 16 no. 1 (Spring 1997): 110–125.

14. Raymond De Young, "Some Psychological Aspects of Living Lightly: Desired Lifestyle Patterns and Conservation Behavior," *Journal of Environment Systems* vol. 20 no. 3 (1990–1991): 215–227.

15. Ronald Inglehart and Paul Abramson, "Economic Security and Value Change," *American Political Science Review* vol. 88 no. 2 (June 1994): 336–354; Frank Andrews and Stephen Withey, *Social Indicators of Well-Being: American Perceptions of Life Quality* (New York: Plenum, 1976).

16. Tibor Scitovsky, *The Joyless Economy: The Psychology of Human Satisfaction* (Oxford: Oxford University Press, [1976] 1992); Schor, "A New Economic Critique of Consumer Society."

17. Robert Costanza, ed., *Ecological Economics: The Science and Management of Sustainability* (New York: Columbia University Press, 1991); Daly and Townsend, eds., *Valuing the Earth*.

18. James J. Kay, "A Nonequilibrium Thermodynamic Framework for Discussing Ecosystem Integrity," *Environmental Management* vol. 15 no. 4 (1991): 483–495; Robert E. Ulanowicz, *Ecology: The Ascendent Perspective* (New York: Columbia University Press, 1997); Costanza et al., *Institutions, Ecosystems, and Sustainability*.

19. To characterize cultivation as degrading is not to judge it as wrong. I use "degrade" primarily in the thermodynamic sense of increased entropy but also in the ecological sense of decreased autonomous functioning over long periods of time. Thus, a corn field may generate more calories than its grassland predecessor but it does so only with continuous external inputs. It likely operates at a net energy loss and without the resilience of a less "productive" yet self-organizing system.

Also, this treatment is not to suggest that there is no value in cultivation. The consumption angle on cultivation merely directs analytic attention to *degradation* and *irreversibility* in a way that the prevailing perspective—the production angle —does not do, or does so only as an add-on where value added is the focus and environmental impacts are unfortunate side effects that can be cleaned up if actors have the funds, the interest, and the political will.

20. John S. Dryzek, *Rational Ecology: Environment and Political Economy* (New York: Blackwell, 1987); Princen, chapter 5, this volume.

21. Georgescu-Roegen, "Energy and Economic Myths"; Boulding, "Spaceship Earth"; and Daly, "Steady-State Economy"; all in Daly and Townsend, eds., *Valuing the Earth*.

22. Bobbi S. Low and Joel T. Heinen, "Population, Resources, and Environment: Implications of Human Behavioral Ecology for Conservation," *Population and Environment* vol. 15 no. 1 (September 1993): 7–41.

23. Raymond De Young, "Some Psychological Aspects of Reduced Consumption Behavior: The Role of Intrinsic Satisfaction and Competence Motivation," *Environment and Behavior* vol. 28 no. 3 (May 1996): 358–409.

24. This scenario, although highly simplified to make the argument, is not unlike that which occurred in the great cutovers of North America (William Cronon, *Nature's Metropolis: Chicago and the Great West* (New York: Norton, 1991)) or that are occurring currently in South America and Southeast Asia (Nancy Peluso, "Coercing Conservation: The Politics of State Resource Control," in Ronnie

Lipshcutz and Ken Conca, eds., *The State and Social Power in Global Environmental Politics* (New York: Columbia University Press, 1993).

25. More efficient use of a tree may appear to be a logical response to increasing demand. Certainly getting more usable wood per tree would, all else being equal, accommodate at least some of the excess demand. But, in general, such an efficiency always makes sense, regardless of demand. The issue raised here is what a producer must do outside of production efficiencies to deal with excess demand and still ensure long-term production from the resource.

26. Robert U. Ayres, "Limits to the Growth Paradigm," *Ecological Economics* vol. 19 no. 2 (November 1996): 117–134; Daly, *Beyond Growth*; Scitovsky, *The Joyless Economy*; Fred Hirsh, *Social Limits to Growth* (Cambridge, MA: Harvard University Press, [1976] 1995).

27. Princen, chapter 5, this volume.

28. For evidence of the role credit plays in creating excess demand for commodities, see Fred Gale, *The Tropical Timber Trade Regime* (Basingstoke, UK and New York: Macmillan and St. Martin's Presses, 1998).

29. James R. McGoodwin, *Crisis in the World's Fisheries: People, Problems, and Policies* (Stanford, CA: Stanford University Press, 1990); Marcus Colchester and Larry Lohman, eds., *The Struggle for Land and the Fate of the Forests* (London: Zed Books, 1993); Janis B. Alcorn and Victor M. Toledo, "The Role of Tenurial Shells in Ecological Sustainability: Property Rights and Natural Resource Management in Mexico," in Susan Hanna and Mohan Munasinghe, eds., *Property Rights in a Social and Ecological Context: Case Studies and Design Applications* (Washington, DC: Beijer International Institute of Ecological Economics and the World Bank, 1995).

30. Some evidence does exist that extractors who attempt to maximize their long-term economic security rather than respond to extant demand pursue strategies of diversified production. When either demand or the resource declines, they shift to other pursuits. Fishers in the Norwegian Arctic and many independent farmers follow this model: Svein Jentoft and Trond Kristoffersen, "Fishermen's Co-Management: The Case of the Lofoten Fishery," *Human Organization* vol. 48 no. 4 (Winter 1989): 355–365; Tracy Clunies-Ross and Nicholas Hildyard, "The Politics of Industrial Agriculture," *The Ecologist* vol. 22 no. 2 (March/April 1992): 65–71; McGoodwin, *Crisis in the World's Fisheries*, 1990. To my knowledge, however, no systematic research has been done on such work strategies and their impact on natural resources.

31. Empirical support for restraint does exist, especially in the common-property literature. See Daniel W. Bromley, ed., *Making the Commons Work: Theory, Practice, and Policy* (San Francisco: Institute for Contemporary Studies Press, 1992); Elinor Ostrom, *Governing the Commons: The Evolution of Institutions for Collective Action* (Cambridge, England: Cambridge University Press, 1990); James M. Acheson and James A. Wilson, "Order Out of Chaos: The Case of Parametric Fisheries Management," *American Anthropologist*, vol. 98 no. 3 (September 1996): 579–594. Other instances in private property are beginning to develop. See, for example, the cases being developed by the MacArthur Founda-

tion's Sustainable Forestry program. On resistance, see Madhav Gadgil and Ramachandra Guha, *This Fissured Land: An Ecological History of India* (Berkeley: University of California Press, 1992); Peluso, "Coercing Conservation"; and special sections of *The Ecologist.*

Chapter 3

I thank Thomas Princen, Benjamin Slote, Richard Bowden, and Brian Hill for their comments on earlier versions of this chapter.

1. Theodore Seuss Geisel (Dr. Seuss), *The Lorax* (New York: Random House, 1971).

2. Saul Alinksy, *Reveille for Radicals* (New York: Vintage Books, 1969), p. 94.

3. Jonathan Larsen, "Holy Seuss!", *Amicus Journal* vol. 20 no. 2 (Summer 1998), p. 39.

4. See in particular the "Introduction: We Live Through Institutions" in Robert Bellah, Richard Madsen, William Sullivan, Ann Swidler, and Steven Tipton, *The Good Society* (New York: Alfred A. Knopf, 1991).

5. For example, John Tierney, "Recycling Is Garbage," *New York Times Magazine* (June 30, 1996): 24–30, 48–53.

6. Green Mountain Energy (greenmountain.com), for example, aggressively markets "clean" and "green" energy at premium prices to consumers in states with deregulated electricity markets. Green Mountain urges customers to become part of a "renewables revolution" by buying its "green" electricity, but critics argue that Green Mountain is only repackaging existing power (principally hydroelectricity) and that only a small portion of its electricity comes from new renewables, predominantly from polluting combustion sources like biomass combustion and landfill gas (see http://www.boycottgreenmountain.com/). Green Mountain, which has entered into collaboration with BP Petroleum and Amoco, blunts such criticism by pointing to ambitious plans to build additional renewable energy capacity and entering into alliances with major environmental groups like Environmental Defense.

7. See for example "Plant Trees in Dr. Seuss's Honor," *Instructor: Primary Edition* vol. 108 no. 1 (August 1998): 9.

8. Mark Dowie, *Losing Ground: American Environmentalism at the Close of the Twentieth Century* (Cambridge, MA: MIT Press, 1996).

9. Real Goods (at http://www.realgoods.com) offers an example: it markets cute contraptions, upscale consumer goods, and renewable energy systems that it says promote "the transition to a sustainable society." But one strains to imagine its gadget-heavy catalog (solar-powered radios, reconfigured showerheads, and "healthy, recycled, tasteful furnishings for your natural home") as anything approaching "anticonsumerist" or "pro-frugality."

10. The Environmental Defense Fund has since undergone a name change to "Environmental Defense."

11. Peter Montague, "1996 Review: More Straight Talk," *Rachel's Environmental and Health Weekly* (December 19, 1996, no. 525), at http://www.rachel.org

12. At work here is a fundamental suspicion of collective action and a shift toward less ambiguous, "quick-payback" community-service and "service-learning" initiatives. On students' suspicion of politics as conventionally practiced, see Ted Halstead, "A Politics for Generation X," *Atlantic Monthly*, vol. 284 no. 2 (August 1999): 33–42; Paul Loeb, *Generation at the Crossroads* (New Brunswick, NJ: Rutgers University Press, 1994); Rene Sanchez, "College Freshmen Have the Blahs, Survey Indicates," *Washington Post* (January 12, 1998): 1. On the attractiveness of community service and service learning to students yearning to make a difference, see Halstead, "A Politics for Generation X," and Janet Eyler and Dwight Giles, *Where's the Learning in Service Learning?* (San Francisco: Jossey-Bass, 1999). For a mild corrective to the claim that undergraduates are predisposed against collective activism for institutional reform, see Arthur Levine and Jeanette S. Cureton, *When Hope and Fear Collide: A Portrait of Today's College Student* (San Francisco: Jossey-Bass, 1998). For a stronger counterargument see Liza Featherstone, "The New Student Movement," *The Nation*, vol. 270 no. 19 (May 15, 2000): 11–15.

13. The words to "The Pollution Solution" go like this:

All of this mess on the ocean floor could have been avoided if we thought about it more. 'Cuz sometimes folks are lazy, and sometimes we forget that if you don't pick up your garbage, this is what we get. And I think there's something that we can do, and the answer's in a song I'll sing for you. CHORUS: The solution to pollution is to think before you act. Re-use, recycle, remember to say, it's a beautiful world, let's keep it that way. Here's a bottle. That's made of glass. Here's a can. That's made of tin. Here's a box made from cardboard. We can use it all again. (Repeat Chorus.) Metal, paper, plastic, a whole lot of stuff. Save it, don't toss it, the earth has had enough! We're all in this together, so let's all lend a hand and recycle all we can. (Repeat Chorus.) Just stop before you drop it. Think before you act. The earth is not a garbage can, and that's a fact. We're all in this together, so let's all lend a hand and recycle all we can. (Repeat Chorus.) It's a beautiful world, let's all keep it that way. Re-use, recycle, remember to say, it's a beautiful world, let's keep it that way.

14. See The Earthworks Group, *50 Simple Things You Can Do to Save the Earth* (Berkeley, CA: Earthworks Press, 1989). The failings of a "50 Simple Things" approach to social change are discussed by J. Robert Hunter, *Simple Things Won't Save the Earth* (Austin: University of Texas Press, 1997). Nevertheless, guidebooks to "easy" social change through small modifications in consumer choice continue to see publication. See, for example, Judith Getis, *You Can Make a Difference*, 2nd ed. (Boston: WCB McGraw-Hill, 1999), which like so many such guides appeals to the reader's sense of political disempowerment.

15. For more on the distinction between consumer and citizen within the context of environmental struggles and processes of consumerism and commodification see Daniel Coleman, *Ecopolitics: Building a Green Society* (New Brunswick, NJ: Rutgers University Press, 1994). An analysis of the forces driving "environmental

consumerism" as the primary popular response to a perceived environmental crisis is provided by Toby Smith, *The Myth of Green Marketing: Tending our Goats at the Edge of Apocalypse* (Toronto: University of Toronto Press, 1998).

16. For a lively review of the commodification of dissent, see Thomas Frank, *The Conquest of Cool* (Chicago: University of Chicago Press, 1997) and Thomas Frank and Matt Weiland, eds., *Commodify Your Dissent: Salvos from the Baffler* (New York: Norton, 1997).

17. Benjamin Barber, "Democracy at Risk: American Culture in a Global Culture," *World Policy Journal*, vol. 15 no. 2 (Summer 1998): 29–41; John Freie, *Counterfeit Community: The Exploitation of Our Longing for Connectedness* (Lanham, MD: Rowman & Littlefield, 1998); Christopher Lasch, *The Revolt of the Elites and the Betrayal of Democracy* (New York: Norton, 1995); Richard Norgaard, *Development Betrayed* (New York: Routledge, 1994); Robert Putnam, *Bowling Alone: The Collapse and Revival of American Community* (New York: Simon and Schuster, 2000); Michael Sandel, *Democracy's Discontent* (Cambridge, MA: Belknap Press, 1996).

18. Murray Bookchin, "Death of a Small Planet," *The Progressive* (August 1989), p. 22, emphasis in the original.

19. Paul Hawken, *The Ecology of Commerce* (New York: Harper Business, 1993), pp. xii, emphasis added.

20. A good entry point into the discussion of the "Reagan years" as they relate to a growing distrust of government and declining faith in individual capacity to "make a difference" is Garry Wills, "It's His Party," *New York Times Magazine* (August 11, 1996): 31–37, 52–59.

21. For example, Mark Dowie, *Losing Ground*; and Robert Gottlieb, *Forcing the Spring: The Transformation of the American Environmental Movement* (Washington, DC: Island Press, 1995).

22. For instance, Kirtpatrick Sale, *The Green Revolution: The American Environment Movement 1962–1992* (New York: Hill and Wang, 1993); Robert Mitchell, Angela G. Mertig, and Riley E. Dunlap, "Twenty Years of Environmental Mobilization: Trends among National Environmental Organizations," in Riley E. Dunlap and Angela G. Mertig, eds., *American Environmentalism: The U.S. Environmental Movement, 1970–1990* (Philadelphia: Taylor and Francis, 1992), pp. 11–26.

23. Especially useful is Matthew Cahn's *Environmental Deceptions: The Tension between Liberalism and Environmental Policymaking in the United States* (Albany: State University of New York Press, 1995).

24. Paul Wapner, "Toward a Meaningful Ecological Politics," *Tikkun* vol. 11 no. 3 (May/June 1996): 21–22, emphasis in the original.

25. Samuel P. Hays, *Conservation and the Gospel of Efficiency* (New York: Atheneum, 1980).

26. *Time*, for example, ran a five-part "Visions of the 21st Century" series covering the following topics: Health and the Environment (vol. 154 no. 19,

November 8, 1999); How We Will Live (vol. 155 no. 7, February 21, 2000); Science (vol. 155 no. 14, April 10, 2000); Our World, Our Work (vol. 155 no. 21, May 22, 2000); and Technology (vol. 155 no. 25, June 19, 2000). Each delivered a large dose of technological utopianism.

27. Langdon Winner, *The Whale and the Reactor* (Chicago: University of Chicago Press, 1986), pp. 79–80.

28. See for example Robert Reich, *The Work of Nations* (New York: Knopf, 1991).

29. From 1997 interviews by the author with individuals who worked as "bottle bill" activists in Colorado and California during the late 1970s and early 1980s (Ross Pumfrey and William Shireman, who led the California initiative process at different times, and Terrence Bensel, an activist in Colorado). Simon Fairlie provides a similar account for the United Kingdom in "Long Distance, Short Life: Why Big Business Favours Recycling," *The Ecologist* vol. 22 no. 5 (September/October 1992): 276–283.

30. Pratap Chatterjee and Matthias Finger, *The Earth Brokers* (New York: Routledge, 1994).

31. Gustavo Esteva and Madhu Suri Prakash, "From Global to Local Thinking," *The Ecologist* vol. 24 no. 5 (September/October 1994): 162–163.

32. These myths (e.g., "public life involves ugly conflict," "power is a dirty word," or "public life is what someone else—a celebrity or big shot—has") are well articulated by Frances Moore Lappé and Paul DuBois, *The Quickening of America* (San Francisco: Jossey-Bass, 1994).

33. See for example Robert Kates, "Population and Consumption: What We Know, What We Need to Know," *Environment* vol. 42 no. 3 (April 2000): 10–19; John Holdren, "Environmental Degradation: Population, Affluence, Technology, and Sociopolitical Factors," *Environment* vol. 42 no. 6 (July/August 2000): 4–5.

34. Robert Costanza and Laura Cornwell, "The 4P Approach to Dealing with Scientific Uncertainty," *Environment* vol. 34 no. 9 (November 1992): 12–20, 42.

35. Donella Meadows, "Who Causes Environmental Pollution?", *International Society of Ecological Economics Newsletter* (July 1995): 1, 8.

36. United Nations Development Programme, *Human Development Report 1998* (New York: Oxford University Press, 1998).

37. Clifford Geertz, "Plenary Session," *Daedalus* (A World to Make: Development in Perspective) vol. 118 no. 1 (Winter 1989): 238.

38. Jack Beatty, "The Year of Talking Radically," *Atlantic Monthly* vol. 277 no. 6 (June 1996): 20.

39. Though only to a point. Pronounced or continuous workplace frustration can, under some conditions, lead to workplace withdrawal and radical frugality, as chapter 9 in this book explains.

40. For discussions of the connections between workplace satisfaction and inclination toward consumerist behaviors, see Murray Bookchin, *Toward an Eco-*

logical Society (Montreal: Black Rose Books, 1980); André Gorz, *Ecology as Politics* (Boston: South End Press, 1980); Thomas Princen, "Consumption, Work, and Ecological Constraint," paper presented at the annual meeting of the International Studies Association, 1997; Jeremy Rifkin, *The End of Work* (New York: Putnam, 1995); Juliet Schor, *The Overspent American: Upscaling, Downshifting, and the New Consumer* (New York: HarperCollins, 1999); Juliet Schor, Joshua Cohen (ed.), and Joel Rogers (ed.), *Do Americans Shop Too Much?* (Boston: Beacon Press, 2000). For an illustration of the environmentalist characterization of the "cultivation of needs," see Alan Durning, *How Much Is Enough?* (New York: Norton, 1992).

41. See for example David Roodman, *Paying the Piper: Subsidies, Politics, and the Environment*, Worldwatch Paper 133 (Washington DC: Worldwatch Institute, 1996).

42. A question driven home by the discussion on renewable energy options in Nancy Cole and P. J. Skerrett, *Renewables Are Ready* (White River Junction, VT: Chelsea Green Publishing Company, 1995).

43. See for example John K. Galbraith, *The New Industrial State*, 4th ed. (Boston: Houghton Mifflin, 1985); David Noble, *Forces of Production: A Social History of Industrial Automation* (New York: Oxford University Press, 1986); Langdon Winner, *Autonomous Technology: Technics-out-of-Control as a Theme in Political Thought* (Cambridge, MA: MIT Press, 1977).

44. See for example Richard Sclove, Madeleine Scammell, and Breena Holland, *Community-Based Research in the United States: An Introductory Reconnaissance Including Twelve Organizational Case Studies and Comparison with the Dutch Science Shops and with the Mainstream American Research System* (Amherst, MA: Loka Institute, July 1998); Langdon Winner, "Citizen Virtues in a Technological Order," in Andrew Feenberg and Alastair Hannay, eds., *The Politics of Knowledge* (Bloomington: University of Indiana Press, 1995).

45. See Paul Wapner, "Toward a Meaningful Ecological Politics."

46. Karen Litfin, *Ozone Discourses* (New York: Columbia University Press, 1994).

47. Susan Griffin, "To Love the Marigold: The Politics of Imagination," *Whole Earth Review* no. 89 (Spring 1996): 65, 67.

48. Saul Alinsky, *Reveille for Radicals*; Theodore Roszak, Mary E. Gomes, and Allen D. Kanner, *Ecopsychology* (San Francisco: Sierra Club Books, 1995).

49. Christopher Plant and Judith Plant, eds., *Green Business: Hope or Hoax?* (Philadelphia: New Society Publishers, 1991); Toby Smith, *The Myth of Green Marketing*. Green-marketing consultants are not shy about trumpeting the profits awaiting firms that recast their product lines as environmentally beneficial. See Patrick Carson and Julia Moulden, *Green Is Gold: Business Talking to Business about the Environmental Revolution* (New York: Harper Business, 1991); Jacquelyn Ottman, *Green Marketing: Opportunities for Innovation in the New Marketing Age*, 2nd ed. (Lincolnwood, IL: NTC/Contemporary Publishing Company, 1997); Walter Coddington, *Environmental Marketing: Positive Strategies for Reaching the Green Consumer* (New York: McGraw-Hill, 1993).

Chapter 4

1. Quantifying consumption efficiency will need to wait for future research. A rough cut at a measure of consumption efficiency at the country level may be made by taking some measure such as the Index of Sustainable Economic Welfare, or the Human Development Index, and dividing by per-capita energy consumption.

2. C. Morey, R. Hwang, J. Kliesch, and J. DeCicco, *Pollution Lineup: An Environmental Ranking of Automakers* (Cambridge, MA: Union of Concerned Scientists, 2000).

3. J.-Y. Ko, C. A. S. Hall, and L. G. Lopez Lemus, "Resource Use Rates and Efficiency as Indicators of Regional Sustainability: An Examination of Five Countries," *Environmental Monitoring and Assessment* vol. 51 (1998): 589.

4. Ko, Hall, and Lopez Lemus, "Resource Use Rates and Efficiency as Indicators of Regional Sustainability," 574.

5. I adopt the term *commoditization* rather than the more familiar *commodification* to stress the active quality of the process. Something is *commodified*, whereas economic forces act to *commoditize* things. For a more in-depth discussion of commoditization, see Jack Manno, *Privileged Goods: Commoditization and Its Impacts on Environment and Society* (Boca Raton, FL: Lewis, 2000).

6. M. Cogoy, "The Consumer as Social and Environmental Actor," *Ecological Economics* vol. 28 (1999): 385–398.

7. T. A. Lyson and R. Welsh, "The Production Function, Crop Diversity, and the Debate between Conventional and Sustainable Agriculture," *Rural Sociology* vol. 58 no. 3 (1993): 424–439.

8. K. De Selincourt, "Intensifying Agriculture—The Organic Way," *The Ecologist* vol. 26 no. 2 (1996): 271.

9. H. Wohlmeyer, "Agro-Eco-Restructuring: Potential for Sustainability," in R. U. Ayres, ed., *Eco-Restructuring: Implications for Sustainable Development* (Tokyo: United Nations University Press, 1998), p. 291.

10. K. De Selincourt, "Intensifying Agriculture—The Organic Way," 271.

11. W. Murdoch, "World Hunger and Population," in C. R. Carroll, J. H. Vandermeer, and P. M. Rossett, eds., *Agroecology* (New York: McGraw-Hill, 1990), pp. 3–20.

12. J. M. Weatherford, *Indian Givers: How the Indians of the Americas Transformed the World* (New York: Crown, 1988), p. 83.

13. D. Pimentel, C. Harvey, P. Resosudarmo, K. Sinclair, D. Kurtz, M. McNair, S. Crist, L. Spritz, L. Fitton, R. Saffouri, and R. Blair, "Environmental and Economic Costs of Soil Erosion and Conservation Benefits," *Science* vol. 267 (1995): 111.

14. P. A. Matson, W. J. Parton, A. G. Power, and M. J. Swift, "Agricultural Intensification and Ecosystem Properties," *Science* vol. 277 (August 25, 1997): 504–509.

15. G. C. Gallopin, "Eco-Restructuring Tropical Land Use," in R. U. Ayres, ed., *Eco-Restructuring*, p. 317.

16. D. Barkin, *Wealth, Poverty, and Sustainable Development* (Mexico: Centro de Ecologia y Desarollo, A., C., 1998).

17. M. A. Mallin, "Impacts of Industrial Animal Production on Rivers and Estuaries," *American Scientist* vol. 88 (January/February 2000): 26–37.

18. U.S. Department of Agriculture Economic Research Service, "Change in Livestock Production, 1969–1992," AER-754 (August 1997), retrieved May 11, 1999, from the World Wide Web at http://www.econ.ag.gov/epubs/htmlsum/are754.htm

19. D. Pimentel and M. Pimentel, eds., *Food, Energy, and Society* (Niwot: University Press of Colorado, 1996), p. 168.

20. Pimentel and Pimentel, eds., *Food, Energy, and Society*, p. 8.

21. C. A. S. Hall and M. Hall, "The Efficiency of Land and Energy Use in Tropical Economies and Agriculture," *Agriculture, Ecosystems, and Environment* vol. 46 (1993): 1–30.

22. William Ophuls, *Requiem for Modern Politics* (Boulder: Westview Press, 1997), p. 100.

23. H. Wohlmeyer, "Agro-Eco-Restructuring," p. 292.

24. H. Wohlmeyer, "Agro-Eco-Restructuring"; 292.

25. M. Lipson, *Searching for the O-Word* (Santa Cruz, CA: Organic Farming Research Foundation, 1997).

26. M. Anderson, "The Life Cycle of Alternative Agricultural Research," p. 4.

27. H. Wohlmeyer, "Agro-Eco-Restructuring," p. 292.

28. J. Warick, "Cultivating Farms to Soak up a Greenhouse Gas," *Washington Post*, December 23, 1998, p. A3.

29. L. R. Brown, G. Gardner, and B. Halwell, *Beyond Malthus: Sixteen Dimensions of the Population Problem, Worldwatch Paper 143* (Washington, DC: Worldwatch Institute, 1998), p. 31.

30. P. R. Ehrlich, A. H. Ehrlich, and G. C. Daily, *The Stork and the Plow* (New York: Putnam, 1995), p. 171.

31. World Health Organization, "Micronutrient Malnutrition—Half the World's Population Affected," *World Health Organization* vol. 78 no. 1 (November 13, 1996).

32. U.S. Environmental Protection Agency, "Extended Producer Responsibility, General EPA Resources," retrieved November 6, 2000 from the World Wide Web at http://www.epa.gov/epr/r-genres.htm

33. W. Stahel and T. Jackson, "Optimal Utilization and Durability," in T. Jackson, ed., *Clean Production Strategies* (London: Lewis (for the Stockholm Environment Institute), 1993).

34. K. A. Gould, A. Schnaiberg, and A. S. Weinberg, *Local Environmental Struggles: Citizen Activism in the Treadmill of Production* (New York: Cambridge University Press, 1996), p. 146.

35. International Institute for Sustainable Development, "Reduction, Reuse, Recycling & Recovery," retrieved August 24, 2000 from the World Wide Web at http://iisd1.iisd.ca/business/4rs.htm

36. W. McDonough and M. Braungart, "The Next Industrial Revolution," *Atlantic Monthly* (October 1998).

37. P. K. Rohatgi and R. Ayres, "Materials Futures: Pollution Prevention, Recycling, and Improved Functionality," in R. U. Ayres, ed., *Eco-Restructuring*, p. 111.

38. Franklin Associates, "Characterization of Municipal Solid Waste in the United States: 1998 Update" (Washington, DC: U.S. Environmental Protection Agency), retrieved on August 24, 2000 from the World Wide Web at http://www.wastenews.com/library/execsumm.txt

39. J. DePinto, J. Manno, M. Milligan, R. Guzman, J. Honan, and L. Lopez Lemus, *Proposed Chlorine Sunsetting in the Great Lakes* (Donald W. Rennie Memorial Monograph Series, No. 11, Buffalo: State University of New York at Buffalo, 1998).

40. M. Walzer, *Spheres of Justice: A Defense of Pluralism and Equality* (New York: Basic Books, 1983); also Thomas Princen and Matthias Finger, *Environmental NGOs in the World Politics: Linking the Local and the Global* (London: Routledge, 1994); Ronnie Lipschutz, *Global Civil Society and Global Environmental Governance: The Politics of Nature from Place to Planet* (Albany: State University of New York Press, 1996); Paul Wapner, *Environmental Activism and World Civic Politics* (Albany: State University of New York Press, 1996).

Chapter 5

1. Paul H. Templet "Grazing the Commons: An Empirical Analysis of Externalities, Subsidies, and Sustainability," *Ecological Economics* vol. 12 no. 2 (February 1995): 141–159.

2. Jim MacNeill, Pieter Winsemius, and Taizo Yakushiji, *Beyond Interdependence: The Meshing of the World's Economy and the Earth's Ecology* (New York: Oxford University Press, 1991).

3. Mathis Wackernagel and William Rees, *Our Ecological Footprint: Reducing Human Impact on the Earth* (Gabriola Island, BC: New Society Publishers, 1996).

4. Elinor Ostrom, *Governing the Commons: The Evolution of Institutions for Collective Action* (Cambridge, England: Cambridge University Press, 1990).

5. I am purposely vague about the meaning of *local*, which has yet to be assigned an ecologically and socially useful definition in the literatures that use the term.

In *Governing the Commons*, Ostrom seems to equate it with small-scale, self-contained resource systems that are embedded in a larger, sometimes national and even international, institutional setting. Daly and Cobb seem to equate it with community and yet include everything from small towns to nations as communities; see Herman E. Daly and John B. Cobb, *For the Common Good: Redirecting the Economy toward Community, the Environment, and a Sustainable Future* (Boston: Beacon Press, 1989). My purpose here is merely to highlight the extremes in jurisdictional discontinuity as one moves from a tightly integrated, densely networked community to a loose, amorphous, anarchic international setting.

6. Lynton K. Caldwell, "Beyond Environmental Diplomacy: The Changing Institutional Structure of International Cooperation," in John E. Carroll, ed., *International Environmental Diplomacy* (Cambridge, England: Cambridge University Press, 1988), pp. 13–28; James Rosenau, *Turbulence in World Politics: A Theory of Change and Continuity* (Princeton, NJ: Princeton University Press, 1990).

7. Karl Polanyi, *The Great Transformation* (Boston: Beacon Press, 1944).

8. A. M. Spence, "Entry, Capacity, Investment, and Oligopolistic Pricing," *Bell Journal of Economics* vol. 8 (1977): 534–544; Michael E. Porter, *Competitive Strategy: Techniques for Analyzing Industries and Competitors* (New York: Free Press, 1980); Oliver E. Williamson, *Markets and Hierarchies* (New York: Free Press, 1975).

9. *Popular Science* magazine's view of DDT was typical: "At last science has found the weapons for total victory on the insect front.... It will unlock the doors to vast areas where disease-bearing insects have barred the way to development and progress" (Alfred H. Sinks, 1944, "Another Enemy Surrenders," *Popular Science* (June 1944): 56A–D, in D. L. Murray, *Cultivating Crisis: The Human Cost of Pesticides in Latin America* (Austin: University of Texas Press, 1944), p. 14).

10. A. M. Spence, "Investment Strategy and Growth in a New Market," *Bell Journal of Economics* vol. 10 (1979): 1–19; O. E. Williamson, "Predatory Pricing: A Strategic and Welfare Analysis," *Yale Law Journal* vol. 87 (1977): 284–340.

11. William Cronon, *Nature's Metropolis: Chicago and the Great West* (New York: Norton, 1991), p. 253.

12. This distinction is not confined to huge multinational corporations and quaint natural resource firms. The owner of a highly profitable professional football team in the United States, the Cleveland Browns, moved the team to Baltimore to get an even higher return on his investment. By contrast, the owners of the Green Bay Packers, namely, several thousand resident fans, have a secure hold on their "community asset," what is also a highly profitable, yet locally grounded, business. See R. Sandomir, "America's Small-Town Team," *New York Times*, Business Day (1996), pp. 17–18.

13. Ostrom, *Governing the Commons*.

14. Jentoft Svein and Trond Kristoffersen, "Fishermen's Co-Management: The Case of the Lofoten Fishery," *Human Organization* vol. 48 (1989): 355–365.

15. Douglas Murray, "Export Agriculture, Ecological Disruption, and Social Inequity: Some Effects of Pesticides in Southern Honduras," *Agriculture and Human Values* (Fall 1991): 19–29, 26.

16. Cronon, *Nature's Metropolis.*

17. Ivette Perfecto, "Pesticide Exposure of Farm Workers and the International Connection," in Bunyan Bryant and Paul Mohai, eds., *Race and the Incidence of Environmental Hazards* (Boulder: Westview Press, 1992): 177–203.

18. Stephan Schmidheiny with the Business Council for Sustainable Development, *Changing Course: A Global Business Perspective on Development and the Environment* (Cambridge, MA: MIT Press, 1992).

19. Thomas Princen, "Toward a Theory of Restraint," *Population and Environment* vol. 18 no. 3 (January 1997): 233–254.

20. In some cultures, a special name is given to sellers one buys from regularly and from whom one can expect good quality at a reasonable price. In the Philippines, it is *suki* (personal communication, Carmencita Cui, 1995). These relationships may typify low-distance commerce and suggest a direction for research on more sustainable forms of commerce.

21. The pesticide industry provides one of the most egregious illustrations of the impact of such cultural distance. According to agricultural economist C. F. Runge, "Modern chemical inputs require substantially more information to use safely and effectively, and such standards are complicated both to develop and to apply. Especially in the South, the inputs themselves are aggressively marketed and subsidized, yet farm-level education (including the basic literacy necessary to read package instruction) is seldom given comparable attention.... Per capita pesticide poisonings in the seven countries of Central America are 1,800 times higher than in the United States" (C. F. Runge, "Trade Protectionism and Environmental Regulations: The New Nontariff Barriers," *Northwestern Journal of International Law and Business* vol. 11 no. 1 (Spring 1990): 47–61, 55).

22. John S. Dryzek, *Rational Ecology: Environment and Political Economy* (New York: Blackwell, 1987).

23. Richard P. Tucker, "The Depletion of India's Forests under British Imperialism: Planters, Foresters, and Peasants in Assam and Kerala," in Donald Worster, ed., *The Ends of the Earth: Perspectives on Modern Environmental History* (Cambridge, England: Cambridge University Press, 1988), pp. 118–140.

24. There is a danger in this outsider-insider argument of romanticizing the "local" or the indigenous or resident peoples, of ascribing to them "natural" propensities of sustainability. The advantage of the distance notion in general and cultural distance in particular is that it can be more analytic than such romanticized notions.

25. Daniel W. Bromley, "The Commons, Property, and Common-Property Regimes," in Daniel W. Bromley, ed., *Making the Commons Work: Theory,*

Practice, and Policy (San Francisco: Institute for Contemporary Studies Press, 1992), pp. 3–15.

26. Outsiders can narrow cultural distance by minimizing mobility and by accumulating local knowledge. For example, John Browder describes a migrant to the Amazon who, in his first year, had to live off the land. He could not sell timber or clear the forest for pasture until a road gave him access to markets and supplies. As a result, in his first year he learned from native peoples hundreds of plants and animals that could be used for food and medicine. He then did clear some land but now grows crops much as the indigenous peoples do, mimicking natural succession patterns. See John Browder, "Redemptive Communities: Indigenous Knowledge, Colonist Farming Systems, and Conservation of Tropical Forests," *Agriculture and Human Values* vol. 12 (1995): 17–30.

27. Simon Fairlie, ed., "Overfishing: Its Causes and Consequences," *The Ecologist* (special issue) vol. 25 (1995); M. J. Peterson, "International Fisheries Management," in Peter Haas, Robert Keohane, and Marc Levy, eds., *Institutions for the Earth: Sources of Effective International Environmental Protection* (Cambridge, MA: MIT Press, 1993): 249–305.

28. Howard Raiffa, *The Art and Science of Negotiation* (Cambridge, MA: Harvard University Press, 1982).

29. Madhav Gadgil and Ramachandra Guha, *This Fissured Land: An Ecological History of India* (Berkeley: University of California Press, 1992).

30. Runge, "Trade Protectionism and Environmental Regulations."

31. The production chain beginning with a highly resource dependent primary producer is a highly stylized construction of actual production systems. The autonomy of decision making and dependency on the resource of the producers in the system can vary considerably. To the extent primary producers are autonomous yet not dependent on a given resource, a frontier economy once again is simulated. Thus the construction offered here represents one extreme, or "ideal type," at least at the point of resource extraction, namely, stewardship.

32. To clarify the argument to this point, regarding the distance effect of multiple agency I am employing the most generous assumptions about the grower and consumer, namely, that both desire long-term production and maintenance of the resource. Regarding farmers, this assumption is reasonable if they are not desperate. They want a productive farm over their lifetime and they want to pass it on to their children as a productive farm. They can be described, in short, as having low discount rates with respect to the soil, water, and other resources they depend on and that they expect their descendants to depend on. Regarding consumers, the long-term assumption is reasonable because they too want dependable supplies. We cannot say with precision whether the producer or the consumer has lower discount rates since those rates are in part a function of the alternatives each faces. But in general, the farmer will be most dependent and hence, to the extent interpersonal utility comparisons can be made, have the lowest discount rate.

33. Raymond De Young, personal communication, 1995.

34. Larue Hosmer, personal communication, 1995.

35. Tracey Clunies-Ross and Nicholas Hildyard, "The Politics of Industrial Agriculture," *The Ecologist* vol. 22 no. 2 (March/April 1992): 65–71.

36. Although I use the zero-distance case for illustration here, such production is not insignificant. In the United States, by one estimate, some $18 billion of home produce is generated by households every year (Kenneth A. Dahlberg, "Regenerative Food Systems: Broadening the Scope and Agenda of Sustainability," p. 82, in P. Allen, ed., *Food for the Future: Conditions and Contradictions of Sustainability* (New York: Wiley, 1993): 75–102).

37. James J. Kay and Eric Schneider, "Embracing Complexity: The Challenge of the Ecosystem Approach," *Alternatives* vol. 20 no. 3 (1994): 32–39.

38. Cronon, *Nature's Metropolis*, pp. 114–119.

39. The easy case is to explain degradation with greed, shortsightedness, ignorance, and villainy. Not only is this analytically inadequate, it leads to hopeless prescriptions: rid the world of greed and ignorance and we solve the environmental problem. The hoped-for effect of this chapter is to provide a preliminary guide to turning the most compelling arguments regarding economic progress on their head. The chapter does this largely by considering firms and states just what they themselves claim to be—rational calculators of strategic advantage.

Chapter 6

1. Timothy W. Luke, "The (UnWise) (Ab)(Use) of Nature: Environmentalism as Globalized Consumerism," *Alternatives* vol. 23 no. 2 (April/June 1998): 175–212.

2. Luke, "The (UnWise) (Ab)Use of Nature," p. 175.

3. World Commission on Environment and Development, *Our Common Future* (New York: Oxford University Press, 1987).

4. See for example A. Fuat Firat and Nikhilesh Dholakia, *Consuming People: From Political Economy to Theaters of Consumption* (London: Routledge, 1998).

5. See for instance Michael Veseth, *Selling Globalization: The Myth of the Global Economy* (Boulder, CO: Lynne Rienner, 1998).

6. The classic statement of the interdependence thesis is found in Robert Keohane and Joseph Nye, *Power and Interdependence: World Politics in Transition* (Boston: Little, Brown, 1977). For popular interpretations of globalization as the deepening of the economic and technological linkages of interdependence, see Benjamin Barber, *Jihad vs. McWorld* (New York: Times Books, 1995); Thomas L. Friedman, *The Lexus and the Olive Tree* (New York: Farrar Straus & Giroux, 1999).

7. The technological dimension is a particularly strong theme in popular accounts of globalization; see for example Friedman's paean to late-twentieth-century American consumer capitalism, *The Lexus and the Olive Tree*.

8. Gary Gereffi, Miguel Korzeniewicz, and Roberto P. Korzeniewicz, "Introduction: Global Commodity Chains," in Gary Gereffi and Miguel Korzeniewicz, eds., *Commodity Chains and Global Capitalism* (Westport, CT: Praeger, 1994), p. 2.

9. Gary Gereffi, "The Organization of Buyer-Driven Global Commodity Chains: How U.S. Retailers Shape Overseas Production Networks," in Gereffi and Korzeniewicz, *Commodity Chains and Global Capitalism*, pp. 96–97.

10. Gereffi and Korzeniewicz, *Commodity Chains and Global Capitalism*, p. 96. See also Gary Gerreffi and Miguel Korzeniewicz, "Commodity Chains and Footwear Exports in the Semi-Periphery," in William Martin, ed., *Semiperipheral States in the World Economy* (Westport, CT: Greenwood Press, 1990).

11. See for example Kenneth Maxwell, "Brazil in Meltdown," *World Policy Journal* vol. 16 no. 1 (Spring 1999), pp. 25–33. On the problem of capital flight see James E. Mahon, Jr., *Mobile Capital and Latin American Development* (University Park: Pennsylvania State University Press, 1996).

12. UN Economic Commission for Latin America and the Caribbean, "Capital Flows to Latin America and the Caribbean 1997," *CEPAL News* 18 no. 3 (March 1998), p. 3.

13. UN Economic Commission for Latin America and the Caribbean, *Foreign Investment in Latin America and the Caribbean* (Santiago, Chile: United Nations, 1998), p. 17.

14. ECLAC, *Foreign Investment in Latin America and the Caribbean*, p. 21.

15. ECLAC, *Foreign Investment in Latin America and the Caribbean*, p. 247.

16. ECLAC, *Foreign Investment in Latin America and the Caribbean*, part IV.

17. ECLAC, *Foreign Investment in Latin America and the Caribbean*.

18. ECLAC, *Foreign Investment in Latin America and the Caribbean*, p. 262.

19. Laura T. Reynolds, "Institutionalizing Flexibility: A Comparative Analysis of Fordist and Post-Fordist Models of Third World Agro-Export Production," in Gereffi and Korzeniewicz, *Commodity Chains and Global Capitalism*, p. 145.

20. Ibid.

21. For an overview of these debates, see Ash Amin, "Post-Fordism: Models, Fantasies, and Phantoms of Transition," in Ash Amin, ed., *Post-Fordism: A Reader* (Oxford: Blackwell, 1994).

22. Thomas Princen, "The Shading and Distancing of Commerce: When Internalization Is Not Enough," *Ecological Economics* 20 (1997): 244.

23. Princen, "The Shading and Distancing of Commerce," p. 250.

24. On this point see Herman Daly, "The Perils of Free Trade," in Ken Conca and Geoffrey D. Dabelko, *Green Planet Blues: Environmental Politics from Stockholm to Kyoto* (Boulder, CO: Westview Press, 1998).

25. Princen, "The Shading and Distancing of Commerce," p. 246.

26. Susan Strange, *The Retreat of the State: The Diffusion of Power in the World Economy* (Cambridge, England: Cambridge University Press, 1996).

27. On export-credit agencies, see Jenny Hsieh, Doris McDonald, Militsa Plavsic, and Andrew Spejewski, "An Analysis of the Environmental Standards of Export and Overseas Private Investment Support Agencies," Yale Environmental Protection Clinic, May 14, 1998. See also *A Race to the Bottom: Creating Risk, Generating Debt and Guaranteeing Environmental Destruction: A Compilation of Export Credit Agencies & Investment Insurance Agency Case Studies*, a report of the Berne Declaration, Environmental Defense Fund, and other organizations, March 1999.

28. Gereffi, "The Organization of Buyer-Driven Global Commodity Chains," p. 97.

29. See Gary Gereffi, "Commodity Chains and Regional Divisions of Labor in East Asia," in Eun Mee Kim, ed., *The Four Asian Tigers: Economic Development and the Global Political Economy* (San Diego: Academic Press, 1998).

30. Piore and Sabel's work on post-Fordism as flexible specialization has been criticized for "being too naive in imagining the likelihood or possibility of a large-scale return to a craft industrial paradigm, on the grounds that the embedded structures of Fordism will persist and adapt to new circumstances rather than disappear." See Amin, "Post-Fordism," p. 16. See also Michael J. Piore and Charles F. Sabel, *The Second Industrial Divide: Possibilities for Prosperity* (New York: Basic Books, 1984).

31. Jan Mazurek, *Making Microchips: Policy, Globalization, and Economic Restructuring in the Semiconductor Industry* (Cambridge, MA: MIT Press, 1999).

32. Alan Durning, *How Much Is Enough? The Consumer Society and the Fate of the Earth* (New York: Norton, 1992).

Chapter 7

I am deeply indebted to Thomas Princen and Michael Maniates for helpful comments on earlier drafts. I also would like to thank Noba Anderson and Matt Griem for their superb research assistance.

1. See, for example, Louis Blumberg and Robert Gottlieb, *War on Waste: Can American Win Its Battle with Garbage?* (Washington, DC: Island Press, 1989); Robert Allen, *Waste Not, Want Not: The Production and Dumping of Toxic Waste* (London: Earthscan, 1992); K. A. Gourlay, *World of Waste: Dilemmas of Industrial Development* (London: Zed Books, 1992); John Young, *Discarding the Throwaway Society* (Washington, DC: Worldwatch, 1991).

2. See World Wildlife Fund, *Getting at the Source: Strategies for Reducing Municipal Solid Waste* (Washington, DC: World Wildlife Fund, 1991); Homer Neal and J. R. Schubel, *Solid Waste Management and the Environment: The Mounting Garbage and Trash Crisis* (Englewood Cliffs, NJ: Prentice Hall, 1987). The waste-management hierarchy is outlined in Michael Carley and Philippe Spapens, *Sharing the World: Sustainable Living and Global Equity in the 21st Century* (London: Earthscan, 1998), p. 121.

3. Susan Strasser, *Waste and Want: A Social History of Trash* (New York: Metropolitan Books, 1999), pp. 13–14.

4. For an overview of the globalization project and its impact on developing countries, see Philip McMichael, *Development and Social Change: A Global Perspective* (Thousand Oaks, CA: Pine Forge Press, 2000).

5. Harold Crooks, *Giants of Garbage: The Rise of the Global Waste Industry and the Politics of Pollution Control* (Toronto: Lorimer, 1993); Kate O'Neill, *Waste Trading among Rich Nations: Building a New Theory of Environmental Regulation* (Cambridge, MA: MIT Press, 2000); Jennifer Clapp, *Toxic Exports: The Transfer of Hazardous Wastes from Rich to Poor Countries* (Ithaca, NY: Cornell University Press, 2001).

6. United Nations Development Program (UNDP), *Human Development Report 1992* (Oxford: Oxford University Press, 1992), pp. 34–36.

7. Robert Wade, "Global Inequality: Winners and Losers," *The Economist* (April 28, 2001), pp. 72–74.

8. See Robert Bullard, "Overcoming Racism in Environmental Decisionmaking," *Environment* vol. 36 no. 4 (1994): 11–20, 39–44; Timothy Maher, "Environmental Oppression," *Journal of Black Studies* vol. 28 no. 3 (1998): 357–367.

9. OECD, retrieved May 14, 2000, from the World Wide Web at http://www.oecd.org/env/indicators/an1e.pdf

10. U.S. Environmental Protection Agency (USEPA), "Municipal Solid Waste Generation, Recycling and Disposal in the United States: Facts and Figures for 1998," Washington, DC: USEPA, 2000. Retrieved May 11, 2001, from the World Wide Web at http://www.epa.gov/epaoswer/non-hw/muncpl/99charac.pdf

11. UNDP, *Human Development Report 1998* (Oxford: Oxford University Press, 1998), p. 56.

12. UNDP, *Human Development Report 1998*, p. 71. On Nigeria, see Adepoju G. Onibokun and A. J. Kumuyi, "Governance and Waste Management in Africa," in Adepoju G. Onibokun, ed., *Managing the Monster: Urban Waste and Governance in Africa* (Ottawa: International Development Research Centre, 1999), p. 3.

13. Strasser, *Waste and Want*, p. 12.

14. Young, *Discarding the Throwaway Society*, pp. 20–21.

15. Zero Waste America claims that municipal waste is between 2 and 20 percent of the waste disposed of in landfills in the United States. Retrieved May 11, 2001, from the World Wide Web at http://www.zerowasteamerica.org

16. See Juliet Schor, *The Overspent American: Upscaling, Downshifting, and the New Consumer* (New York: Basic Books, 1998).

17. Carly and Spapens, *Sharing the World*, pp. 134–138.

18. Gourlay, *World of Waste*, pp. 33–34; Young, *Discarding the Throwaway Society*, p. 12.

19. Young, *Discarding the Throwaway Society*, p. 21.

20. Strasser, *Waste and Want*, p. 10.

21. Conrad Lodziak, "On Explaining Consumption," *Capital and Class*, no. 72 (2000), pp. 124–125.

22. Kim Nash, "Millions of Obsolete PCs Enter Waste Stream," *Computerworld* vol. 34 no. 15 (April 10, 2000), p. 20.

23. Danielle Knight, "U.S. Fights EU Initiative against Electronics Pollution," *Inter Press Service*, May 25, 2000.

24. Knight, "U.S. Fights EU Initiative against Electronics Pollution."

25. Nash, "Millions of Obsolete PCs Enter Waste Stream," p. 20.

26. On the packaging issue and policies directed toward it, see Bette Fishbein, *Germany, Garbage, and the Green Dot: Challenging the Throwaway Society* (New York: INFORM, 1994); Sally Eden, "The Politics of Packaging in the UK: Business, Government, and Self-Regulation in Environmental Policy," *Environmental Politics* vol. 5 no. 4 (1996): 632–653.

27. USEPA, "Municipal Solid Waste Generation, Recycling and Disposal in the United States."

28. Blumberg and Gottlieb, *War on Waste*, pp. 260–262.

29. On this point, see World Wildlife Fund, *Getting at the Source*, pp. 9–10, and the discussion of consumer sovereignty in chapter 14 of this book.

30. Nora Goldstein and Celeste Madtes, "The State of Garbage in America," *BioCycle* vol. 41 no. 11 (November 2000): 41; USEPA, "Municipal Solid Waste Generation, Recycling and Disposal in the United States."

31. Strasser, *Waste and Want*, pp. 271–272.

32. UNDP, *Human Development Report 1998* (Oxford: Oxford University Press, 1998).

33. See "Barging into a Trashy Saga," *Newsday*, retrieved May 10, 2001, from the World Wide Web at http://www.lihistory.com/9/hs9garb.htm

34. See Bullard, "Overcoming Racism in Environmental Decisionmaking," for an overview of this literature. See also Daniel Brook, "Environmental Genocide: Native Americans and Toxic Waste," *American Journal of Economics and Sociology* vol. 57 no. 1 (1998): 105–113.

35. On this point, see David Pellow, Allan Schnaiberg, and Adam Weinberg, "Putting the Ecological Modernisation Thesis to the Test: The Promises and Performances of Urban Recycling," *Environmental Politics* vol. 9 no. 1 (2000): 114.

36. Nancy Linn, Joanne Vining, and Patricia Ann Feeley, "Toward a Sustainable Society: Waste Minimization through Environmentally Conscious Consuming," *Journal of Applied Social Psychology* vol. 24 no. 17 (1994): 1551.

37. See Clapp, *Toxic Exports*, pp. 64–66.

38. I. Bokerman and J. Vorfelder, *Plastics Waste to Indonesia: The Invasion of the Little Green Dots* (Hamburg: Greenpeace Germany, 1993).

39. Greenpeace Canada, *We've Been Had! Montreal's Plastics Dumped Overseas* (Montreal: Greenpeace, 1993); "German Wastes Flood Latvia," *Toxic Trade Update* vol. 6 no. 3 (1993): 11–13.

40. Andrew Noone, "PET Recycling in Asia Pacific," in *Resource Recycling* vol. 16 no. 2 (February 1999): 37–40.

41. Blumberg and Gottlieb, *War on Waste*, pp. 35–42.

42. See Paul Connett and Ellen Connett, "Municipal Waste Incineration: Wrong Question, Wrong Answer," *The Ecologist* vol. 24 no. 1 (1994): 14–20.

43. Alicia Marie Belchak, "Arctic Dioxin Tracked to North American Sources," *Environmental News Service* (October 4, 2000), retrieved May 29, 2001, from the World Wide Web at http://ens.lycos.com/ens/oct2000/2000l%2D10%2D04%2D06.html

44. See Jim Vallette and Heather Spalding, eds., *The International Trade in Wastes: A Greenpeace Inventory* (Washington, DC: Greenpeace, 1990), pp. 21–25.

45. See Ramona Smith, "New Ship Hauls Haitian Ash," *Philadelphia Daily News*, October 30, 1998; "Haiti: No Welcome Mat for Return of U.S. Toxic Waste," *Inter Press Service*, June 13, 1999; Victor Fiorillo and Liz Spiol, "Ashes to Ashes, Dust to Dust," *Philadelphia Weekly*, January 18, 2001; Melissa E. Holsman, "2,000 Tons of Ash Still on Barge," *Sun Sentinel* (Fort Lauderdale, FL), April 5, 2001. All sources available online at www.ban.org

46. Preface to the 1999 version of the Basel Convention (Geneva: Secretariat of the Basel Convention, 1999).

47. Robert Lucas, David Wheeler, and Hemamala Hettige, "Economic Development, Environmental Regulation, and the International Migration of Toxic Industrial Pollution 1960–88," *Policy Research Working Papers*, WPS 1602 (Washington, DC: World Bank, December 1992), p. 14.

48. Young, *Discarding the Throwaway Society*, pp. 7–12; Gourlay, *World of Waste*, pp. 43–49.

49. Environmental Mining Council of British Columbia, "Mining and the Environment Primer," retrieved May 22, 2001, from the World Wide Web at http://www.miningwatch.org/emcbc/primer/default.htm

50. Gareth Porter, "Trade Competition and Pollution Standards: 'Race to the Bottom' or 'Stuck at the Bottom,'" *Journal of Environment and Development* vol. 8 no. 2 (1999): 133–151.

51. Joshua Karliner, "The Environmental Industry," *The Ecologist* vol. 24 no. 2 (1994): 59–63.

52. On plastic in particular, see Blumberg and Gottlieb, *War on Waste*, pp. 265–274; on the chemical industry, see Gourlay, *World of Waste*, pp. 36–43.

53. O'Neill, *Waste Trading among Rich Nations*, pp. 35–36.

54. Francis Adeola, "Environmental Hazards, Health, and Racial Inequity in Hazardous Waste Distribution," *Environment and Behavior* vol. 26 no. 1 (1994): 99–126.

55. David Allen, "Social Class, Race, and Toxic Release in American Counties, 1995," *Social Science Journal* vol. 38 no. 1 (2001): 13–25.

56. See Clapp, *Toxic Exports*, p. 35.

57. Vallette and Spalding, eds., *The International Trade in Wastes*.

58. The Basel Action Network (BAN) tracks these incidents at http://www.ban.org

59. "Illegal Dumping," *Mainichi Daily News*, Niigata, Japan, January 13, 2000 (posted online at: http://www.ban.org/ban_news/illegal.html).

60. "Japan: Philippines Case Bares Inadequacy of Waste Rules," *The Yomiuri Shimbun*, Tokyo, Japan, January 12, 2000 (posted online at: http://www.ban.org/ban_news/japan2.html).

61. Jim Puckett, "Disposing of the Waste Trade: Closing the Recycling Loophole," *The Ecologist* vol. 24 no. 2 (1994): 53–58.

62. Susan Young, "Feds Refuse HoltraChem Mercury, Company May Send Chemical to India," *Bangor Daily News*, November 17, 2000 (posted online at: http://www.ban.org/ban_news/feds.html); Danielle Knight, "Outcry over U.S. Toxic Chemical Shipment to India," *Inter Press Service*, December 11, 2000 (posted online at: http://www.ban.org/ban_news/outcry.html).

63. For a discussion of this case, see Clapp, *Toxic Exports*, pp. 62–64.

64. Danielle Knight, "Controversy Around Mercury Shipment from U.S. to India," *Inter Press Service*, January 25, 2001 (posted online at: http://www.ban.org/ban_news/controversy).

65. Susan Young, "New Home for Mercury Hard to Find," *Bangor Daily News*, March 28, 2001 (posted online at: http://www.ban.org/ban_news/new_home.html).

66. Patrick Low, "The International Location of Polluting Industries and the Harmonization of Environmental Standards," in H. Muñoz and R. Rosenberg, eds., *Difficult Liaison: Trade and the Environment in the Americas* (London: Transaction, 1993), p. 25.

67. Edward Williams, "The Maquiladora Industry and Environmental Degradation in the United States–Mexico Borderlands," *St. Mary's Law Journal* vol. 27 no. 4 (1996): 777–779.

68. Leslie Sklair, *Assembling for Development* (San Diego: University of California Center for U.S.-Mexican Studies, 1993), pp. 79–80.

69. Sklair, *Assembling for Development*, pp. 253–254; Diane Perry, Roberto Sanchez, William Glaze, and Marisa Mazari, "Binational Management of Hazardous Waste: The Maquiladora Industry at the U.S.-Mexico Border," *Environmental Management* vol. 14 no. 4 (1990): 442.

70. John Harbison and Taunya McLarty, "A Move away from the Moral Arbitrariness of Maquila and NAFTA-Related Toxic Harms," *UCLA Journal of Environmental Law and Policy* vol. 14 no. 1 (1995–1996): 6.

71. Cyrus Reed, "Hazardous Waste Management on the Border: Problems with Practices and Oversight Continue," *Borderlines* vol. 6 no. 5 (1998), retrieved May 23, 2001, from the World Wide Web at http://www.us-mex.org/borderlines/

1998/bl46/bl46haz.html. See also HAZTRAKS on the World Wide Web at http://www.epa.gov/earth1r6/6en/h/haztraks/haztraks.htm

72. Enrique Medina, "Overview of Transboundary Pollution Issues along the Mexico-U.S. Border," in Thomas La Point, Fred Price, and Edward Little, eds., *Environmental Toxicology and Risk Assessment: Fourth Volume* (West Conshohocken, PA: American Society for Testing and Materials, 1996), p. 9.

73. Lawrence Speer, "Environmentalists Assail Taiwan's Plans to Ship Waste to French Treatment Facility," *International Environment Reporter* vol. 22 no. 21 (October 13, 1999): 830.

74. Glen Perkinson, "Company Agrees to Take Back Mercury-Laden Waste Sent to Cambodia," *International Environment Reporter* vol. 22 no. 5 (March 3, 1999).

75. For a full account of the incident, see Basel Action Network (BAN), "Victory for Global Environmental Justice: Toxic Waste Dumped on Cambodia Will Finally Be Treated by Producer," press release, Seattle, June 9, 2000, retrieved April 10, 2001 from the World Wide Web at http://www.ban.org

76. See John Nicol, "Down in the Dumps," *Macleans*, Feb. 26, 2001, p. 16; Amy Cameron, "Garbage North," *Macleans*, Aug. 21, 2000, p. 38.

77. "Michigan Bill Seeks Ban on City's Garbage," *Toronto Star*, March 8, 2001, p. GT1.

78. Gary Davis and Catherine Wilt, "Extended Product Responsibility," *Environment* vol. 39 no. 7 (September 1997): 10. See also Paul Hawkin, Amory Lovins, and L. Hunter Lovins, *Natural Capitalism* (Boston: Little, Brown, 1999), who report (p. 79) that by 1998 there were some 28 countries with take-back laws for packaging.

79. Davis and Wilt, "Extended Product Responsibility."

Chapter 8

1. Sidney W. Mintz, *Sweetness and Power: The Place of Sugar in Modern History* (New York: Penguin, 1985), p. xvii.

2. Robert D. Sack, "The Realm of Meaning: The Inadequacy of Human-Nature Theory and the View of Mass Consumption," in B. L. Turner II, John F. Richards, and William B. Meyer, eds., *The Earth as Transformed by Human Action: Global and Regional Changes in the Biosphere over the Past 300 Years* (Cambridge, England: Cambridge University Press, 1990), pp. 659–672.

3. Grant McCracken, *Culture and Consumption: New Approaches to the Symbolic Character of Consumer Goods and Activities* (Bloomington: University of Indiana Press, 1990), p. 28.

4. For example, see Arjun Appadorai, ed., *The Social Life of Things: Commodities in Cultural Perspective* (Cambridge, England: Cambridge University Press, 1986); David Howes, ed., *Cross-Cultural Consumption: Global Markets, Local Realities* (London: Routledge, 1996); Daniel Miller, ed., *Acknowledging Con-*

sumption: A Review of New Studies (London: Routledge, 1995); Henry J. Rutz and Benjamin S. Orlove, eds., *The Social Economy of Consumption* (Lanham, MD: University Press of America, 1989); Richard Wilk, *Economies and Cultures: Foundations of Economic Anthropology* (Boulder, CO: Westview Press, 1996).

5. Paul Glennie, "Consumption within Historical Studies," in Miller, ed., *Acknowledging Consumption*, pp. 164–203.

6. Mark H. Lytle, "An Environmental Approach to American Diplomatic History," *Diplomatic History* vol. 20 no. 2 (1996): 279–300.

7. Thomas Princen, personal communication, October 1998. For further elaboration of the analysis, see Princen, "Consumption and Environment: Some Conceptual Issues," *Ecological Economics* 31 (1999): 347–363.

8. Bananas, coffee, and timber, as well as cane sugar, rubber, and beef, are discussed at chapter length each in Richard P. Tucker, *Insatiable Appetite: The United States and the Ecological Degradation of the Tropical World* (Berkeley: University of California Press, 2000).

9. Neva R. Goodwin, Frank Ackerman, and David Kiron, eds., *The Consumer Society* (Washington, DC: Island Press, 1997); Lisa Jardine, *Worldly Goods: A New History of the Renaissance* (New York: Doubleday, 1998); Mintz, *Sweetness and Power*; John E. Wills, Jr., "European Consumption and Asian Production in the Seventeenth and Eighteenth Centuries," in J. Brewer and R. Porter, eds., *Consumption and the World of Goods* (London: Routledge, 1993).

10. K. N. Chaudhuri, *Trade and Civilization in the Indian Ocean: An Economic History from the Rise of Islam to 1750* (Cambridge, England: Cambridge University Press, 1985).

11. For a parallel and competitive trend centered in Amsterdam, see Simon Schama, *The Embarrassment of Riches: An Interpretation of Dutch Culture in the Golden Age* (New York: Knopf, 1987); Brewer and Porter, *Consumption and the World of Goods*, 85–132.

12. For an analysis of the subject in the nineteenth century, see J. F. Richards and Michelle B. McAlpin, "Cotton Cultivating and Land Clearing in the Bombay Deccan and Karnatak: 1818–1920," in Richard P. Tucker and J. F. Richards, eds., *Global Deforestation and the Nineteenth-Century World Economy* (Durham, NC: Duke University Press, 1983), pp. 68–94.

13. Neil McKendrick, John Brewer, and J. H. Plumb, *The Birth of a Consumer Society: The Commercialization of Eighteenth-Century England* (Bloomington: University of Indiana Press, 1982), p. 9.

14. Richard S. Tedlow, *New and Improved: The Story of Mass Marketing in America* (New York: Basic Books, 1990), pp. 3–4.

15. Tedlow, *New and Improved*, chap. 4. See also William Cronon, *Nature's Metropolis: Chicago and the Great West* (New York: Norton, 1991), chap. 7; William Leach, *Land of Desire: Merchants, Power, and the Rise of a New American Culture* (New York: Pantheon Books, 1993), parts I–II; Susan Strasser, *Satisfaction Guaranteed: The Making of a Mass Market* (New York: Pantheon Books, 1989), chaps. 3–4.

16. Charles Goodrum and Helen Dalrymple, *Advertising in America: The First 200 Years* (New York: Abrams, 1990), chaps. 1–2; Jackson Lears, *Fables of Abundance: A Cultural History of Advertising in America* (New York: Basic Books, 1994); Daniel Pope, *The Making of Modern Advertising* (New York: Basic Books, 1983).

17. James D. Norris, *Advertising and the Transformation of American Society, 1865–1920* (New York: Greenwood Press, 1990), pp. 97–98.

18. Norris, *Advertising and the Transformation of American Society, 1865–1920*, pp. 167–168.

19. Harvey A. Levenstein, *Paradox of Plenty: A Social History of Eating in Modern America* (Oxford, England: Oxford University Press, 1983), chap. 3.

20. Kenneth T. Jackson, *Crabgrass Frontier: The Suburbanization of the United States* (Oxford, England: Oxford University Press, 1985); Stanley Lebergott, *Pursuing Happiness: American Consumers in the Twentieth Century* (Princeton, NJ: Princeton University Press, 1993); William E. Leuchtenburg, *A Troubled Feast: American Society Since 1945* (Boston: Little, Brown, 1973).

21. Richard W. Fox and T. J. Jackson Lears, eds., *The Culture of Consumption: Critical Essays in American History, 1880–1980* (New York: Pantheon Books, 1983), p. ix.

22. Mira Wilkins, *The Maturing of Multinational Enterprise: American Business Abroad from 1914 to 1970* (Cambridge, MA: Harvard University Press, 1974).

23. Emily S. Rosenberg, *Spreading the American Dream: American Economic and Cultural Expansion, 1890–1945* (New York: Hill and Wang, 1982), p. 27.

24. David Burner, *Herbert Hoover: A Public Life* (New York: Viking Press, 1978).

25. John Soluri, *Landscape and Livelihood: An Agroecological History of Export Banana Growing in Honduras, 1870–1970*, Ph.D. dissertation, University of Michigan, Ann Arbor, 1997.

26. Frederick Upham Adams, *Conquest of the Tropics: The Story of the Creative Enterprises Conducted by the United Fruit Company* (Garden City, NY: Doubleday Page, 1914).

27. John Soluri, "Banana Cultures: Linking the Production and Consumption of Export Bananas, 1800–1980," unpublished paper, 1999, p. 22.

28. Soluri, "Banana Cultures," p. 11.

29. Soluri, "Banana Cultures," p. 21.

30. Virginia Scott Jenkins, *Bananas: An American History* (Washington, DC: Smithsonian Institution Press, 2000).

31. Cynthia Enloe, *Bananas, Beaches, and Bases: Making Feminist Sense of International Politics* (Berkeley: University of California Press, 1989), pp. 124–132.

32. Joseph Grunwald and Philip Musgrove, *Natural Resources in Latin American Development* (Baltimore, MD: Johns Hopkins University Press, 1970), pp. 375–376.

33. Lori Ann Thrupp, "Sterilization of Workers from Pesticide Exposure: The Causes and Consequences of DBCP-Induced Damage in Costa Rica and Beyond," *International Journal of Health Services* vol. 21 no. 4 (1991): 731–757.

34. Steven C. Topik and Allen Wells, eds., *The Second Conquest of Latin America: Coffee, Henequen, and Oil during the Export Boom 1850–1930* (Austin: University of Texas Press, 1998), chap. 2.

35. Joseph E. Sweigart, *Coffee Factorage and the Emergence of a Brazilian Capital Market, 1850–1888* (New York: Garland, 1987).

36. For further detail, see Tucker, *Insatiable Appetite*, chap. 4.

37. William H. Ukers, *All about Coffee* (New York: Tea and Coffee Trade Journal Company, 1922).

38. Michael Jimenez, "'From Plantation to Cup': Coffee and Capitalism in the United States, 1830–1930," in William Roseberry, Lowell Gudmundson, and Mario Samper Kutschbach, eds., *Coffee, Society, and Power in Latin America* (Baltimore, MD: Johns Hopkins University Press, 1995), pp. 38–64; Mark Pendergrast, *Uncommon Grounds: The History of Coffee and How It Transformed Our World* (New York: Basic Books, 1999).

39. Marco Palacios, *Coffee in Colombia, 1850–1970: An Economic, Social, and Political History* (Cambridge, England: Cambridge University Press, 1980), p. 212.

40. Grunwald and Musgrove, *Natural Resources in Latin American Development*, p. 328.

41. Pendergrast, *Uncommon Grounds*, pp. 285–287.

42. Warren Dean, *Rio Claro: A Brazilian Plantation System, 1820–1920* (Stanford, CA: Stanford University Press, 1976); Stanley J. Stein, *Vassouras: A Brazilian Coffee County, 1850–1900* (Cambridge, MA: Harvard University Press, 1957).

43. United Nations, Food and Agriculture Organization, *Coffee in Latin America*, vol. 1 (New York: United Nations, 1958).

44. Catherine LeGrand, *Frontier Expansion and Peasant Protest in Colombia, 1850–1936* (Albuquerque: University of New Mexico Press, 1986); Palacios, *Coffee in Colombia, 1850–1970.*

45. T. Lynn Smith, "Land Tenure and Soil Erosion in Colombia," in *Proceedings of the Inter-American Conference on Conservation of Renewable Natural Resources* (Washington, DC: Department of State, 1948), pp. 155–160.

46. Jeffery Paige, *Coffee and Power: Revolution and the Rise of Democracy in Central America* (Cambridge, MA: Harvard University Press, 1997).

47. For an example of recent field research, see Liana I. Babbar and Donald R. Zak, "Nitrogen Cycling in Coffee Agroecosystems: Net N Mineralization and Nitrification in the Presence and Absence of Shade Trees," *Agriculture, Ecosystems & Environment* 48 (1994): 107–113.

48. F. Bruce Lamb, *Mahogany of Tropical America: Its Ecology and Management* (Ann Arbor: University of Michigan Press, 1966).

49. William F. Payson, *Mahogany: Antique and Modern* (New York: Dutton, 1926); Arthur M. Wilson, "The Logwood Trade in the Seventeenth and Eighteenth Centuries," in D. C. McKay, ed., *Essays in the History of Modern Europe* (New York: Harper Brothers, 1936).

50. Lamb, *Mahogany of Tropical America*, p. 13.

51. P. Macquoid, *History of English Furniture* (New York: Putnam, 1904); J. H. Pollen, *Ancient and Modern Furniture and Woodwork* (London: Chapman and Hall, 1875).

52. John C. Callahan, *The Fine Hardwood Veneer Industry in the United States: 1838–1990* (Lake Ann, MI: National Woodlands Publishing Company, 1990).

53. Richard P. Tucker, "Foreign Investors, Timber Extraction, and Forest Depletion in Central America before 1941," in Harold K. Steen and Richard P. Tucker, eds., *Changing Tropical Forests: Historical Perspectives on Today's Challenges in Central and South America* (Durham, NC: Forest History Society, 1992), pp. 265–276.

54. John C. Callahan, "The Mahogany Empire of Ichabod T. Williams & Sons, 1838–1973," *Journal of Forest History* vol. 29 no. 3 (July 1985): 120–130.

55. Lamb, *Mahogany of Tropical America*, pp. 13–21.

56. Tom Gill, *Tropical Forests of the Caribbean* (Washington, DC: Charles Lathrup Pack Foundation, 1931); Lamb, *Mahogany of Tropical America*; Samuel J. Record and Clayton D. Mell, *Timbers of Tropical America* (New Haven, CT: Yale University Press, 1924).

57. See Gill, *Tropical Forests of the Caribbean.*

58. Eugene F. Horn, "The Lumber Industry of the Lower Amazon Valley," *Caribbean Forester* vol. 18 no. 3–4 (1957): 56–67; John Browder, "Lumber Production and Economic Development in the Brazilian Amazon: Regional Trends and a Case Study," *Journal of World Forest Resource Management* vol. 4 no. 2 (1989): 1–19.

59. Richard P. Tucker, "Managing Subsistence Use of the Forest: The Philippine Bureau of Forestry, 1904–1960," in John Dargavel and Richard P. Tucker, eds., *Changing Pacific Forests: Historical Perspectives on the Forest Economy of the Pacific Basin* (Durham, NC: Forest History Society, 1992), pp. 105–115.

60. David Kummer, *Deforestation in the Postwar Philippines* (Chicago: University of Chicago Press, 1992).

Chapter 9

I thank Allegheny College for its financial support of field research.

1. Linda Breen Pierce, *Choosing Simplicity: Real People Finding Peace and Fulfillment in a Complex World* (Carmel, CA: Gallagher Press, 2000).

2. Rich Hayes, *A Survey of People Attending a Conference on Voluntary Simplicity*, unpublished manuscript (Berkeley: Energy and Resources Group, University of California, 1999).

3. David Shi, *The Simple Life: Plain Living and High Thinking in American Culture* (New York: Oxford University Press, 1985), p. 3.

4. Alan Durning, *How Much Is Enough?* (New York: Norton, 1992), p. 141.

5. Amitai Etzioni, "Voluntary Simplicity: Characterization, Select Psychological Implications, and Societal Consequences," *Journal of Economic Psychology* vol. 19 no. 5 (1998): 619–643.

6. Cecile Andrews, *The Circle of Simplicity: Return to the Good Life* (New York: HarperPerennial, 1997). Andrews's book spawned the "The Simplicity Circles Project," which fosters small-group discussions of simplicity and links more than 500 "simplicity circles" across the United States.

7. Andrews, *The Circle of Simplicity*, p. xiv.

8. Daniel Elgin, *Voluntary Simplicity: An Ecological Lifestyle That Promotes Personal and Social Renewal* (New York: Bantam Books, 1981).

9. For example, Alan Durning, *How Much Is Enough?*, p. 139.

10. Harwood Group, *Yearning for Balance: Views of Americans on Consumption, Materialism, and Environment* (Tacoma, MD: Merck Family Fund, 1995).

11. Juliet Schor, *The Overspent American: Upscaling, Downshifting, and the New Consumer* (New York: Basic Books, 1998).

12. See Gerald Celente, *Trends 2000* (New York: Warner Books, 1997), and later reports from the Trends Research Institute at http://www.trendsresearch.com/

13. For the purposes of this chapter, indexes covering the following major newspapers were reviewed: *New York Times*, *Washington Post*, *Los Angeles Times*, *San Jose Mercury News*, *Chicago Tribune*, and *Atlanta Journal-Constitution*.

14. The first was *Affluenza*, which characterized consumerism as a sickness and explored ways of opposing it, both individually and collectively. The follow-up production was *Escape from Affluenza*, which focused in detail on the voluntary simplicity movement. John de Graaf and Vivia Boe, *Affluenza*, a coproduction of KCTS/Seattle and Oregon Public Broadcasting (Oley, PA: Bullfrog Films, 1997); John de Graaf and Vivia Boe, *Escape from Affluenza*, a coproduction of KCTS/Seattle and John de Graaf (Oley, PA: Bullfrog Films, 1998).

15. In two separate communications with John Hoskyns-Abrahall of Bullfrog Films, he describes *Affluenza*, as "by far our biggest hit in years." Hoskyns-Abrahall notes that "the interesting thing is that the video (*Affluenza*) is that rarity that cuts across all boundaries—geographical, political, age, you name it. We've had a lot of calls from the South (which is rare for us); a lot from older folks; a lot from congregations of all kinds including fundamentalists; and of course the simple living types. *Affluenza* is being used by Consumer Credit Counseling service nationwide. Even some enterprising Deans have used it as part of freshman orientation."

16. As, for example, James Twitchell does in his *Lead Us into Temptation: The Triumph of American Materialism* (New York: Columbia University Press, 1999).

17. As Duane Elgin sometimes does in his *Promise Ahead: A Vision of Hope and Action for Humanity's Future* (New York: Morrow, 2000).

18. See for example, Lester Brown, Alan Durning, Christopher Flavin, Hilary French, Jodi Jacobson, Marcia Lowe, Sandra Postel, Michael Renner, Linda Starke, and John Young, *State of the World* (New York: W.W. Norton and Company, 1990); and The World Commission on Environment and Development, *Our Common Future* (New York: Oxford University Press, 1987).

19. Robert Frank, *Luxury Fever: Money and Happiness in an Era of Excess* (Princeton, NJ: Princeton University Press, 1999). Though Gyawali's views on issues of environment, development, and democracy in South Asia are well documented, both in his own writings and from varied conference proceedings and speeches, his published work has not explored the tyranny of expectations. I owe my use of the term to a series of 1987 conversations with Gyawali in Kathmandu.

20. Frank, *Luxury Fever*, p. 4.

21. See for example John Freie, *Counterfeit Community: The Exploitation of Our Longings for Connectedness* (Lanham, MD: Rowman & Littlefield, 1998), especially chap. 5.

22. See Frank, *Luxury Fever*, and Juliet Schor, *The Overworked American: The Unexpected Decline of Leisure* (New York: Basic Books, 1991).

23. See Schor, *The Overspent American*, especially chap. 6. The "upward creep of desire" in the following sentence is Schor's phrase.

24. Robert Edwards Lane, *The Loss of Happiness in Market Democracies* (New Haven: Yale University Press, 2000).

25. This is not Gyawali's argument alone; scholars of South Asia have drawn special attention to this phenomenon. See for example Ponna Wignaraja and Akmal Hussain, eds., *The Challenge in South Asia: Development, Democracy, and Regional Cooperation* (New Delhi: Sage, 1989), especially their "Editorial Overview: The Crisis and Promise of South Asia pp. 18–24."

26. National Film Board of Canada, *The Bomb under the World* (videorecording), produced by Green Lion Productions in association with the Canadian Broadcasting Corporation; director, Werner Volkmer; writer, Gwynne Dyer; producers, Catherine Mullins and Marrin Cannelo (Oley, PA: Bullfrog Films, 1994).

27. Alan Durning, *How Much Is Enough?*, pp. 117–118.

28. For example, see Constance Hays, "Be True to Your Cola, Rah! Rah! (Battle for Soft-Drink Loyalties Moves to Public Schools)," *New York Times* (March 10, 1998): C1.

29. Ronald Sider, *Christians in an Age of Hunger: Moving from Affluence to Generosity* (Dallas: Word Publishing, 1997), emphasis in the original. Also see Kim Thomas, *Simplicity: Finding Peace by Uncluttering Your Life* (Nashville: Broadman and Holman, 1999).

30. Scott Russell Sanders, "Simplicity," *Audubon* vol. 100 no. 4 (July/August 1998): 70–78.

31. For an example of UK longings for simplicity, see Maureen Rice, "Dejunk Your Life," *The Observer* (January 23, 2000). A broader reconnaissance of European simplicity sentiment is found in Pat Kane, "Play for Today," *The Observer* (October 22, 2000), and Walter Schwarz and Dorothy Schwarz, "Seeing the Light," *Guardian Unlimited* (February 3, 1999). All available at http://www.guardian.co.uk Simplicity, of course, weaves its way throughout Asian traditions. Those in South and Southeast Asia promoting the capacity for consumer restraint do not generally view themselves as part of a "voluntary simplicity movement," but rather draw on Buddhism and the teachings of Gandhi. See for example Ishwar Harris, *Gandhians in Contemporary India: The Vision and the Visionaries* (Lewiston, NY: Edwin Mellen Press, 1998); Christopher Queen and Sallie King, eds., *Engaged Buddhism: Buddhist Liberation Movements in Asia* (Albany: State University of New York Press, 1996); Tom Walz and Heather Ritchie, "Gandhian Principles in Social Work Practice: Ethics Revisited," *Social Work* vol. 45 no. 3 (May 2000): 213–222; Simon Zadek, "The Practice of Buddhist Economics?", *American Journal of Economics and Sociology* vol. 52 no. 4 (October 1993): 433–445.

32. Beth Brophy, "Stressless—and Simple—in Seattle," *U.S. News and World Report* vol. 119 no. 23 (December 11, 1995): 96.

33. Hayes, *A Survey of People Attending a Conference on Voluntary Simplicity*, p. 1.

34. It is an interesting list, one suggestive of political coalitions to come. The cosponsors were Adbusters Media Foundation, Ashley Montagu Institute, Center for the New American Dream, Earth Service, Eco-Home Network, Environmental Ministries, HopeDance, the Jane Goodall Institute's Roots and Shoots Program, Kushi Institute, New Road Map Institute, Plain Magazine, Sierra Club, *Simple Living Journal*, Simple Living Network, and the University of Northern Iowa's Center for Energy and Environmental Education.

35. See Amitai Etzioni, "Voluntary Simplicity"; Osamu Iwata, "Attitudinal and Behavioral Correlates of Voluntary Simplicity Lifestyles," *Social Behavior and Personality* vol. 25 no. 3 (1997): 233–240. Elements of this perspective also are found in Frank, *Luxury Fever*.

36. An accessible account of this dynamic is provided by Kevin Phillips's *The Politics of Rich and Poor: Wealth and the American Electorate in the Reagan Aftermath* (New York: Random House, 1990), and in his *Boiling Point: Republicans, Democrats, and the Decline of Middle-Class Prosperity* (New York: Random House, 1993). In recent years the *New York Times* has brought unusual scrutiny to bear on the stagnating fortunes of the middle class. One article among many is Louis Uchitelle's "The Middle Class: Winning in Politics, Losing in Life," *New York Times* (July 19, 1998), sec. 4: 1. Doug Henwood maintains sharp watch on the distributional effects of U.S. macroeconomic policy. A frequent contributor to *The Nation* (see, for example, "The Boom Year," *The Nation* (March 10, 1999): 10), his *Wall Street: How It Works and for Whom* (New York: Verso,

1998), is a goldmine of data, as is his Web page at http://www.panix.com/~dhenwood/LBO_home.html See too Ruy Teixeira and Joel Rogers's *America's Forgotten Majority: Why the White Working Class Still Matters* (New York: Basic Books, 2000), which reacts in a different way to the seeming marginalization of the middle class with respect to broad macroeconomic policy. The dynamic of "overwork" may not have begun with the compensatory, two-income strategies of *1980s* Americans, however. Schor, *The Overworked American*, argues that, from 1970 to 1990, the average worker has added an extra 164 hours (another month) to his or her work year.

37. See for example J. D. Biersdorfer, "Staying Wired on the Go and Doing More Work," *New York Times* (November 16, 2000): D9, G9; and Arlie Hochschild's *The Time Bind: When Work Becomes Home and Home Becomes Work* (New York: Holt, 1997). For insight into the "time crunch" for households in the 1980s, see Sylvia Ann Hewlett, *When the Bough Breaks: The Costs of Neglecting Our Children* (New York: Basic Books, 1991). In an extended endnote (chap. 1, note 3), Hochschild argues that those who claim that Americans are *not* spending more time at work fail to recognize the increasing invasiveness of work, in which out-of-office errands or coffee breaks are coopted by work-related tasks.

38. Kristen Davis, "Downshifters," *Kiplinger's Personal Finance Magazine* vol. 50 (August 1996): 32–35.

39. As Schor observes in *The Overspent American*, p. 114.

40. Thompson seems to have settled on a hopeful "maybe." See Bob Thompson, "Consumed," *Washington Post Magazine* (December 20, 1998): W10–15, W25–30.

41. Marcy Ann Darnovsky, *The Green Challenge to Consumer Culture: The Movement, the Marketers, and the Environmental Imagination* (Ann Arbor: UMI Dissertation Services, 1996); David Shi, *The Simple Life*, and Shi, ed., *In Search of the Simple Life: American Voices, Past and Present* (Salt Lake City, UT: Peregrine Smith Books, 1986); Peter Gould, *Early Green Politics: Back to Nature, Back to the Land, and Socialism in Britain, 1880–1900* (New York: St. Martin's Press, 1988). One example of 1930s essay writing on simplicity is Richard Gregg's *The Value of Voluntary Simplicity*, Pendle Hill essays, no. 3 (Wallingford, PA: Pendle Hill, 1936).

42. See Shi, *The Simple Life*; Tom Vanderbilt, "It's a Wonderful (Simplified) Life: Is the Voluntary Simplicity Movement True Liberation, or Diminished Expectations under a Different Name?," *The Nation* vol. 262 no. 3 (January 22, 1996): 20.

43. Though one should not underestimate the struggle, negotiation, and compromise bound up in this "household politics." On this score, see de Graaf and Boe, *Escape from Affluenza*, or Joe Dominguez and Vicki Robin, *Your Money or Your Life: Transforming Your Relationship with Money and Achieving Financial Independence* (New York: Viking Penguin, 1999).

44. Activity on this front appears to be accelerating. See for example the resources of the Families and Work Institute at http://www.familiesandwork.org/,

and the activities, agenda, and certification strategies of the Alliance of Work/Life Professionals, at http://www.awlp.org/

45. Vanderbilt, "It's a Wonderful (Simplified) Life," p. 21.

46. Vanderbilt, "It's a Wonderful (Simplified) Life," p. 23.

47. Marci McDonald, "Cashing Out," *Maclean's* vol. 109 (October 28, 1996): 44–50.

48. Household income data are from the U.S. Bureau of the Census (http://www.census.gov/hhes/income/histinc/index.html). Hayes's estimates are from his *A Survey of People Attending a Conference on Voluntary Simplicity*, while Schor's numbers are drawn from her *The Overspent American*. For summary data from Pierce's study, see Pierce, *Choosing Simplicity*, p. 291, or http://www.gallagherpress.com/pierce/stats.htm

49. Credit-card debt among the middle 20 percent of American households has increased greatly over the past decade, as consumers in this class in particular have sought to increase their consumption levels on stagnant incomes. It perhaps is not surprising, then, that the VSM appears to be largely a "white-collar" movement. See Hayes, *A Survey of People Attending a Conference on Voluntary Simplicity*; also Schor, *The Overspent American*, p. 137.

50. U.S. Census Bureau, "Educational Attainment" (http://www.census.gov/population/www/socdemo/educ-attn.html)

51. Harwood Group, *Yearning for Balance*, p. 15.

52. Schor, *The Overspent American*, p. 137, emphasis in the original.

53. Hochschild, *The Time Bind*, p. 245, emphasis added.

54. Seeds of Simplicity maintains a Web page at http://www.seedsofsimplicity.org/ The Simplicity Circles Project is found at http://www.simplicitycircles.com/ For more on the "online" revolution in voluntary simplicity, see Dolores Kong, "The Path to a Simpler Life Winds Its Way to the Web," *Boston Globe* (June 6, 2000): J7.

55. Andrews, *The Circle of Simplicity*, p. 194.

56. See Joshua Glenn, "Thriftzines," *Utne Reader* (March/April 1996): 26, 28.

57. At http://frugalliving.about.com/

58. So says Remarq/Critical Path, a major USENET access provider.

59. These organizations are the New Road Map Foundation, Institute for Earth Education, Financial Integrity Associates, Seeds of Simplicity, *Simple Living Journal*, and Center for the New American Dream.

60. http://simpleliving.net/ Also see Kong, "The Path to a Simpler Life Winds Its Way to the Web," and Andrews, *The Circle of Simplicity*.

61. Paul Elkins, "A Subversive Idea," *UNESCO Courier* (January 1998): 6–9.

62. Robert Putnam, *Bowling Alone: The Collapse and Revival of American Community* (New York: Simon and Schuster, 2000).

63. See for example Benjamin Barber, "Democracy at Risk: American Culture in a Global Culture," *World Policy Journal* vol. 15 no. 2 (Summer 1998): 29–41; National Commission on Civic Renewal, *A Nation of Spectators: How Civic Engagement Weakens America and What We Can Do about It* (College Park: University of Maryland, 1998); Michael Sandel, *Democracy's Discontent: America in Search of a Public Philosophy* (Cambridge, MA: Belknap Press of Harvard University Press, 1996); Freie, *Counterfeit Community*.

64. Durning, *How Much Is Enough?*, p. 149.

65. Durning, *How Much Is Enough?*, p. 142.

66. Thomas Frank, *The Conquest of Cool: Business Culture, Counterculture, and the Rise of Hip Consumerism* (Chicago: University of Chicago Press, 1997).

67. "Doing Well, Not Doing Without," *U.S. News and World Report* (December 14, 1998): 60.

68. Maria Blackburn, "Simplicity Has Its Price," *Baltimore Sun* (March 26, 2000): 2G; Nichole Torres, "Simplify, Simplify: A New Magazine and Web Site Seek to Help People Enjoy the Simple Pleasures of Life," *Business Start-Ups* (June 2000); Dave Wampler, "Faux Simplicity," *Simple Living Newsletter* no. 29 (April/June 2000), http://www.simpleliving.net/; John Zebrowski, "Selling Simplicity," *Seattle Times* (March 31, 2000): C1.

69. See chapter 3 of this book.

70. Schor, *The Overspent American*, pp. 169–173.

Chapter 10

1. See http://www.adbusters.org

2. The term *culture jamming* is not exclusive to Adbusters. It was coined in 1984 by San Francisco band Negativland, and refers to various undercurrents of billboard liberation graffiti, activist art, and resistant pranks. See Naomi Klein, *No Logo: Taking Aim at the Brand Bullies* (New York: Picador, 1999), chap. 12.

3. Rick Poynor, "Design Is Advertising #2," *Eye* vol. 8 no. 30 (Winter 1998), p. 37.

4. James MacKinnon, "Marketing: The Language of Consumerism," address given at "Versus: International Symposium on Marketing," Mexico City Campus of the Monterrey Institute of Technology and Higher Education, Mexico City, Mexico (March 11, 2000). Courtesy of author.

5. The Gap is to be featured as a case study in an introductory textbook at UC Berkeley's Haas School of Business, to which Gap owner Don Fisher is a major financial contributor. See Eyal Press and Jennifer Washburn, "The Kept University," *Atlantic Monthly* vol. 285 no. 3 (March 2000): 39–54.

6. See Anthony J. Cortese, *Provocateur: Images of Women and Minorities in Advertising* (Lanham, MD: Rowman & Littlefield, 1999), p. 4.

7. Stephen Sawicki, "Stopping the Insanity," *E: The Environmental Magazine* vol. 9 no. 6 (November/December 1998): 14–15.

8. Current estimates of total advertising expenditures in the United States range from $160 billion (Hayes) to $180 billion (Cortese). Robert Coen of McCann-Erickson predicted that national ad spending (excluding local advertising) would grow by 9.1 percent in 2000, to $139.2 billion (Teresa Ennis and Jennifer Steil, "Publishers Are Bullish about Y2K Ad Prospects," *Folio: The Magazine for Magazine Management* vol. 29 no. 1 (January 2000): 12–13). See Cortese, *Provocateur*; Cassandra Hayes, "Dogfight on Madison Avenue," *Black Enterprise* vol. 29 no. 11 (June 1999): 171–174.

9. Michael McCarthy and Theresa Howard, "Big-Name Marketers Return to Lineup," *USA Today* (January 16, 2001). Retrieved May 29, 2001, from the World Wide Web: http://www.usatoday.com/money/advertising/ad0069.htm#readmore

10. For example, national advertisers' spending in national newspapers rose 20.6 percent in 1999 (Nancy Coltun Webster, "Ad Spending in Dailies Rises 9.4% in 1999," *Advertising Age* vol. 71 no. 20 (May 8, 2000): S10).

11. John Consoli, "The 11.8-Hour Daily Diet," *Mediaweek* vol. 8 no. 16 (20 April 1998), p. 10.

12. Chuck Ross, "Who Wants to Watch More Commercials? 'Millionaire' Crams in Ads But Keeps Viewers" *Advertising Age* (September 9, 1999): 3.

13. Robert W. McChesney, *Rich Media, Poor Democracy: Communication Politics in Dubious Times* (Urbana, IL: University of Illinois Press, 1999), p. 45.

14. McChesney, *Rich Media, Poor Democracy*, p. 45.

15. See Juliet Schor, *The Overspent American: Upscaling, Downshifting, and the New Consumer* (New York: Basic Books, 1998).

16. Lasn, quoted in John Haynes, "Straight Ahead," *Illusions* vol. 2 no. 1 (Winter 1997), p. 14.

17. Lasn, interview with author, August 3, 2000.

18. Lasn, *Culture Jam: The Uncooling of America™* (New York: Eagle Brook, 1999), p. 31.

19. A recent PowerShift project was the Coca-Cola Climate Change campaign for Greenpeace Australia, which succeeded in forcing the corporate giant to promise to redesign the cooling elements in its soft drink machines to eliminate the use of HFCs.

20. Lasn, quoted in Haynes, "Straight Ahead," p. 14.

21. Lasn, quoted in Haynes, "Straight Ahead," p. 15.

22. Lasn, interview with author, August 3, 2000.

23. MacKinnon, "Marketing: The Language of Consumerism."

24. McChesney, *Rich Media, Poor Democracy*, p. 19.

25. Ben Bagdikian, *The Media Monopoly*, 6th ed. (Boston: Beacon Press, 2000), pp. x–xi.

26. McChesney, *Rich Media, Poor Democracy*, p. 35.

27. McChesney, *Rich Media, Poor Democracy*, p. 77.

28. Lasn, quoted in David Edwards, "The Millennial Moment of Truth," *The Ecologist* vol. 28 no. 6 (November/December 1998): 338–342.

29. Lasn, *Culture Jam*, pp. 32–33.

30. Robert Berner, "A Holiday Greeting Networks Won't Air: Shoppers Are 'Pigs,'" *Wall Street Journal* (November 19, 1997), sec. A, p. 1.

31. Lasn, interview with author, August 3, 2000.

32. Mark Leiren-Young, "Madison Avenue's Worst Nightmare," *Utne Reader* no. 79 (January/February 1997): 77.

33. Allan McDonald, interview with author, August 4, 2000.

34. Kalle Lasn, quoted in John Haynes, "Straight Ahead," p. 9.

35. I borrow the phrase "perspective by incongruity" from Kenneth Burke, *Permanence and Change: An Anatomy of Purpose*, 2nd ed. (Indianapolis, IN: Bobbs-Merrill, 1954), esp. part II, pp. 69–70.

36. "Buy Nothing Day '99: Call for Organizers," Posting on September 3, 1999, by the Culture Jammers Network, ⟨buynothingday@adbusters.org⟩.

37. Lasn, interview with author, August 3, 2000.

38. Sawicki, "Stopping the Insanity," p. 14.

39. Ferdinand M. De Leon, "Buy? Humbug!—Movement Afoot to Delay the Bell Opening Holiday Shopping Season," *Seattle Times* (November 25, 1997), p. E1.

40. Lasn, interview with author, August 3, 2000.

41. *Adbusters* vol. 6 no. 2 (Autumn 1998), p. 7.

42. Lasn, *Culture Jam*, p. 124.

43. Lasn, *Culture Jam*, pp. 114–118.

44. Lasn, interview with author, August 3, 2000.

45. Todd Gitlin, *The Twilight of Common Dreams: Why America Is Wracked by Culture Wars* (New York: Holt, 1995).

46. Lasn, interview with author, August 3, 2000.

47. Here I differ from Naomi Klein, who critiques *Adbusters* for being "repetitive and obvious," and for "crossing the fine line between information-age civil disobedience and puritanical finger-waving" (Klein, *No Logo*, p. 293).

48. Bruce Grierson, Kalle Lasn, and James MacKinnon, "Flashpoints," *Adbusters* vol. 7 no. 4 (Winter 2000): 57–66.

49. Klein, *No Logo*, p. 296.

Chapter 11

For research help, I am very grateful to Andrea Harrington, Derek Hall, Laura Chrabolowsky, and the Social Sciences and Humanities Research Council of Canada. Thanks also to Jennifer Clapp, Benjamin Cohen, Gita Sud, and one anonymous reviewer for their very helpful comments. The author wishes to acknowledge that this chapter was first written for a conference panel convened by the coeditors of this book and then published in *Global Society: Journal of Interdisciplinary International Relations* vol. 14 no. 1 (January 2000)—http://www.tandf.co.uk

1. See for example Stephen Gill, "Globalisation, Market Civilisation and Disciplinary Neoliberalism," *Millennium* vol. 23 no. 3 (1995): 399–423.

2. The largest LETS network is in one region of Australia with over 4,000 members. See Lawrence Scanlan, "LETS Make a Deal," *Harrowsmith* vol. 19 no. 1 (June 1994): 37–43.

3. At least one million U.S. citizens were involved in local currency or "scrip" networks in a brief one- or two-year period during the early 1930s, according to Wayne Weishaar and Wayne Parrish, *Men without Money* (New York: Putnam's, 1933).

4. See for example Thomas Greco, *New Money for Healthy Communities* (Tucson, AZ: Thomas Greco, 1994); Susan Meeker-Lowry, "The Potential of Local Currency," *Z Magazine* vol. 8 no. 7/8 (July/August 1995): 16–23; Ross Dobson, *Bringing the Economy Home from the Market* (Montreal: Black Rose Books, 1993); Paul Ekins, ed., *The Living Economy* (London: Routledge and Kegan Paul, 1986); Edgar Cahn and Jonathan Rowe, *Time Dollars* (Emmaus, PA: Rodale Press, 1992); Margrit Kennedy, *Interest and Inflation Free Money* (Gabriola Island, BC: New Society Publishers, 1995); Shann Turnbull, "Creating a Community Currency," in Ward Morehouse, ed., *Building Sustainable Communities: Tools and Concepts for Self-Reliant Economic Change* (New York: Bootstrap Press, 1989).

5. See for example Makoto Maruyama, "The Socio-Economic Role of LETS in Canada," Working Paper no. 66, University of Tokyo, Komaba, Department of Social and International Relations, February 1996; L. Thorne, "Local Exchange Trading Systems in the UK: A Case of Re-embedding," *Environment and Planning A* vol. 28 no. 8 (1996): 1361–1376; C. Williams, "Local Exchange and Trading Systems: A New Source of Work and Credit for the Poor and Unemployed?," *Environment and Planning A* vol. 28 no. 8 (1996): 1395–1415; C. Williams, "The New Barter Economy," *Journal of Public Policy* vol. 16 no. 1 (1996): 85–101; R. Lee, "Moral Money? LETS and the Social Construction of Local Economic Geographies in Southeast England," *Environment and Planning A* vol. 28 no. 8 (1996): 1377–1394. Some interesting, more theoretical analyses can be found in Abraham Rotstein and Colin Duncan, "For a Second Economy," in D. Drache and M. Gertler, eds., *The New Era of Global Competition* (Montreal: McGill-Queen's University Press, 1991); L. D. Solomon, *Rethinking Our Centralized Monetary System: The Case for a System of Local Currencies*

(Westport, CT: Praeger, 1996); Rachel Tibbett, "Alternative Currencies: A Challenge to Globalisation?," *New Political Economy* vol. 2 no. 1 (1996): 127–135. My focus on the use of consumption as a political challenge to neoliberal priorities is, however, somewhat different than the focus of these writings.

6. The broad contours of "green" political-economic thought are outlined below in this chapter. For a more detailed discussion, see Andrew Dobson, *Green Political Thought: An Introduction* (London: Unwin Hyman, 1990); Herman Daly and John Cobb, *For the Common Good: Redirecting the Economy Toward Community, the Environment, and a Sustainable Future* (Boston: Beacon Press, 1989); Eric Helleiner, "International Political Economy and the Greens," *New Political Economy* vol. 1 no. 1 (1996): 59–78.

7. See for example Gill, "Globalisation, Market Civilisation and Disciplinary Neoliberalism"; Eric Helleiner, *States and the Reemergence of Global Finance* (Ithaca, NY: Cornell University Press, 1994).

8. Kennedy, *Interest and Inflation Free Money*, p. 119.

9. Paul Glover, "Ithaca HOURS," in Susan Meeker-Lowry, ed., *Invested in the Common Good* (Philadelphia: New Society Publishers, 1995), p. 156.

10. Guy Dauncey, *After the Crash: The Emergence of the Rainbow Economy* (Basingstoke, UK: Green Print, 1988), p. 62.

11. Eric Helleiner, "Historicizing Territorial Currencies: Monetary Space and the Nation-State in North America," *Political Geography* vol. 18 (1999): 309–339; Helleiner, "One Money, One Nation: Territorial Currencies and the Nation-State," ARENA Working Paper no. 17 (Oslo: ARENA, 1997).

12. See for example Helleiner, *States and the Reemergence of Global Finance*.

13. Vandana Shiva, "The Greening of Global Reach," in W. Sachs, ed., *Global Ecology* (London: Zed Books, 1993).

14. See for example Rotstein and Duncan, "For a Second Economy." For Polanyi, see Karl Polanyi, *The Great Transformation* (Boston: Beacon Press, 1944).

15. See for example Rotstein and Duncan, "For a Second Economy"; Dobson, *Bringing the Economy Home from the Market*. For Braudel, see Fernand Braudel, *Civilization and Capitalism: 15th–18th Centuries* (3 vols.), translated by Sian Reynolds (London: Fontana, 1985).

16. Rotstein and Duncan, "For a Second Economy." This historical world was also one in which "bad" money often pushed out "good" according to Gresham's Law, a point that local currency advocates have not explored in much depth.

17. James Robertson, *Future Wealth* (London: Cassell, 1990), p. 125.

18. Dauncey, *After the Crash*, p. 58.

19. See for example Friedrich A. von Hayek, *The Road to Serfdom* (Chicago: University of Chicago Press, 1944).

20. See for example Helleiner, *States and the Reemergence of Global Finance*.

21. See for example Lawrence White, *Competition and Currency* (New York: New York University Press, 1989); George Selgin, *The Theory of Free Banking*

(Totowa, NJ: Rowman & Littlefield, 1988); Steve Hanke, *Currency Boards for Developing Countries* (San Francisco: International Center for Economic Growth, 1994).

22. Kennedy, *Interest and Inflation Free Money.*

23. See for example Greco, *New Money for Healthy Communities.*

24. For these arguments, see for example Jane Jacobs, *Cities and the Wealth of Nations* (New York: Random House, 1984); David Weston, "Green Economics: The Community Use of Currency," paper presented at The Other Economic Summit (mimeo, 1985).

25. Kennedy, *Interest and Inflation Free Money.*

26. Williams, "Local Exchange and Trading Systems."

27. Kennedy, *Interest and Inflation Free Money*, p. 128.

28. Silvio Gesell, *The Natural Economic Order* (San Antonio: Free Economy Publishing Company, 1934).

29. Glover, "Ithaca HOURS," p. 156. Schuman reports that the Barter Potluck has given Ithaca HOURS to groups such as the Stop Wal-Mart Campaign, Ithaca Rape Center, Senior Citizens Center, and the Committee on US-Latin American Relations. See Michael Schuman, *Going Local: Creating Self-Reliant Communities in a Global Age* (New York: Free Press, 1998).

30. Glover, "Ithaca HOURS," p. 155.

31. Thomas Greco, "The Essential Nature of Money," *Earth Island Journal* (Winter 1995), p. 36. See also Rotstein and Duncan, "For a Second Economy."

32. Quoted in Lee, "Moral Money," p. 1383.

33. Dobson, *Bringing the Economy Home from the Market*, p. 163.

34. Andrew Leyshon and Nigel Thrift, "Moral Geographies of Money," in Emily Gilbert and Eric Helleiner, eds., *Nation-States and Money: The Past, Present and Future of National Currencies* (London: Routledge, 1999).

35. See for example Alexander Shand, *Free Market Morality: The Political Economy of the Austrian School* (London: Routledge, 1990).

36. Daly and Cobb, *For the Common Good.*

37. Thorne, "Local Exchange Trading Systems in the UK," p. 1365. But Williams's data contradict this, in that they suggest that economic reasons were most important in explaining interest in participating in the Manchester local currency network; Williams, "Local Exchange and Trading Systems," p. 1404.

38. Karl Marx, *Capital,* vol. 1 (London: Lawrence and Wishart, 1974), p. 132.

39. George Simmel, *The Philosophy of Money*, translated by Tom Bottomore and David Frisby (London: Routledge and Kegan Paul, [1900] 1978), pp. 377, 373, 128, 440.

40. Viviana Zelizer, *The Social Meaning of Money* (New York: Basic Books, 1994).

41. Eric Helleiner, "National Identities and National Currencies," *American Behavioral Scientist* vol. 41 no. 10 (1998), pp. 1409–1436.

42. Virginia Hewitt, *Beauty and the Banknote* (London: British Museum Press, 1994), p. 11.

43. Glover, "Ithaca HOURS," p. 156.

44. Indeed, Lee ("Moral Money," p. 1382) reports that the naming process is often the subject of "fierce local debate."

45. LETS Toronto, "What Is LETS?" (mimeo, undated).

46. Dauncey, *After the Crash*, p. 61.

47. Glover, "Ithaca HOURS," p. 156.

48. Dobson, *Bringing the Economy Home from the Market*, p. 189.

49. Williams, "Local Exchange and Trading Systems," p. 1396.

50. Makoto Maruyama, "The Social Contexts of Money Uses," in Robert Albritton and Thomas Sekine, eds., *A Japanese Approach to Political Economy* (London: Macmillan, 1995), p. 102.

51. Quoted in Scanlan, "LETS Make a Deal," p. 40.

52. Dauncey, *After the Crash*, pp. 63–64.

53. Eric Helleiner, "International Political Economy and the Greens," *New Political Economy* vol. 1 no. 1 (1996): 59–78.

54. E. F. Schumacher, *Small Is Beautiful: Economics as If People Mattered* (New York: Harper and Row, 1973).

55. Benjamin Cohen, *The Geography of Money* (Ithaca, NY: Cornell University Press, 1998).

56. The Australian government invited Michael Linton to Australia in 1992 to discuss LETS and subsequently promoted the establishment of LETS networks with computer equipment, education, publicity, postage, and stationery; see Patricia Knox, "A New Green Economy? LETS Do It!," *Earth Island Journal* (Winter 1994). The Department of Social Services actively encourages people on benefits to join local LETS, and a 1995 Act declared that LETS earnings would be exempt from income tests for social security, as long as those unemployed were still looking for work. The European Commission has also incorporated promotion of LETS into its urban and regional policy; see Gill Seyfang and Colin Williams, "Give DIY Economies a Break," *New Statesman* (March 27, 1998).

57. Michael Schuman notes that the first grant to E. F. Schumacher's newsletter *Local Currency News* came from the conservative Bradley Foundation, which has also funded the U.S.-based Heritage Foundation and American Enterprise Institute (Schuman, *Going Local*, p. 137).

58. Friedrich Hayek, *The Denationalisation of Money: The Argument Refined* (London: Institute for Economic Affairs, 1990). His ideas are cited in Ekins, *The Living Economy*, p. 197; Robertson, *Future Wealth*, pp. 127–128; Solomon, *Rethinking Our Centralized Monetary System*. Hayek's goals are in fact very different from those of the local currency advocates; he favors private corporate

issuers of money competing in the marketplace as a means of promoting the values of laissez-faire economics and individualism.

59. Even in a country such as Britain, where the local currency movement has been relatively successful, only approximately 20,000 people are members of local currency networks (of which there are about 300) and often no more than 10 percent of each individual's economic activity is involved in the local currency system. See Williams, "The New Barter Economy." Similarly, Maruyama ("The Socio-Economic Role of LETS in Canada") notes that in Canada, participants in LETS networks conduct, on average, less than 10 percent of their economic activity within the network.

Chapter 12

1. United Nations Food and Agricultural Association (FAO), "Organic Farming," retrieved May 2001 from the World Wide Web at www.fao.org/WAICENT/FAOINFO/AGRICULT/magazine/9901sp3.htm

2. See, for example, Laure Waridel, *Coffee with Pleasure: Just Java and World Trade* (Montreal: Black Rose Books, 2001).

3. Nigel Dudley, Jean-Paul Jeanrenaud, and Francis Sullivan, *Bad Harvest? The Timber Trade and the Degradation of the World's Forests* (London: Earthscan, 1995).

4. Fred Gale, *The Tropical Timber Trade Regime* (Basingstoke, UK, and New York: Macmillan and St. Martin's Presses, 1998), p. 160.

5. For a detailed account of ECLs at the ITTO, see Gale, *The Tropical Timber Trade Regime*, pp. 155–177.

6. V. Viana, J. Ervin, R. Donovan, C. Elliott, and H. Gholz (eds.), *Certification of Forest Products: Issues and Perspectives* (Covelo, CA: Island Press, 1996), p. 5.

7. Government of Canada, Standing Committee on Natural Resources and Government Operations, *Forest Management Practices in Canada as an International Trade Issue* (Final Report) (Ottawa: Government of Canada), retrieved May 2001 from the World Wide Web at www.parl.gc.ca/infocomdoc/36/2/nrgo/studies/reports/nrgo01/07-toc-e.html

8. Daniel Botkin, *Discordant Harmonies: A New Ecology for the Twenty-First Century* (New York: Oxford University Press, 1990); R. Grumbine, "What Is Ecosystem Management?," *Conservation Biology* vol. 8 no. 1 (1994): 27–38; Norman L. Christensen, Ann M. Bertuska, James H. Brown, Stephen Carpenter, Carla D'Antonio, Rober Francis, Jerry F. Franklin, James A. MacMahon, Reed F. Noss, David J. Parsons, Charles H. Peterson, Monica G. Turner, Robert G. Woodmansee. "The Report of the Ecological Society of America Committee on the Scientific Basis for Ecosystem Management," *Ecological Applications* vol. 6 no. 3 (1996): 665–691.

9. J. Wilson, J. Acheson, M. Metcalfe, and P. Kleban, "Chaos, Complexity, and Community Management," *Marine Policy* vol. 18 no. 4 (1994): 291–305.

10. Fred Gale and Cheri Burda, "Attitudes towards Eco-Certification in the BC Forest Products Industry" (Victoria: Eco-Research Chair Report, University of Victoria, 1996).

11. American Forest and Paper Association (AFPA), *5th Annual Sustainable Forestry Initiative Progress Report* (Washington, DC: American Forest and Paper Association, 2000), retrieved May 2001 from the World Wide Web at www.afandpa.org/forestry/sfi/menu.html

12. AFPA, *5th Annual Sustainable Forestry Initiative Progress Report*, p. ii.

13. AFPA, "Sustainable Forestry Initiative," retrieved May 2001 from the World Wide Web at www.afandpa.org/forestry/sfi/Final_Standard.pdf, p. 3.

14. AFPA, "Sustainable Forestry Initiative," p. 24.

15. AFPA, "Independent Expert Review Panel," retrieved October 2001 from the World Wide Web at http://www.afandpa.org/forestry/sfi_frame.html

16. AFPA, *5th Annual Sustainable Forestry Initative Progress Report*, p. 1.

17. AFPA, "Organizations Supporting the Goals of the SFI Program," retrieved May 2001 from the World Wide Web at www.afandpa.org/forestry/sfi/menu.html

18. AFPA, "Sustainable Forestry Initiative," p. 5.

19. AFPA, "Sustainable Forestry Initiative," p. 5.

20. AFPA, "Sustainable Forestry Initiative," p. 15.

21. AFPA, *5th Annual Sustainable Forestry Initiative Progress Report*, p. 16.

22. AFPA, *5th Annual Sustainable Forestry Initiative Progress Report*, p. 5.

23. AFPA, *5th Annual Sustainable Forestry Initiative Progress Report*, p. 1.

24. AFPA, *5th Annual Sustainable Forestry Initiative Progress Report*, p. 21.

25. Canadian Standards Association (CSA), *A Sustainable Forest Management System: Guidance Document*, CAN/CSA-Z808-96 (Ottawa: CSA, 1996).

26. Canadian Sustainable Forestry Certification Coalition, "The CSA Standards for Canada's Forests," retrieved May 2001 from the World Wide Web at www.sfms.com/2s_2.htm

27. Chris Elliott, "Forest Certification: Analysis from a Policy Network Perspective," Thèse No. 1965, École Polytechnique Fédérale de Lausanne, Lausanne, Switzerland, 1998, pp. 305–306.

28. PriceWaterhouseCoopers, "Small Business Forest Enterprise Program (SBFEP) Certification Pre-Assessment" (Victoria: BC Ministry of Forests, October 1999), retrieved May 2001 from the World Wide Web at http://www.for.gov.bc.ca/hfe/pwcreport.pdf, p. iii.

29. Jennifer Clapp, "The Privatization of Global Environmental Governance: ISO 14000 and the Developing World," *Global Governance* vol. 4 no. 3 (July/September 1998): 295–316, 306.

30. Forest Stewardship Council (FSC), "FSC Principles and Criteria" (Oaxaca, Mexico: FSC, February 1999), retrieved May 2001 from the World Wide Web at http://www.fscoax.org/principal.htm

31. FSC, "Forest Stewardship Council A.C. By-Laws" (Oaxaca, Mexico: FSC, February 1999), retrieved May 2001 from the World Wide Web at www.fscoax.org/principal.htm

32. FSC, "Forest Stewardship Council A.C. By-Laws."

33. FSC, "Forest Stewardship Council A.C. By-Laws."

34. See FSC, "II Annual Conference," retrieved May 2001 from the World Wide Web at www.fscoax.org/html/social_issues_index.htm

35. Charles Restino, "J. D. Irving Exerts Pressure on a Malleable Forest Stewardship Council," *Northern Forest Forum* vol. 8 no. 2 (Spring 2000): 22–25.

36. Restino, "J. D. Irving Exerts Pressure," p. 25.

37. FSC, "FSC Commission of Enquiry on Maritime Regional Standards" (Oaxaca, Mexico: FSC, February 28, 2000).

38. FSC, *Group Certification: FSC Guidelines for Certification Bodies* (Oaxaca, Mexico: FSC, July 31, 1998), retrieved May 2001 from the World Wide Web at http://www.fscoax.org/html/groupp.html

39. FSC, "Forest Stewardship Council A.C. By-Laws," para 13.

Chapter 13

1. Notably, this estimate of the scale of activity must be derived from earlier estimates in the light of continuing vigorous growth in mostly mail-order, home power equipment sales. In a nation that seems to count everything, it is surprising that a census of home power activity has not been taken and that no substantial government or policy-relevant attention has yet been focused on the home power movement. Here as elsewhere, however, actual human experience may yet prove to count in ways that the merely countable never will. See "Here Comes the Sun," *Time* (October 18, 1993), p. 84; Jesse Tatum, "The Home Power Movement and the Assumptions of Energy Policy Analysis," *Energy: The International Journal* vol. 17 no. 2 (February 1992): 99–107; Jesse Tatum, *Muted Voices: The Recovery of Democracy in the Shaping of Technology* (Bethlehem, PA: Lehigh University Press, 2000).

2. I use the term *movement* as shorthand for what Neil Smelser has termed a "value-oriented movement"—that is, a pattern of collective behavior that seeks "to restore, protect, modify, or create values in the name of a generalized [value-oriented] belief." Such a belief "envisions a modification of those conceptions concerning 'nature, man's place in it, man's relation to man, and the desirable and nondesirable as they may relate to man-environment and inter-human relations.' This kind of belief involves a basic reconstitution of self and society." See Neil Smelser, *Theory of Collective Behavior* (New York: Free Press, 1962), p. 120.

3. Anecdotal evidence and occasional articles in *Home Power Magazine* indicate that home power efforts in Europe and other parts of the industrialized world parallel those in the United States. Third World applications of home power

developments have also been vigorous, ranging from large-scale government-sponsored or U.S.-utility-led programs to provide power without a grid in Indonesia and elsewhere, to the more locally tailored efforts of nongovernmental organizations like the Solar Electric Light Fund (headquartered in Washington, DC). This chapter focuses, however, on the home power movement in the United States.

4. See, for example, Tatum, "The Home Power Movement and the Assumptions of Energy Policy Analysis"; Tatum, "Technology and Values: Getting Beyond the 'Device Paradigm' Impasse," *Science, Technology, and Human Values* vol. 19 no. 1 (Winter 1994): 70–87; Tatum, *Muted Voices*.

5. I use the term *participatory research* somewhat loosely to refer to research undertaken and pursued by ordinary people, which they will themselves primarily apply and from which they will primarily benefit, in response to needs and questions they themselves have defined. See Peter Park, M. Brydon-Miller, B. Hall, and T. Jackson, eds., *Voices of Change: Participatory Research in the United States and Canada* (Westport, CT: Bergin & Garvey, 1993).

6. Mick Sagrillo, "Applies and Oranges," *Home Power Magazine*, no. 65 (June/July 1998): 18–32.

7. Jesse Tatum, *Energy Possibilities: Rethinking Alternatives and the Choice-Making Process* (Albany, NY: SUNY Press, 1995).

8. Connie Said, personal communication, November 6, 2000 (*Home Power Magazine*, P.O. Box 520, Ashland, OR 97520).

9. A more detailed account of home power activities around Amherst, Wisconsin, is provided in Tatum, *Muted Voices*, pp. 118–123.

10. Langdon Winner, *The Whale and the Reactor: A Search for Limits in an Age of High Technology* (Chicago: University of Chicago Press, 1986).

11. Tatum, *Muted Voices*, pp. 127–131.

12. Susan Berlin, *Ways We Live: Exploring Community* (Gabriola Island, BC: New Society Publishers, 1997).

13. Winner, *The Whale and the Reactor*, p. 12.

14. These arguments, tied closely to Borgmann's theoretical constructs, are developed more fully in Tatum, "Technology and Values."

15. Nigel Whiteley, *Design for Society* (London: Reaktion Books, 1993).

16. Whiteley, *Design for Society*.

17. The model here is of the child who enters a room in which one other child stands holding a single block, with a hundred identical blocks spread about the floor. The first child, taking a cue from the second's possession, decides that he or she must have the particular block held by the second child. Jim Grote, "Rene Girard's Theory of Violence: An Introduction," *Research in Philosophy and Technology* vol. 12 (1992): 261–270. See also Reginald Luyf and Pieter Tijmes, "Modern Immaterialism," *Research in Philosophy and Technology* vol. 12 (1992): 271–284.

18. Wiebe E. Bijker, "The Social Construction of Fluorescent Lighting, or How an Artifact Was Invented in Its Diffusion Stage," in John Law and Wiebe Bijker, eds., *Shaping Technology/Building Society* (Cambridge, MA: MIT Press, 1992): 75–102.

19. Stanton A. Glantz, John Slade, Lisa A. Bero, Peter Hanauer, and Deborah E. Barnes, *The Cigarette Papers* (Berkeley: University of California Press, 1996).

20. Herbert Marcuse, *One Dimensional Man* (Boston: Beacon Press, 1964).

21. "Equal to the world" is the apt phrase of philosopher of technology Albert Borgmann in *Technology and the Character of Contemporary Life* (Chicago: University of Chicago Press, 1984).

22. Lewis Mumford, *Technics and Human Development* (New York: Harcourt Brace Jovanovich, 1967.

23. Lewis Mumford, *The Pentagon of Power* (New York: Harcourt Brace Jovanovich, 1970.

Chapter 14

1. William Cronon, *Nature's Metropolis: Chicago and the Great West* (New York: Norton, 1991).

2. See especially Mark Edmundson, "On the Uses of a Liberal Education: 1. As Lite Entertainment for Bored College Students," *Harper's Magazine* vol. 295 no. 1768 (September 1997): 39–49.

3. "The marketplace," observed economist E. F. Schumacher, "is the institutionalization of nonresponsibility." See E. F. Schumacher, *Small Is Beautiful: Economics as If People Mattered* (New York: Harper and Row, 1973).

4. Matthew Wald, "Clinton Energy-Saving Rules Are Getting a Second Look," *New York Times* (March 31, 2001), p. C1.

5. See for example Jim Klein and Marth Olson, *Taken for a Ride* (Hohokus, NJ: New Day Films, 1996, videorecording), or see http://www.newday.com/films/Taken_for_a_Ride.html

6. This was a central argument of John Kenneth Galbraith's *The New Industrial State* (Boston: Houghton Mifflin, 1967). Galbraith observed that in an increasingly complex economy with growing scales of production, corporate decisions about new-product launches, investments in new productive capacity and final networks of product distribution must be made months if not years before products are marketed to consumers. Firms, in Galbraith's view, are thus compelled to invest in the production of new products based on what they think the consumer will want (or can be induced to want), and then invest heavily in advertising when the product hits the shelves. In this "planning-economy" world, consumer sovereignty is turned on its head.

7. David Leonhardt, "Economy's Growth Proves Steadfast Although Modest," *New York Times* (April 28, 2001), pp. A1, B3.

Contributors

Marilyn Bordwell is Assistant Professor of Communication Arts at Allegheny College. She has published articles in *Text and Performance Quarterly, Theatre Annual*, and *JASHM: Journal for the Anthropological Study of Human Movement.*

Jennifer Clapp is Associate Professor in the Comparative Development and Environmental and Resource Studies Programs at Trent University. She is the author of a number of articles on the international movement of hazardous wastes and dirty production technologies. Her most recent book is *Toxic Exports: The Transfer of Hazardous Wastes from Rich to Poor Countries.*

Ken Conca (editor) is Associate Professor of Government and Politics at the University of Maryland, where he also directs the Harrison Program on the Future Global Agenda. He is the author of *Manufacturing Insecurity: The Rise and Fall of Brazil's Military-Industrial Complex*; the editor of *Green Planet Blues: Environmental Politics from Stockholm to Kyoto*; and the coeditor, with Ronnie D. Lipschutz, of *The State and Social Power in Global Environmental Politics.*

Fred Gale is Lecturer at the School of Government, University of Tasmania, Australia. He is coeditor of *Nature, Production, Power: Towards an Ecological Political Economy* and author of *The Tropical Timber Trade Regime.* His current research focuses on the trade/environment interface, specifically on the implications of ecocertification and ecolabeling for trade theory and policy.

Eric Helleiner is Canada Research Chair in International Political Economy at Trent University, Peterborough, Ontario, Canada. He is the author of *States and the Reemergence of Global Finance; One State, One Money: A History of Territorial Currencies*; and coeditor of *Nation-States and Money.*

Michael Maniates (editor) is Associate Professor of Political Science and Environmental Science at Allegheny College. He also codirects the Meadville Community Energy Project, a student-centered program fostering regional energy efficiency and sustainability, and administers the International Project on Teaching Global Environmental Politics, an electronic network of college and university professors exploring solutions to global environmental ills. His work on the nature and future of undergraduate environmental studies programs has appeared in *BioScience*, and he is editor of the forthcoming volume *Empowering*

Knowledge: Teaching and Learning Global Environmental Politics as if Education Mattered.

Jack Manno is Executive Director of the New York Great Lakes Research Consortium and adjunct Associate Professor in the Faculty of Environmental Studies at the State University of New York College of Environmental Science and Forestry in Syracuse. His most recent book is *Privileged Goods: Commoditization and Its Impact on Environment and Society.*

Thomas Princen (editor) is Associate Professor of Natural Resources and Environmental Policy in the School of Natural Resources and Environment at the University of Michigan, where he also codirects the Workshop on Consumption and Environment. He is coauthor, with Matthias Finger, of *Environmental NGOs in World Politics: Linking the Local and the Global* and author of *Intermediaries in International Conflict.*

Jesse Tatum has taught for many years in the field of science, technology, and society, most recently at Rensselaer Polytechnic Institute. He is the author of two books, *Energy Possibilities: Rethinking Alternatives and the Choice-Making Process* and *Muted Voices: The Recovery of Democracy in the Shaping of Technology.*

Richard Tucker is Adjunct Professor of Natural Resources at the University of Michigan. He is an environmental historian, specializing in the ecological impacts of colonial regimes. His latest book is *Insatiable Appetite: The United States and the Ecological Degradation of the Tropical World.*

Index